Physics and Chemistry in Space Vol. 24
Planetology

Edited by M. C. E. Huber, Noordwijk, L. J. Lanzerotti, Murray Hill, and D. Stöffler, Münster

A. D. M. Walker

Plasma Waves in the Magnetosphere

With 119 Figures

Springer-Verlag
Berlin Heidelberg New York
London Paris Tokyo Hong Kong
Barcelona Budapest

Professor Dr. A. D. M. WALKER
Department of Physics, University of Natal
Durban 4000, South Africa

Series Editors:

Professor Dr. M. C. E. HUBER
European Space Research and Technology Centre
Keplerlaan 1, NL-2200 AG Noordwijk, The Netherlands

Dr. L. J. LANZEROTTI
AT&T Bell Laboratories, 600 Mountain Avenue
Murray Hill, NJ 07974-2070, USA

Professor D. STÖFFLER
Institut für Planetologie, Universität Münster
Wilhelm-Klemm-Str. 10, W-4400 Münster, FRG

ISBN-13: 978-3-642-77869-8 e-ISBN-13: 978-3-642-77867-4
DOI: 10.1007/978-3-642-77867-4

Softcover reprint of the hardcover 1st edition 1993

Production Editor: Renate Münzenmayer
Typesetting: Camera-ready by author
32/3145 - 5 4 3 2 1 0 – Printed on acid-free paper

Preface

This book is a study of plasma waves which are observed in the earth's magnetosphere. The emphasis is on a thorough, but concise, treatment of the necessary theory and the use of this theory to understand the manifold varieties of waves which are observed by ground-based instruments and by satellites. We restrict our treatment to waves with wavelengths short compared with the spatial scales of the background plasma in the magnetosphere. By so doing we exclude large scale magnetohydrodynamic phenomena such as ULF pulsations in the Pc2–5 ranges. The field is an active one and we cannot hope to discuss every wave phenomenon ever observed in the magnetosphere! We try instead to give a good treatment of phenomena which are well understood, and which illustrate as many different parts of the theory as possible. It is thus hoped to put the reader in a position to understand the current literature.

The treatment is aimed at a beginning graduate student in the field but it is hoped that it will also be of use as a reference to established workers. A knowledge of electromagnetic theory and some elementary plasma physics is assumed. The mathematical background required includes a knowledge of vector calculus, linear algebra, and Fourier transform theory encountered in standard undergraduate physics curricula. A reasonable acquaintance with the theory of functions of a complex variable including contour integration and the residue theorem is assumed. Other mathematical techniques which are needed are summarized in Appendices. These are: (1) Cartesian tensor notation, a brief summary of which is given in Appendix A, (2) Some elementary properties of Bessel functions and other related results which are summarized in Appendix B.

The book is divided into two parts. In Part I the theory of wave propagation and generation in a plasma is introduced. All the major techniques needed in Part II are discussed. Part II is devoted to the application of these techniques to the menagerie of plasma waves encountered in the magnetosphere. Some readers may prefer to start with Part II and refer back to Part I when they encounter an unfamiliar procedure.

The scope of this book is inevitably limited by the space available. There are numerous books which treat the various topics introduced here in greater depth. Those listed here have greatly influenced the author and are particularly recommended for further reading.

The propagation of waves in a cold plasma without the effect of ions is treated in the classic work by Budden [17]. In particular, the mathematical techniques are lucidly treated and waves in varying media receive much attention. This subject has been brought up to date in a new work [19]. Another treatment which emphasizes waves in a uniform medium is that by Booker [14]. Fundamental books on plasma physics which give detailed treatments of waves in hot plasmas include the well-known book by Stix [100] and the very thorough treatise by Clemmow and Dougherty [27]. While this book was in preparation Swanson [106] published an excellent book on plasma waves. While its main thrust is towards thermo-nuclear applications its approach is similar to ours. Dungey [35] has given a masterly exposition of the physical ideas behind quasi-linear theory. Our approach to this topic is influenced by this approach but attempts to include more of the detail.

A comprehensive collection of reviews of various topics in solar system plasma physics has been edited by Kennel, Lanzerotti and Parker [69]. Plasma instabilities with particular reference to the magnetosphere are covered concisely in Hasegawa's book in this series [49]. Hargreaves [47] has written a useful student text on the magnetosphere. Lyons and Williams [76] emphasize those parts of magnetospheric physics which are well understood quantitatively. Treatments of magnetospheric physics can be found in many review articles. References to some of these are made in the text.

Observations of VLF waves are treated in the well-known book by Helliwell [53]. This is largely limited to ground-based observations. While it was published many years ago it contains a large number of examples of ground-based data as well as discussion of the theory as then understood. Much is still relevant. Other types of magnetospheric wave, particularly those which can only be observed *in situ* by satellites, are discussed in a number of excellent reviews e. g. [67, 94, 45].

The question of citation of references is a thorny one. In a book of this sort, if full credit were to be given to all the work which has been done, the bibliography would be unmanageable. We have cited works which we believe will be helpful to the reader in amplifying or clarifying the material in the text. No attempt has been made to cite every relevant reference and thus many workers who have made significant contributions to the field may not get due credit. This applies particularly to work which has not been published in English.

Modern spreadsheet and graphics software has made it feasible for almost all the calculations originally made by other authors to be recalculated. This has provided a useful check on the results but, of course, may have introduced errors.

I am very grateful to numerous friends and colleagues who have read and criticised parts of this book. Particular mention should be made of Prof. M. A. Hellberg, Prof. M. W. J. Scourfield, Mr Peter Krumm and several students. The errors and omissions which doubtless remain are my own. I am also grateful to Dr A. R. W. Hughes for selecting suitable data to illustrate a number of points and to Gavin Hough and June Watson for processing it in an appropriate form. Terry Holloway provided exemplary computer support. Reiner Friedel and Pravesh Ramlall were generous in supplying me with the unpublished results of some of their computations. My wife and family deserve thanks for their tolerance of my penchant for using the small hours of the morning as the best time for uninterrupted work.

Note added in proof: I am grateful to Dr S. Sazhin for supplying me with a copy of the list of contents of his book *Whistler-mode waves in a hot plasma* (Cambridge University Press, 1992, in press). Its subject matter is very relevant to the material in this book.

September 1992 A. D. M. Walker

Contents

I Propagation and Generation of Plasma Waves

1 Basic Equations 3
1.1 Introduction . . . 3
1.2 Electromagnetic Equations . . . 4
1.3 Fluid Equations . . . 6
1.4 The Kinetic Equation . . . 7
1.5 Poynting's Theorem . . . 11
1.6 Harmonic Oscillations . . . 11
1.7 The Wave Equation . . . 13
1.8 Summary . . . 13

2 Waves in a Uniform Cold Magnetoplasma – 1. Infinite Plane Waves
2.1 Introduction . . . 15
2.2 Characteristic Frequencies and Speeds . . . 16
2.3 Linearization of the Equation of Motion . . . 18
2.4 Constitutive Relations . . . 18
2.5 Plane Waves . . . 24
2.6 Polarization . . . 30
2.7 Properties of the Refractive Index . . . 34
2.8 Energy Flux in a Plane Wave . . . 40
2.9 Summary . . . 41

3 Waves in a Uniform Cold Magnetoplasma – 2. Rays and Wave Packets
3.1 Introduction . . . 43
3.2 Wave Packets and Rays . . . 43
3.3 Classification of Waves in a Cold Plasma . . . 53
3.4 Refractive Index and Dispersion Relation . . . 66
3.5 Summary . . . 72

4 Propagation of Electromagnetic Waves in a Non-Uniform Cold Magnetoplasma
4.1 Introduction . . . 74
4.2 Plane Stratified Media . . . 74
4.3 Ray Tracing in General Media . . . 79
4.4 Summary . . . 81

5 Waves in a Uniform Warm Magnetoplasma

5.1 Introduction . . . 83
5.2 Characteristic Speeds . . . 84
5.3 The Constitutive Relation . . . 84
5.4 Dispersion Relations and Refractive Index . . . 86
5.5 Polarization . . . 97
5.6 Summary . . . 99

6 Waves in a Hot Plasma – 1. General Features

6.1 Introduction . . . 101
6.2 Unperturbed Particle Orbits . . . 101
6.3 Electrostatic Approximation . . . 102
6.4 Propagation Parallel to the Magnetic Field . . . 108
6.5 Growth and Decay of Waves . . . 112
6.6 The Equilibrium Distribution Function – The Maxwellian 114
6.7 Non-Equilibrium Distribution Functions . . . 115
6.8 Summary . . . 121

7 Waves in a Hot Plasma – 2. Equilibrium and Non-Equilibrium Distributions

7.1 Introduction . . . 123
7.2 Waves in Plasmas in Thermal Equilibrium . . . 124
7.3 Longitudinal Waves Excited by a Particle Beam . . . 130
7.4 Electrostatic Waves Associated with Anisotropic Distributions . . . 135
7.5 Summary . . . 150

8 The Effect of Wave Fields on Energetic Particles

8.1 Introduction . . . 151
8.2 Particle Resonance . . . 152
8.3 Trajectories of Resonant Particles in Velocity Space . . . 156
8.4 Diffusion in Velocity Space . . . 169
8.5 Some Non-Linear Effects . . . 176
8.6 Waves Resonant with a Test Particle . . . 180
8.7 Summary . . . 185

II Applications of the Theory to Plasma Wave Observations

9 Magnetospheric Plasmas 189

9.1 Introduction . . . 189
9.2 Structure of the Earth's Magnetosphere . . . 189
9.3 Cold Plasma Populations in the Magnetosphere . . . 195

9.4 Hot Plasma Populations 198
9.5 Waves in the Magnetosphere 202
9.6 Summary 207

10 Waves in the Plasmasphere – 1. Whistler Observations and Basic Theory
10.1 Introduction 208
10.2 Properties of the Whistler Mode 208
10.3 Observations of Whistlers 213
10.4 Elementary Theory of Whistlers 216
10.5 Use of Whistlers as a Magnetospheric Probe 221
10.6 Summary 222

11 Waves in the Plasmasphere – 2. Details of Whistler Propagation
11.1 Introduction 223
11.2 Lightning as a Source of Electromagnetic Radiation . . . 223
11.3 Propagation in the Earth-Ionosphere Waveguide 223
11.4 Transmission of Whistlers Through the Ionosphere 226
11.5 Propagation of Unducted Whistlers 227
11.6 Propagation in Ducts 233
11.7 Summary 247

12 Waves in the Plasmasphere – 3. Ion Cyclotron Whistlers
12.1 Introduction 249
12.2 Observations 249
12.3 Ion Cyclotron Whistler Propagation in a Uniform Medium 251
12.4 Nature of Ion Cyclotron Whistler Generation 255
12.5 The Effect of Collisions on Coupling 260
12.6 Summary 264

13 Waves in the Plasmasphere – 4. Doppler Shifted Cyclotron Resonance of Electrons with Whistlers
13.1 Introduction 265
13.2 Some Relevant Observations 266
13.3 Whistler Wave–Particle Interaction in a Uniform Medium 268
13.4 Whistler-Mode Noise in a Non-Uniform Medium 277
13.5 Whistler-Mode Signals Generated by Energetic Particles . 283
13.6 Summary 287

14 Waves in the Auroral Region 289
14.1 Introduction 289
14.2 Observations 289
14.3 Propagation of Whistler Mode Hiss 293

14.4 Z-Mode Radiation 294
14.5 Terrestrial Myriametric Radiation 304
14.6 Auroral Kilometric Radiation 307
14.7 Summary 315

15 Some Final Words
15.1 Introduction 317
15.2 Man-Made Plasma Waves 317
15.3 Ultra Low Frequency Pulsations 318
15.4 Waves in the Magnetotail, Magnetosheath and Solar Wind 318
15.5 Future Work 318

A The Essence of Cartesian Tensors 321

B Some Mathematical Results 323
B.1 Properties of Bessel Functions and Related Results 323
B.2 The Plasma Dispersion Function 327

C Properties of the Earth's Dipole Field 330

D Definition of Symbols 331

Bibliography 335

Subject Index 343

Part I

Propagation and Generation of Plasma Waves

Chapter 1

Basic Equations

1.1 Introduction

A plasma is a gas in which a substantial number of the molecules are ionized and show collective behavior. This collective behavior occurs provided that there are a sufficient number of particles within a sphere of radius equal to the Debye length,

$$\lambda_D = (\epsilon_0 KT/Nq^2)^{1/2} \tag{1.1}$$

All symbols are defined in Appendix D. On scales small compared with λ_D the individual particle behavior is important. On scales larger than λ_D the Coulomb potential of a single particle is screened by charges of the opposite sign. The unperturbed plasma is electrically neutral over length scales greater than λ_D. Although neutral, because of the ionization, a plasma is a conductor, and this leads to behavior which is far more complicated than that of an un-ionized gas. A plasma permeated by a background magnetic field is called a magnetoplasma.

A wide variety of wave phenomena can occur in plasmas. This book deals with the propagation and generation of such waves in the magnetosphere. In Part I we discuss the theory of such waves. In Part II the theory is applied to observations of magnetospheric plasma waves.

The equations describing the waves in a plasma are essentially the equations of motion of the charged plasma particles under the action of the Lorentz force, together with Maxwell's equations. When studying such waves the equations of motion must be suitably averaged over the particles. The nature of this averaging process depends on the nature of the problem.

Frequently the waves turn out to have speeds near the speed of light. In this case the thermal velocities of the particles are usually very much less than the wave velocity. In the averaging process we can ignore the thermal motions of the particles and treat each species of particle as a continuous charged fluid. This is the cold plasma approximation which is introduced in the next chapter.

When the waves have speeds comparable with the thermal speeds of the particles the averaging process is more complicated and use must be made of the distribution function as is done in later chapters.

The waves are often classified as (1) electromagnetic waves which have electric and magnetic fields approximately perpendicular to the wave normal, and (2) electrostatic waves which have negligible magnetic field and electric field approximately parallel to the wave normal. It should be emphasized that this terminology simply refers to particular approximations which can be made in certain circumstances. There is no fundamental difference between electrostatic and electromagnetic waves; as conditions change one type can deform continuously into another and they cannot always conveniently be decoupled. Both types of waves can be regarded as plasma waves.

In this chapter we introduce the basic equations which we shall use later in the book. They are Maxwell's equations, the Lorentz force law and the kinetic equation which determines how the particle distribution function develops in time. The basic ideas of electromagnetism are assumed familiar. The nature of particle distribution functions and their evolution will be discussed in more detail.

1.2 Electromagnetic Equations

1.2.1 Maxwell's Equations

We shall take as our basic electromagnetic equations Maxwell's equations in their free space form. All interaction with matter will be through the charge and current densities. The equations are then:

$$\nabla\times\mathbf{E} = -\frac{\partial\mathbf{B}}{\partial t} \tag{1.2}$$

$$\nabla\times\mathbf{B} = \mu_0\left\{\mathbf{J}+\epsilon_0\frac{\partial\mathbf{E}}{\partial t}\right\} \tag{1.3}$$

$$\nabla.\mathbf{E} = \rho/\epsilon_0 \tag{1.4}$$

$$\nabla.\mathbf{B} = 0 \tag{1.5}$$

These equations are not sufficient in themselves to describe the electromagnetic fields in a plasma. If **J** is given then **E** and **B** may be found; what we need, however is a self-consistent solution for **E**, **B** and **J** in the plasma. This requires the introduction of another equation, the *constitutive relation* for the medium, which relates two of the variables. In a plasma the most straightforward procedure is to find a conductivity relationship between **J** and **E**. This is done by considering the force exerted on a single particle by the electromagnetic field and averaging it over all the particles to find a generalized version of Ohm's law. For a cold plasma this averaging process is quite simple. For plasmas in which temperature effects are important the averaging process is more difficult. The principle is, however, the same in each case. An alternative procedure for finding constitutive relations is to treat the plasma as a dielectric. Then the constitutive relation relates the polarization **P** to **E** through a susceptibility relation or the displacement **D** to **E** through a permittivity relation.

The problem of wave propagation is generally only tractable if the equations to be solved are linear. This is so if the wave disturbance is small enough. In this case the constitutive relation between **J** and **E** must be linear. The simplest linear relation is isotropic:

$$\mathbf{J} = \sigma \mathbf{E}$$

where σ may be a function of position but not of direction of wave propagation. The plasmas we shall deal with are usually anisotropic because the background magnetic field introduces a preferred direction. The most general anisotropic linear relationship is a tensor one. In the notation of Cartesian tensors (Appendix A) it can be written

$$J_j = \sigma_{jk} E_k \tag{1.6}$$

The problem of finding what waves can be propagated by the plasma is then the problem of finding the conductivity tensor σ_{jk}. It is worth noting that, in the absence of loss mechanisms such as collisions the tensor is expected to be Hermitian

$$\sigma_{jk} = \sigma_{kj}^*$$

1.2.2 The Lorentz Force Law

Whatever the model of the plasma it is necessary to perform an averaging procedure over all the particles. Each particle moves under the action of the electromagnetic field; it is this motion which is suitably averaged. The force on a single charged particle, mass m, is given by the Lorentz force law. The equation of motion is then

$$m\dot{\mathbf{v}} = q\{\mathbf{E} + \mathbf{v} \times \mathbf{B}\} \tag{1.7}$$

We adopt the convention that q is a charge which may be positive or negative. The value of the electron charge is

$$e = 1.602 \times 10^{-19}\ C$$

Thus $q_e = -e$ and $q_i = +pe$ where p is the degree of ionization. Ions are usually singly charged so that we shall almost always take $p = 1$.

When the thermal speed is negligible compared with the wave speed the unperturbed particles may be assumed to be at rest. Each particle of a given species then moves according to (1.7). If the particle density is large enough we can consider a volume small compared with the spatial scale on which **E** and **B** vary, and large compared with the interparticle distance. Then we can perform the averaging by replacing each species of particle within such a volume by a continuous fluid with mass and charge densities equal to the average values for the particle species. Each fluid element δV for a particular species experiences a force (1.7) with $m = \rho_m \delta V$ and $q = \rho_q \delta V$. When the thermal speed is not negligible the Lorentz force must be used in combination with the kinetic equation as described in the next section.

1.2.3 Collisions with Neutral Particles

In most of the magnetosphere the plasma is fully ionized and we shall not need to take account of collisions. In the ionosphere the plasma is not fully ionized and collision processes may be important. A full treatment is complicated. A good account is given by Budden [19, §3.12]. For our purposes we can consider a simple theory in which a charged particle is accelerated by the electromagnetic fields until it collides with a neutral particle when, on average, it is brought to rest. If the collision frequency is ν its acceleration is $-\nu\mathbf{v}$, where $\mathbf{v}$ is the average velocity. We approximate the effect of collisions by adding an extra force term to the Lorentz force, getting

$$m\dot{\mathbf{v}} = q\{\mathbf{E} + \mathbf{v} \times \mathbf{B}\} - m\nu\mathbf{v} \tag{1.8}$$

The approximation is crude but adequate. We shall seldom need to make use of this formula.

1.3 Fluid Equations

When the thermal speed of the particles is not small compared with the wavelength a possible approach is again to treat each species of particle as a continuous charged fluid but include the effects of temperature and pressure. In this case we write the equation of motion of each species in the form

$$\rho_m \frac{D\mathbf{v}}{Dt} = Nq\mathbf{E} + \mathbf{J} \times \mathbf{B} - \nabla P \tag{1.9}$$

where N is the number density of the particles and the mass density is $\rho_m = Nm$. The derivative with respect to time is taken following an element of the fluid. In a coordinate system in which the fluid moves with velocity $\mathbf{v}(\mathbf{r})$ it is given by

$$\left\{\frac{D}{Dt}\right\} \equiv \left\{\frac{\partial}{\partial t} + \mathbf{v}.\nabla\right\} \tag{1.10}$$

Now that the pressure term is included it is necessary to relate p to the fluid velocity. The continuity equation requires that

$$\frac{\partial \rho_m}{\partial t} + \nabla.\rho_m\mathbf{v} = 0 \tag{1.11}$$

or

$$\frac{D\rho_m}{Dt} + \rho_m \nabla . \mathbf{v} = 0 \tag{1.12}$$

Finally an equation of state relating p to ρ is required. In many circumstances changes occur adiabatically and the equation of state is

$$\frac{D}{Dt}\left(\frac{P}{\rho_m^\gamma}\right) = 0 \tag{1.13}$$

where the ratio of specific heats $\gamma = (n+2)/n$ and n is the number of degrees of freedom. There are some circumstances in which the frequency is so low that the electrons respond fast enough for isothermal conditions to hold so that P is proportional to ρ_m. In this case we can still use (1.13) with $\gamma = 1$. Another way of writing this equation is

$$\frac{DP}{D\rho_m} = \frac{\gamma P}{\rho_m} = V_S^2 \tag{1.14}$$

where V_S has the dimensions of speed and is the sound speed for the fluid.

1.4 The Kinetic Equation

1.4.1 The Distribution Function

When the thermal speeds of the particles cannot be neglected the problem of finding mean values of quantities like charge or current density involves knowing the distribution of particle velocities. In order to specify this distribution it is convenient to work in a six-dimensional *phase space*. Three of the coordinates are the coordinates x, y, z of *configuration space* and the other three are the coordinates, v_x, v_y, v_z of *velocity space*. When a particle has an x-component between x and $x + dx$ we say for brevity that it lies in the range x, dx. Consider an element of volume, $d^3\mathbf{x}\, d^3\mathbf{v} \equiv dx\, dy\, dz\, dv_x\, dv_y\, dv_z$, located at the position $(\mathbf{r}, \mathbf{v})$ in this phase space as illustrated in Figure 1.1a. At some instant t the volume element contains all those particles whose positions lie in the range x, dx; y, dy; z, dz; v_x, dv_x; v_y, dv_y; v_z, dv_z. We assume that the volume element is large enough to contain many particles. At the same time we assume that the density of particles in phase space varies sufficiently slowly so that it is approximately constant over the volume. In these circumstances we can replace the individual particles by a continuous fluid. Let the density of this fluid be $f(\mathbf{r}, \mathbf{v}, t) \equiv f(x, y, z, v_x, v_y, v_z, t)$. Then the number of particles lying within the volume element is

$$f(\mathbf{r}, \mathbf{v}, t) d^3\mathbf{x}\, d^3\mathbf{v}$$

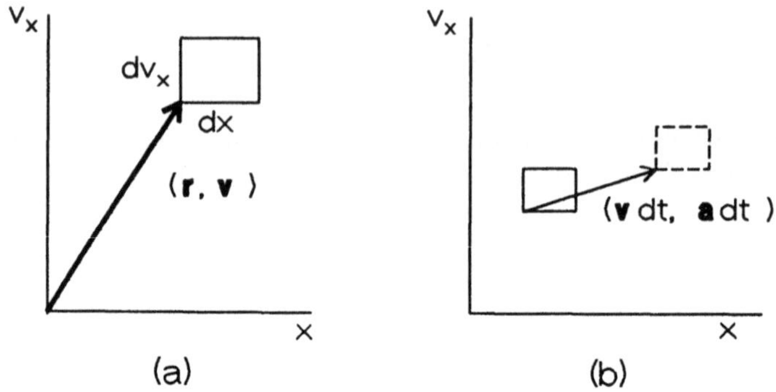

Figure 1.1: (a) Element of volume in phase space. Only the x and v_x components are shown. (b) Motion of an element of plasma in phase space

The number density of particles in configuration space is found by integrating over all velocity space:

$$N(x,y,z,t) = \int_{-\infty}^{\infty}\int_{-\infty}^{\infty}\int_{-\infty}^{\infty} f(x,y,z,v_x,v_y,v_z,t)dv_x\,dv_y\,dv_z \tag{1.15}$$

For brevity this can be written

$$N(\mathbf{r},t) = \int_{-\infty}^{\infty} f(\mathbf{r},\mathbf{v},t)d^3\mathbf{v} \tag{1.16}$$

The function f is called the *velocity distribution function* or simply the *distribution function* of the plasma.

1.4.2 Mean Values of Plasma Properties

To find the mean value of a property of the particle motion such as charge or current density at a point in configuration space, we simply compute the value of the quantity in an element of velocity space and integrate it over velocity. For example the charge density in $\mathbf{v}$, $d^3\mathbf{v}$ is $qf(\mathbf{v},t)\,d^3\mathbf{v}$ and thus

$$\rho(\mathbf{r},t) = \int_{-\infty}^{\infty} qf(\mathbf{r},\mathbf{v},t)\,d^3\mathbf{v} \tag{1.17}$$

where q is the charge on a particle. Similarly the current density is

$$\mathbf{J}(\mathbf{r},t) = \int_{-\infty}^{\infty} q\mathbf{v}f(\mathbf{r},\mathbf{v},t)\,d^3\mathbf{v} \tag{1.18}$$

In general the mean value of any function $G(\mathbf{r},\mathbf{v},t)$ at $\mathbf{r}$ is

$$\langle G(\mathbf{r},t)\rangle = \frac{1}{N}\int_{-\infty}^{\infty} G(\mathbf{r},\mathbf{v},t)f(\mathbf{r},\mathbf{v},t)\,d^3\mathbf{v} \tag{1.19}$$

1.4.3 The Boltzmann and Vlasov Equations

We now derive the kinetic equation which describes how the distribution function f evolves in time. Consider the element of volume in phase space shown in Figure 1.1b. It is occupied by a number of particles, each of which is subject to an acceleration determined by the resultant force acting on it. The forces acting on a particle can be placed in two classes. The first type is a mean macroscopic force arising from external fields or from the integrated effect of all the particles. This force varies on a spatial scale large compared with $d^3\mathbf{v}$ and is approximately constant over $d^3\mathbf{v}$. Each particle within $d^3\mathbf{v}$ experiences the same acceleration due to this force. In the case of a plasma this is generally the Lorentz force (1.7). The second type is a small scale force arising from collisions with nearby particles. In the case of neutral particles this is the result of a series of impulses arising from binary collisions. For charged particles the situation is more complicated as the Coulomb force is a long range force and a particle may interact with a number of other particles simultaneously. Inside a Debye sphere there will be a fluctuating field arising from the interaction with other particles; outside a Debye sphere the particle is effectively screened from the Coulomb force.

If collisions can be neglected each particle within $d^3\mathbf{r}\, d^3\mathbf{v}$ has the same acceleration $\mathbf{a}$. Suppose that the volume element moves with the same acceleration. After time dt its position has changed from $\mathbf{r}$ to

$$\mathbf{r}' = \mathbf{r} + d\mathbf{r} = \mathbf{r} + \mathbf{v}\, dt \tag{1.20}$$

and its velocity from $\mathbf{v}$ to

$$\mathbf{v}' = \mathbf{v} + d\mathbf{v} = \mathbf{v} + \mathbf{a}\, dt = \mathbf{v} + \frac{q}{m}(\mathbf{E} + \mathbf{v} \times \mathbf{B})dt \tag{1.21}$$

The change in f is

$$df = \frac{\partial f}{\partial t}dt + (\nabla f).d\mathbf{r} + (\nabla_v f).d\mathbf{v} = \left\{\frac{\partial f}{\partial t} + \mathbf{v}.\nabla f + \mathbf{a}.\nabla_v f\right\} dt$$

In the absence of collisions all the particles occupying the volume element at time t will also occupy it at time $t + dt$. The number of particles, $f d^3\mathbf{r} d^3\mathbf{v}$, in the volume element is therefore constant. It follows that, if the size of the volume element does not change then the density, f, is constant as it moves through phase space. Since the acceleration $\mathbf{a}$ is velocity-dependent as a result of the magnetic force, it is not obvious that the size of the volume element does not change. Let the volume elements in configuration space and velocity space change from $d^3\mathbf{r}$, $d^3\mathbf{v}$ to $d^3\mathbf{r}'$, $d^3\mathbf{v}'$ as t advances to $t + dt$. Then

$$d^3\mathbf{r}' = \frac{\partial(x', y', z')}{\partial(x, y, z)} d^3\mathbf{r} \text{ and } d^3\mathbf{v}' = \frac{\partial(v_x', v_y', v_z')}{\partial(v_x, v_y, v_z)} d^3\mathbf{v}$$

where

$$\frac{\partial(x',y',z')}{\partial(x,y,z)} = \begin{vmatrix} \frac{\partial x'}{\partial x} & \frac{\partial x'}{\partial y} & \frac{\partial x'}{\partial z} \\ \frac{\partial y'}{\partial x} & \frac{\partial y'}{\partial y} & \frac{\partial y'}{\partial z} \\ \frac{\partial z'}{\partial x} & \frac{\partial z'}{\partial y} & \frac{\partial z'}{\partial z} \end{vmatrix}$$

and

$$\frac{\partial(v'_x,v'_y,v'_z)}{\partial(v_x,v_y,v_z)} = \begin{vmatrix} \frac{\partial v'_x}{\partial v_x} & \frac{\partial v'_x}{\partial v_y} & \frac{\partial v'_x}{\partial v_z} \\ \frac{\partial v'_y}{\partial v_x} & \frac{\partial v'_y}{\partial v_y} & \frac{\partial v'_y}{\partial v_z} \\ \frac{\partial v'_z}{\partial v_x} & \frac{\partial v'_z}{\partial v_y} & \frac{\partial v'_z}{\partial v_z} \end{vmatrix}$$

are the Jacobians of the transformations. From (1.20) and (1.21) we get

$$\frac{\partial(x',y',z')}{\partial(x,y,z)} = \begin{vmatrix} 1 & 0 & 0 \\ 0 & 1 & 0 \\ 0 & 0 & 1 \end{vmatrix} = 1$$

and

$$\frac{\partial(v'_x,v'_y,v'_z)}{\partial(v_x,v_y,v_z)} = \begin{vmatrix} 1 & \frac{qB_z}{m}dt & -\frac{qB_y}{m}dt \\ -\frac{qB_z}{m}dt & 1 & \frac{qB_x}{m}dt \\ \frac{qB_y}{m}dt & -\frac{qB_x}{m}dt & 1 \end{vmatrix} = 1 + O(dt^2)$$

Thus to first order in dt the Jacobians are unity and the size of the volume element in phase space does not change. We conclude that, in the absence of collisions, the density, f, is unchanged as the volume element moves through phase space.

Consider now the effect of a collision. If the particle collides with another particle the momentum of each particle and hence its velocity is abruptly changed. It moves discontinuously from one region of phase space to another.

The change in f as the position of the volume element goes from $\mathbf{r}$ to $\mathbf{r}+d\mathbf{r}$ and its velocity from $\mathbf{v}$ to $\mathbf{v}+d\mathbf{v}$ thus arises entirely from collisions. We represent the effect of collisions by a term $(\partial f/\partial t)_c$. Then

$$\frac{df}{dt} \equiv \frac{\partial f}{\partial t} + \mathbf{v}.\nabla f + \mathbf{a}.\nabla_v f = \left(\frac{\partial f}{\partial t}\right)_c \tag{1.22}$$

This is the Boltzmann equation giving the time evolution of f. In almost all cases of interest in this book the mean free path is so large that collisions can be neglected and the acceleration is given by the Lorentz force. In this case we have the collisionless form of the Boltzmann equation which is known as the Boltzmann-Vlasov or simply the Vlasov equation:

$$\frac{\partial f}{\partial t} + \mathbf{v}.\nabla f + \frac{q}{m}(\mathbf{E} + \mathbf{v} \times \mathbf{B}).\nabla_v f = 0 \tag{1.23}$$

1.5 Poynting's Theorem

We shall need to know how energy is transported by waves. Maxwell's equations can be combined to give the well-known Poynting theorem which expresses the conservation of energy in an electromagnetic field:

$$\nabla.\left(\frac{\mathbf{E} \times \mathbf{B}}{\mu_0}\right) = -\mathbf{J}.\mathbf{E} - \frac{\partial}{\partial t}\left(\frac{\epsilon_0 E^2}{2} + \frac{B^2}{2\mu_0}\right) \tag{1.24}$$

The Poynting vector, $\mathbf{\Pi} \equiv \mathbf{E} \times \mathbf{B}/\mu_0$, is interpreted as the energy flux. When integrated over a closed surface it gives the electromagnetic field energy flowing outwards through the surface per unit time. The Joule heating term, $\mathbf{J}.\mathbf{E}$, represents the rate at which the field does work on matter per unit volume. The field energy density is given by $U = (\epsilon_0 E^2/2) + (B^2/2\mu_0)$.

1.6 Harmonic Oscillations

1.6.1 Complex Notation

We shall often assume that the only time dependence of the field quantities is through a factor $\exp(-i\omega t)$. Normally this factor is assumed in the equations and is not written explicitly. The field amplitudes are then complex quantities. The actual physical quantities may be retrieved by taking the real part of the complex field components. We may only carry out linear operations on these complex fields. Before performing nonlinear operations, such as the evaluation of the Poynting vector or energy density, we must take the real parts of the field quantities.

The operator $\partial/\partial t$ in Maxwell's equations and the equation of motion of the particles can now be replaced using the substitution

$$\partial/\partial t \equiv -i\omega \tag{1.25}$$

The linearized equations then become

$$\nabla\times\mathbf{E} = i\omega\mathbf{B} \tag{1.26}$$

$$\nabla\times\mathbf{B} = \mu_0\mathbf{J} - i\frac{\omega}{c^2}\mathbf{E} \tag{1.27}$$

$$\mathbf{v} = \frac{iq}{m\omega}\{\mathbf{E} + \mathbf{v} \times \mathbf{B}\} \tag{1.28}$$

with **J** being found from (1.6). Equation (1.5) follows by taking the divergence of (1.26). If we take the divergence of (1.27) and use (1.4) we get

$$\nabla.\mathbf{J} = i\omega\rho_q$$

which is the equation of charge continuity.

1.6.2 Time-Averaged Energy Density and Flux

The Poynting vector and field energy density in a harmonically varying electromagnetic field are non-linear combinations of the field variables. If **E** and **B** are the complex fields then we must take their real part before evaluating such non-linear quantities. Denote the complex conjugate of a complex quantity by an asterisk. Then the real part of any complex quantity, F, is given by

$$\Re(F) = \tfrac{1}{2}(F + F^*)$$

Now consider the product of two complex vectors, **F** and **G**:

$$\Re(\mathbf{F}) \circ \Re(\mathbf{G})$$

where $\circ$ represents a scalar or vector product. Each can be written as the sum of a constant real zero-order part and a harmonically varying first order part:

$$\mathbf{F} \equiv \mathbf{F}_0 + \mathbf{F}_1 e^{-i\omega t}$$

Then

$$\begin{aligned}
\Re(\mathbf{F}) \circ \Re(\mathbf{G}) &= \tfrac{1}{2}(\mathbf{F} + \mathbf{F}^*) \circ \tfrac{1}{2}(\mathbf{G} + \mathbf{G}^*) \\
&= \tfrac{1}{4}[\mathbf{F} \circ \mathbf{G} + \mathbf{F}^* \circ \mathbf{G}^* + \mathbf{F} \circ \mathbf{G}^* + \mathbf{F}^* \circ \mathbf{G}] \\
&= \tfrac{1}{2}[\Re(\mathbf{F} \circ \mathbf{G}) + \Re(\mathbf{F} \circ \mathbf{G}^*)] \\
&= \tfrac{1}{2}\Re[\mathbf{F}_0 \circ \mathbf{G}_0 + (\mathbf{F}_1 \circ \mathbf{G}_0 + \mathbf{F}_0 \circ \mathbf{G}_1)e^{-i\omega t} + \mathbf{F}_1 \circ \mathbf{G}_1 e^{-2i\omega t}] \\
&\quad + \tfrac{1}{2}\Re[\mathbf{F}_0 \circ \mathbf{G}_0 + (\mathbf{F}_1 \circ \mathbf{G}_0 e^{-i\omega t} + \mathbf{F}_0 \circ \mathbf{G}_1^* e^{i\omega t}) + \mathbf{F}_1 \circ \mathbf{G}_1^*]
\end{aligned} \quad (1.29)$$

In wave propagation we are interested only in time averaged quantities. The factors $e^{\pm i\omega t}$ and $e^{-2i\omega t}$ average to zero over one complete cycle. Thus

$$\Re(\mathbf{F}) \circ \Re(\mathbf{G}) = \mathbf{F}_0 \circ \mathbf{G}_0 + \tfrac{1}{2}\Re(\mathbf{F}_1 \circ \mathbf{G}_1^*) \quad (1.30)$$

In particular the time-averaged Poynting vector for harmonically varying fields is given by

$$\langle \Pi \rangle = \tfrac{1}{2}\Re[\mathbf{E} \times \mathbf{B}^*/\mu_0] \quad (1.31)$$

the Joule heating by

$$\langle W \rangle = \tfrac{1}{2}\Re[\mathbf{J}.\mathbf{E}^*] \quad (1.32)$$

and the field energy density by

$$\langle U \rangle = \tfrac{1}{2}\Re[\tfrac{1}{2}\epsilon_0|E|^2 + \tfrac{1}{2}|B|^2/\mu_0] \quad (1.33)$$

1.7 The Wave Equation

Assume a linear conductivity relationship of the form (1.6) in Maxwell's curl equations (1.26), (1.27). Take the curl of (1.26) and substitute for $\nabla\times\mathbf{B}$ from (1.27). The result is

$$\nabla\times(\nabla\times\mathbf{E}) = i\omega\mu_0\mathbf{J} + \omega^2/c^2\mathbf{E} \tag{1.34}$$

where $c = 1/\sqrt{\mu_0\epsilon_0}$ is the speed of light in free space. This is a form of the wave equation.

We make use of the vector identity $\nabla\times(\nabla\times\mathbf{E}) = \nabla\nabla.\mathbf{E} - \nabla^2\mathbf{E}$ and write this in Cartesian tensor notation

$$\left(\frac{\partial}{\partial x_j}\frac{\partial}{\partial x_k} - \delta_{jk}\frac{\partial}{\partial x_l}\frac{\partial}{\partial x_l} - (\omega^2/c^2)\delta_{jk}\right) E_k = i\omega\mu_0 J_j \tag{1.35}$$

This is a wave equation for the electric field $\mathbf{E}$. If we use the conductivity relation between J_j and E_j it becomes

$$\left(\frac{\partial}{\partial x_j}\frac{\partial}{\partial x_k} - \delta_{jk}\frac{\partial}{\partial x_l}\frac{\partial}{\partial x_l} - i\omega\mu_0\sigma_{jk} - (\omega^2/c^2)\delta_{jk}\right) E_k = 0 \tag{1.36}$$

1.8 Summary

- A plasma is an ionized, electrically conducting gas, consisting of equal densities of positive and negative ions. It exhibits collective behavior on a length scale greater than the Debye length.

- The fundamental equations describing plasma behavior are (i) Maxwell's equations describing the electromagnetic fields arising from the charge and current density distributions, (ii) the Lorentz force law describing the force exerted on a single particle by an electromagnetic field and (iii) an appropriate constitutive relation giving the current and charge densities in terms of the fields by averaging the Lorentz force over the particles in a manner appropriate to the problem.

- For plasmas with sufficiently small thermal energy the thermal speed of the particles can be neglected. In this cold plasma approximation the averaging process is carried out by assuming that all the particles in a fluid element move with the same velocity.

- For plasmas in which the thermal speed is small but not negligible a warm plasma approximation is possible. In this approximation the cold plasma fluid equations are modified by an additional pressure term.

- For plasmas with significant thermal energy a full treatment using kinetic theory is necessary. The average properties of the particles are expressed in terms of a velocity distribution function which obeys the Boltzmann-Vlasov equation.

- Subsidiary equations which are important are the equations of energy conservation, expressed in the form of Poynting's theorem, and the wave equation.

Chapter 2

Waves in a Uniform Cold Magnetoplasma – 1. Infinite Plane Waves

2.1 Introduction

In this chapter we begin the study of the electromagnetic waves which are propagated by a cold magnetoplasma. Our model of the plasma is a continuous electron fluid and several ion fluids, each with uniform density. Because the thermal speeds are assumed negligible it is unnecessary to use the Boltzmann equation approach; the unperturbed particles are assumed to be at rest. There is a uniform background magnetic field $\mathbf{B}_0$. The number density of electrons is N_e and that of the ith species of ions N_i. The electron charge is q_e and the charge on an ion of the ith species is q_i. Ions are singly charged so that $q_e = -e$ and $q_i = +e$. Then charge neutrality requires that, in equilibrium,

$$N_e = \sum_{species} N_i$$

Generally we shall assume that collisions, whether between charged and neutral particles, electrons and ions, or ions of different species, can be neglected. This assumption is justified over the whole magnetosphere with the exception of the D- and E-regions of the ionosphere where collisions with neutral particles are important.

The continuous fluids are modeled by assuming that all the particles of each species within a small volume δV have the same average perturbation velocity. The fields are averaged over δV which is assumed to be large enough to contain a very large number of particles but to have linear dimensions small compared to the wavelength and other spatial scales in the problem. The mass density of such a fluid is then Nm, the charge density is Nq, and the current density is $Nq\mathbf{v}$.

We introduce important natural frequencies of oscillation and characteristic speeds associated with the medium. By averaging the Lorentz force over the particles we deduce the appropriate constitutive relations connecting current density and electric field in the wave and make appropriate approximations for various limiting cases. This allows us to derive the dispersion relation for plane waves. The properties of this dispersion relation determine the properties of infinite plane waves propagated by the medium.

2.2 Characteristic Frequencies and Speeds

2.2.1 The Plasma Frequency

The plasma frequency is the characteristic frequency with which a fluid consisting of electrons and ions will oscillate when the particles are displaced relative to one another. It occurs naturally in the equations describing electromagnetic wave propagation in plasmas. Consider a simple example. A rectangular "block" of plasma consists of an equal number of positive ions and electrons with number density N, charges $q_i = e$, $q_e = -e$, and masses m_i, m_e. Ions and electrons are displaced distances r_i and r_e relative to their center of mass so that $m_e r_e + m_i r_i = 0$. Their relative displacement is $r = r_e - r_i$. This results in a surface charge density $\sigma = \pm Ner$ and an electric field $E = Ner/\epsilon_0$. As a result there is a restoring force on each particle species and their equations of motion are

$$\begin{aligned} m_e \ddot{r}_e &= -\frac{Ne^2}{\epsilon_0}(r_e - r_i) \\ m_i \ddot{r}_i &= -\frac{Ne^2}{\epsilon_0}(r_i - r_e) \end{aligned}$$

Subtracting these we get

$$\ddot{r} = -\left\{\frac{Ne^2}{\epsilon_0 m_e} + \frac{Ne^2}{\epsilon_0 m_i}\right\} r$$

This is simple harmonic motion with angular frequency given by

$$\omega_p^2 = \omega_{p,e}^2 + \omega_{p,i}^2 = \frac{Ne^2}{\epsilon_0 m_e} + \frac{Ne^2}{\epsilon_0 m_i} \tag{2.1}$$

It is possible to generalize this to a plasma with more than one species of positive ion. The quantities $\omega_{p,e}$ and $\omega_{p,i}$ are called the electron plasma frequency and ion plasma frequency respectively. Of course $m_e \ll m_i$ and thus

$$\omega_p \sim \omega_{p,e} \tag{2.2}$$

In spite of this we cannot neglect the ion plasma frequencies. An external electric field oscillating near an ion plasma frequency will set the corresponding species of ion into oscillation. We expect these natural frequencies to appear in the equations governing wave propagation in the plasma.

2.2.2 The Cyclotron or Gyrofrequency

Consider a particle with charge $q = \pm e$ moving with velocity $\mathbf{v}_\perp$ perpendicular to a uniform magnetic field $\mathbf{B}_0$. It experiences a centripetal force $q\mathbf{v} \times \mathbf{B}_0$ and moves in a circle of radius ρ and period T. Its equation of motion is

$$\frac{mv_\perp^2}{\rho} = qv_\perp B_0$$

Its angular frequency is

$$\Omega = \frac{2\pi}{T} = \frac{v_\perp}{\rho}$$

and this leads to

$$\Omega = \frac{B_0|q|}{m} = \frac{B_0 e}{m} \tag{2.3}$$

Ω is called the angular gyrofrequency or cyclotron frequency. It is independent of $v_\perp$ and hence of the particle energy; all particles of the same type have the same gyrofrequency and it is thus a characteristic frequency of the plasma. It is often convenient to write Ω as a vector:

$$\boldsymbol{\Omega} = -\mathbf{B}_0 q/m \tag{2.4}$$

The minus sign is chosen so that the vector ω represents the angular velocity of the particle; postive particles gyrate in a left-handed and negative in a right-handed sense about $\mathbf{B}_0$. This means that the plane of the orbit of the particle is defined by the direction of $\boldsymbol{\Omega}$. Note that, if the particle is positively charged, $\boldsymbol{\Omega}$ is antiparallel to $\mathbf{B}_0$, and, if it is negatively charged, parallel to $\mathbf{B}_0$.

2.2.3 Alfvén and Åström Speeds

When a conducting fluid moves in a magnetic field currents are generated which modify the field. For sufficiently low frequencies and sufficiently large conductivity the field is modified in such a way that the field lines are "frozen in" to the fluid. In these circumstances waves are propagated with speeds which are of the order of the ratio of the magnetic energy density $B_0^2/2\mu_0$ to the fluid density ρ_m. Define the *normalized Alfvén speed* U_A by

$$U_A^2 = \frac{B_0^2}{\mu_0 \rho_m c^2} \tag{2.5}$$

If we note that, for a plasma, $\rho_m = N(m_e + m_i) \simeq N m_i$ and that $\mu_0 c^2 = 1/\epsilon_0$ then we can write

$$U_A^2 = \frac{B_0^2 e^2}{m_e m_i} \frac{\epsilon_0 m_e}{N e^2} = \frac{\Omega_e \Omega_i}{\omega_{p,e}^2} \tag{2.6}$$

Such waves were first described by Alfvén [2] for conducting fluids with very large conductivity. The case of a collisionless plasma was treated by Åström [6] who obtained the same expression for the characteristic speed. The distinction between Alfvén's and Åström's treatments is discussed in detail by Hines [58]. By analogy we can define a normalized Åström speed for electrons

$$U_{A,e}^2 = \frac{B_0^2 e^2}{m_e m_e} \frac{\epsilon_0 m_e}{N e^2} = \frac{\Omega_e^2}{\omega_{p,e}^2} \tag{2.7}$$

2.3 Linearization of the Equation of Motion

We assume that the disturbance in the plasma is a small perturbation in the background state. The only zero order field quantity is $\mathbf{B}_0$. The zero order number density of each species is taken as $N_{e,i}$. The perturbations in the fields, velocities and number densities are then $\mathbf{E}$, $\mathbf{B}$, $\mathbf{v}_{e,i}$ and $n_{e,i}$. We neglect terms of second order in these quantities.

The Lorentz force for each species may then be written

$$m\dot{\mathbf{v}} = q\{\mathbf{E} + \mathbf{v} \times \mathbf{B}_0\} \tag{2.8}$$

where the second order term $\mathbf{v} \times \mathbf{B}$ has been neglected.

This allows the velocity of each particle species to be found. The total current density is then given by

$$\mathbf{J} = N_e q_e \mathbf{v}_e + \sum_i N_i q_i \mathbf{v}_i \tag{2.9}$$

The contribution of the convection of the charge density perturbations to the current density, $n_{e,i} q_{e,i} \mathbf{v}_{e,i}$, is of second order and may be neglected.

The net charge perturbation is

$$\rho = (N_e + n_e) q_e + \sum_i (N_i + n_i) q_i = n_e q_e + \sum_i n_i q_i$$

This is of first order in n_e and n_i. These quantities, however, do not occur anywhere except in the Maxwell equation (1.4). This equation therefore is not needed when finding the properties of the wave fields. It is only used if the actual charge density perturbation is required.

2.4 Constitutive Relations

2.4.1 Conductivity Tensor

We now average over the particles to obtain the conductivity relation for the medium. The same procedure can be carried out for electrons and each species of ion. We drop the subscripts e and i when we do not wish to distinguish between them. Multiply (1.28) by Nq and use (2.1) and (2.4)

$$-i\omega\mathbf{J} = \epsilon_0\omega_p^2\mathbf{E} - \mathbf{J}\times\boldsymbol{\Omega} \tag{2.10}$$

Take the vector and scalar products of this equation with $\boldsymbol{\Omega}$ and use the results to eliminate $\mathbf{J}\times\boldsymbol{\Omega}$ and $\mathbf{J}.\boldsymbol{\Omega}$. The result is

$$\mathbf{J} = \frac{i\epsilon_0\omega\omega_p^2}{\omega^2-\Omega^2}\left\{\mathbf{E} + \frac{i\boldsymbol{\Omega}\times\mathbf{E}}{\omega} + \frac{\boldsymbol{\Omega}(\boldsymbol{\Omega}.\mathbf{E})}{\omega^2}\right\} \tag{2.11}$$

This is of the form

$$J_j = \sigma_{jk}E_k \tag{2.12}$$

The tensor σ_{jk} is called the conductivity tensor for the particle species. It can be written

$$\sigma_{jk} = \frac{i\epsilon_0\omega\omega_p^2}{\omega^2-\Omega^2}\left(\delta_{jk} + \frac{i\epsilon_{jkl}\Omega_l}{\omega} - \frac{\Omega_j\Omega_k}{\omega^2}\right) \tag{2.13}$$

for the particular particle species. Note that care must be taken with the sign of Ω_j depending on the sign of the charge. We can write (2.12) explicitly as a matrix equation. We choose the z-axis to be parallel to the background magnetic field $\mathbf{B}$. Then for electrons

$$\begin{pmatrix} J_{e,x} \\ J_{e,y} \\ J_{e,z} \end{pmatrix} = -i\omega\epsilon_0\frac{\omega_{p,e}^2}{\omega^2}\begin{pmatrix} -\frac{\omega^2}{\omega^2-\Omega_e^2} & \frac{i\Omega_e\omega}{\omega^2-\Omega_e^2} & 0 \\ -\frac{i\Omega_e\omega}{\omega^2-\Omega_e^2} & -\frac{\omega^2}{\omega^2-\Omega_e^2} & 0 \\ 0 & 0 & -1 \end{pmatrix}\begin{pmatrix} E_x \\ E_y \\ E_z \end{pmatrix} \tag{2.14}$$

A similar equation can be written for ions

$$\begin{pmatrix} J_{i,x} \\ J_{i,y} \\ J_{i,z} \end{pmatrix} = -i\omega\epsilon_0\frac{\omega_{p,i}^2}{\omega^2}\begin{pmatrix} -\frac{\omega^2}{\omega^2-\Omega_i^2} & -\frac{i\Omega_i\omega}{\omega^2-\Omega_i^2} & 0 \\ \frac{i\Omega_i\omega}{\omega^2-\Omega_i^2} & -\frac{\omega^2}{\omega^2-\Omega_i^2} & 0 \\ 0 & 0 & -1 \end{pmatrix}\begin{pmatrix} E_x \\ E_y \\ E_z \end{pmatrix} \tag{2.15}$$

The total conductivity tensor is found by summing the conductivity tensor over the particle species.

$$\sigma_{jk} = -i\omega\epsilon_0\begin{pmatrix} \kappa_1 - 1 & i\kappa_2 & 0 \\ -i\kappa_2 & \kappa_1 - 1 & 0 \\ 0 & 0 & \kappa_3 - 1 \end{pmatrix} \tag{2.16}$$

where

$$\kappa_1 = 1 - \frac{\omega_{p,e}^2}{\omega^2 - \Omega_e^2} - \sum_i \frac{\omega_{p,i}^2}{\omega^2 - \Omega_i^2} \tag{2.17}$$

$$\kappa_2 = \frac{\Omega_e}{\omega} \frac{\omega_{p,e}^2}{\omega^2 - \Omega_e^2} - \sum_i \frac{\Omega_i}{\omega} \frac{\omega_{p,i}^2}{\omega^2 - \Omega_i^2} \tag{2.18}$$

$$\kappa_3 = 1 - \frac{\omega_{p,e}^2}{\omega^2} - \sum_i \frac{\omega_{p,i}^2}{\omega^2} \tag{2.19}$$

Note that $\omega_{p,i} \ll \omega_{p,e}$ and $\Omega_i \ll \Omega_e$. Nevertheless we cannot neglect the ion terms in the expressions for κ_1 and κ_2 unless $\omega^2 \gg \Omega_i^2$. It is an extremely good approximation to write

$$\kappa_3 \simeq 1 - \omega_{p,e}^2/\omega^2 \tag{2.20}$$

It is useful to introduce a notation that is used in ionospheric radio propagation theory. In this notation the plasma frequency and gyrofrequencies are normalized in terms of the wave frequency. Define

$$X = \frac{\omega_{p,e}^2}{\omega^2} \tag{2.21}$$

$$Y = \frac{\Omega_e}{\omega} \tag{2.22}$$

We can also regard X as the normalized electron density and Y as the normalized magnetic field. In the same way quantities X_i, Y_i can be defined for each ion species. In this notation

$$\kappa_1 = 1 - \frac{X}{1 - Y^2} - \sum_i \frac{X_i}{1 - Y_i^2} \tag{2.23}$$

$$\kappa_2 = \frac{XY}{1 - Y^2} - \sum_i \frac{X_i Y_i}{1 - Y_i^2} \tag{2.24}$$

$$\kappa_3 = 1 - X - \sum_i X_i \simeq 1 - X \tag{2.25}$$

It is also useful to write the conductivity tensor in what may be called *circularly polarized coordinates.* Define

$$\begin{array}{rclrcl} J_L & = & J_x + iJ_y & J_R & = & J_x - iJ_y \\ E_L & = & E_x + iE_y & E_R & = & E_x - iE_y \end{array} \tag{2.26}$$

It follows that

$$J_L = -i\omega\epsilon_0(\kappa_L - 1)E_L \quad J_R = -i\omega\epsilon_0(\kappa_R - 1)E_R \tag{2.27}$$

where

$$\kappa_L = \kappa_1 + \kappa_2 \tag{2.28}$$

$$\kappa_R = \kappa_1 - \kappa_2 \tag{2.29}$$

Then

$$\kappa_1 = \tfrac{1}{2}(\kappa_L + \kappa_R), \quad \kappa_2 = \tfrac{1}{2}(\kappa_L - \kappa_R) \tag{2.30}$$

The expressions for κ_L and κ_R can be written explicitly in the form

$$\kappa_L = 1 - \frac{X}{1+Y} - \sum_i \frac{X_i}{1-Y_i} \tag{2.31}$$

$$\kappa_R = 1 - \frac{X}{1-Y} - \sum_i \frac{X_i}{1+Y_i} \tag{2.32}$$

2.4.2 Susceptibility Tensor

So far we have characterized the effect of the medium by a conductivity tensor σ_{jk} which, in (1.3) relates the current density $\mathbf{J}$ to $\mathbf{E}$. Alternatively we can regard $\mathbf{J}$ as a polarization current $\partial\mathbf{P}/\partial t$ where

$$\mathbf{P} = Nq\,\mathbf{r}$$

and $\mathbf{r}$ is the displacement of an element of the particle species from its equilibrium position. The right-hand side of (1.3) becomes

$$\mu_0 \frac{\partial}{\partial t}\{\epsilon_0 \mathbf{E} + \mathbf{P}\} = \mu_0 \frac{\partial \mathbf{D}}{\partial t}$$

Then, for harmonically varying quantities,

$$P_j = \frac{i}{\omega} J_j = \frac{i}{\omega}\sigma_{jk} E_k = \epsilon_0 \chi_{jk} E_k \tag{2.33}$$

where

$$\chi_{jk} = \frac{i}{\omega\epsilon_0}\sigma_{jk} = \begin{pmatrix} \kappa_1 - 1 & i\kappa_2 & 0 \\ -i\kappa_2 & \kappa_1 - 1 & 0 \\ 0 & 0 & \kappa_3 - 1 \end{pmatrix} \tag{2.34}$$

is the *susceptibility tensor.*

2.4.3 Dielectric Tensor

The electric displacement vector, D, is defined by

$$D_j = \epsilon_0 E_j + P_j = \epsilon_0\{\delta_{jk} + \chi_{jk}\} E_k$$

Let the *dielectric tensor*, κ_{jk} be defined by

$$\kappa_{jk} \equiv \delta_{jk} + X_{jk} \equiv \delta_{jk} + \frac{i}{\omega\epsilon_0}\sigma_{jk} \tag{2.35}$$

Then

$$\kappa_{jk} = \begin{pmatrix} \kappa_1 & i\kappa_2 & 0 \\ -i\kappa_2 & \kappa_1 & 0 \\ 0 & 0 & \kappa_3 \end{pmatrix} \tag{2.36}$$

It is a matter of taste whether we choose to characterize the medium as a conductor with $\mathbf{P} = 0$ and $\mathbf{J}$ related to $\mathbf{E}$ through a conductivity tensor, or as a dielectric with $\mathbf{J} = 0$ and $\mathbf{P}$ related to $\mathbf{E}$ through a susceptibility tensor. In elementary electromagnetic theory that part of the current which is in phase or 180^0 out of phase with the electric field is conventionally taken to be conduction current. That part which is $\pm 90^0$ out of phase is taken to be polarization current. This is appropriate in media which can be modeled as containing bound and free charges; in our more complicated medium the conductivity, polarization or dielectric tensor is complex and there is no reason to make such a distinction.

2.4.4 Dielectric Tensor in a More General Coordinate System

It is sometimes convenient to use coordinate systems in which the z-axis is not oriented along the magnetic field. In the most general Cartesian system $\mathbf{B}_0$ has direction cosines (l_x, l_y, l_z). This means that $\mathbf{\Omega}_e$ has components $(-l_x\Omega_e, -l_y\Omega_e, -l_z\Omega_e)$ and $\mathbf{\Omega}_i$ has components $(l_x\Omega_i, l_y\Omega_i, l_z\Omega_i)$. We can then use (2.35) and (2.13) to write out the components of κ_{jk} in this coordinate system. The result is

$$\begin{aligned}
\kappa_{11} &= 1 - \frac{X}{1-Y^2}(1 - l_x^2Y^2) - \sum_i \frac{X_i}{1-Y_i^2}(1 - l_x^2Y_i^2) \\
\kappa_{22} &= 1 - \frac{X}{1-Y^2}(1 - l_y^2Y^2) - \sum_i \frac{X_i}{1-Y_i^2}(1 - l_y^2Y_i^2) \\
\kappa_{33} &= 1 - \frac{X}{1-Y^2}(1 - l_z^2Y^2) - \sum_i \frac{X_i}{1-Y_i^2}(1 - l_z^2Y_i^2) \\
\kappa_{12} = \kappa_{21}^* &= \frac{X}{1-Y^2}(il_zY + l_xl_yY^2) + \sum_i \frac{X_i}{1-Y_i^2}(-il_zY_i + l_xl_yY_i^2) \\
\kappa_{13} = \kappa_{31}^* &= \frac{X}{1-Y^2}(-il_yY + l_xl_zY^2) + \sum_i \frac{X_i}{1-Y_i^2}(il_yY_i + l_xl_zY_i^2) \\
\kappa_{23} = \kappa_{32}^* &= \frac{X}{1-Y^2}(il_xY + l_yl_zY^2) + \sum_i \frac{X_i}{1-Y_i^2}(-il_xY_i + l_yl_zY_i^2)
\end{aligned}$$

Now note that $l_x^2 + l_y^2 + l_z^2 = 1$ and that

$$\kappa_3 - \kappa_1 = \frac{XY^2}{1-Y^2} + \sum_i \frac{X_i Y_i^2}{1-Y_i^2}$$

This allows us to write the general expression for the dielectric tensor κ_{jk}:

$$\begin{pmatrix} \kappa_1(l_y^2+l_z^2)+\kappa_3 l_x^2 & il_z\kappa_2+l_xl_y(\kappa_3-\kappa_1) & -il_y\kappa_2+l_xl_z(\kappa_3-\kappa_1) \\ -il_z\kappa_2+l_xl_y(\kappa_3-\kappa_1) & \kappa_1(l_x^2+l_z^2)+\kappa_3 l_y^2 & il_x\kappa_2+l_yl_z(\kappa_3-\kappa_1) \\ il_y\kappa_2+l_xl_z(\kappa_3-\kappa_1) & -il_x\kappa_2+l_yl_z(\kappa_3-\kappa_1) & \kappa_1(l_x^2+l_y^2)+\kappa_3 l_z^2 \end{pmatrix} \tag{2.37}$$

When $l_z = 1$, $l_x = l_y = 0$ this reduces to the previous expression.

2.4.5 Neglect of Positive Ions: Electron Plasma

If we place the ion and electron terms in (2.17) and (2.18) over a common denominator, note that $\omega_{p,i}^2\Omega_e = \omega_{p,e}^2\Omega_i$ and neglect m_e in comparison with m_i then, when $\omega^2 \gg \Omega_e\Omega_i$, we can neglect the ion terms. This implies that the ions are unaffected by the wave field; they behave as a background "sea" of positive charge which merely serves to preserve neutrality of the unperturbed plasma. In this case we speak of an *electron plasma* and

$$\kappa_1 = 1 - X/(1-Y^2) \tag{2.38}$$

$$\kappa_2 = XY/(1-Y^2) \tag{2.39}$$

This approximation is valid in a very wide set of circumstances in the magnetosphere. In this form the theory is known as *magneto-ionic theory.*

2.4.6 Low Frequency Limit: Hydromagnetic Medium

Consider wave frequencies which are much less than the lowest ion gyrofrequency. Suppose also that $\omega_{p,e}^2 \gg \Omega_e^2$. Then

$$\kappa_1 \simeq \frac{\omega_{p,e}^2}{\Omega_e^2} + \sum_i \frac{\omega_{p,i}^2}{\Omega_i^2} = \frac{N_e m_e + \sum_i N_i m_i}{\epsilon_0 B^2}$$

The mass density of the plasma is $\rho_m = N_e m_e + \sum_i N_i m_i$. Then

$$\kappa_1 \simeq 1/U_A^2 \tag{2.40}$$

where U_A is the normalized Alfvén speed. Also

$$\kappa_2 \simeq -\frac{\omega_{p,e}^2}{\omega\Omega_e} + \sum_i \frac{\omega_{p,i}^2}{\omega\Omega_i} = \frac{|q|}{\epsilon_0\omega B}[-N + \sum_i N_i] = 0 \tag{2.41}$$

It is also easy to show that

$$\kappa_1 \ll \kappa_3 \tag{2.42}$$

In this limit magnetohydrodynamic theory applies.

In the special case where there is only one species of positive ion we can relate the Alfvén speed to X, Y and Y_i through

$$U_A^2 = \frac{YY_i}{X} \tag{2.43}$$

2.4.7 Collisions with Neutral Particles

If the loss process associated with collisions is to be included we can use the modified Lorentz force law (1.8) to give a modified conductivity relation for each species. We obtain

$$-i(\omega - i\nu)\mathbf{J} = \epsilon_0\omega_p^2\mathbf{E} - \mathbf{J} \times \mathbf{\Omega} \tag{2.44}$$

which is identical with (2.10) except that ω is replaced by a complex frequency $\omega - i\nu$. We then follow an identical analysis to obtain the following expressions for the elements of the dielectric tensor:

$$\kappa_1 = 1 - \frac{XU}{U^2 - Y^2} - \sum_i \frac{X_iU_i}{U_i^2 - Y_i^2} \tag{2.45}$$

$$\kappa_2 = \frac{XY}{U^2 - Y^2} - \sum_i \frac{X_iY_i}{U_i^2 - Y_i^2} \tag{2.46}$$

$$\kappa_3 = 1 - X/U - \sum_i X_i/U_i \simeq 1 - X/U \tag{2.47}$$

$$\kappa_L = 1 - \frac{X}{U + Y} - \sum_i \frac{X_i}{U_i - Y_i} \tag{2.48}$$

$$\kappa_R = 1 - \frac{X}{U - Y} - \sum_i \frac{X_i}{U_i + Y_i} \tag{2.49}$$

where

$$U = 1 - i\nu/\omega; \quad U_i = 1 - i\nu_i/\omega \tag{2.50}$$

If it is necessary to include the effects of collisions the elements of the dielectric tensor can be replaced by these values.

2.5 Plane Waves

2.5.1 Wave Normal, Phase Speed, Wave Vector and Refractive Index

A plane wave in a uniform medium has a phase which is a function of distance measured in one direction only. If it varies harmonically we may assume time and space variation of the form $\exp[i(\mathbf{k}.\mathbf{r} - \omega t)]$. The *wave vector* $\mathbf{k}$ is perpendicular to the planes of constant phase; it has the direction of the *wave normal* and magnitude $2\pi/\lambda$. Then the *phase speed*, V_P, is defined as the speed at which the planes of constant phase advance in the wave normal direction. Clearly

$$V_P = \omega/k \tag{2.51}$$

In free space the phase speed is the speed of light, c. If we define $k_0 \equiv \omega\sqrt{\mu_0\epsilon_0} \equiv \omega/c$ then the normalized wave vector $\mathbf{k}/k_0$ is called the *refractive index vector*

$$\mathbf{n} = \mathbf{k}/k_0 = \hat{\mathbf{n}}c/V_P \tag{2.52}$$

where $\hat{\mathbf{n}}$ is a unit vector in the direction of the wave normal.

The operator ∇ in Maxwell's equations (1.26), (1.27) can be replaced by $i\mathbf{k}$. The equations then become

$$\mathbf{k} \times \mathbf{E} = \omega\mathbf{B} \tag{2.53}$$

$$\mathbf{k} \times \mathbf{B} = -i\mu_0\mathbf{J} - \frac{\omega}{c^2}\mathbf{E} \tag{2.54}$$

The wave equation, (1.36), becomes

$$[k_j k_k - (k^2 - k_0^2)\delta_{jk} + i\omega\mu_0\sigma_{jk}]E_k = 0 \tag{2.55}$$

or

$$[k_j k_k - k^2\delta_{jk} + k_0^2\kappa_{jk}]E_k = 0 \tag{2.56}$$

2.5.2 Dispersion Relation

To find the dispersion relation we substitute in (2.56) for the conductivity from (2.16). There is cylindrical symmetry about the z-axis, the direction of the magnetic field. We can, without any loss of generality, choose the directions of the x- and y-axes so that $\mathbf{k}$ lies in the $x - z$ plane. Then $k_x = k\sin\theta$, $k_y = 0$ and $k_z = k\cos\theta$ where θ is the angle measured from $\mathbf{B}$ to $\mathbf{k}$. The result, in matrix form, is

$$\begin{pmatrix} \kappa_1 - n^2\cos^2\theta & i\kappa_2 & n^2\sin\theta\cos\theta \\ -i\kappa_2 & \kappa_1 - n^2 & 0 \\ n^2\sin\theta\cos\theta & 0 & \kappa_3 - n^2\sin^2\theta \end{pmatrix} \begin{pmatrix} E_x \\ E_y \\ E_z \end{pmatrix} = 0 \tag{2.57}$$

This is a set of three homogeneous equations in E_x, E_y, E_z. They are only self-consistent if the determinant of the coefficients is zero. This leads to a quadratic equation for n^2:

$$An^4 + Bn^2 + C = 0 \tag{2.58}$$

where

$$A = \kappa_1 \sin^2\theta + \kappa_3 \cos^2\theta \tag{2.59}$$

$$\begin{aligned} B &= (\kappa_2^2 - \kappa_1^2)\sin^2\theta - \kappa_1\kappa_3(1+\cos^2\theta) \\ &= -\kappa_L\kappa_R \sin^2\theta - \kappa_1\kappa_3(1+\cos^2\theta) \end{aligned} \tag{2.60}$$

$$\begin{aligned} C &= \kappa_3(\kappa_1^2 - \kappa_2^2) \\ &= \kappa_L\kappa_R\kappa_3 \end{aligned} \tag{2.61}$$

$$\begin{aligned} D^2 \equiv B^2 - 4AC &= [\kappa_1(\kappa_1-\kappa_3) - \kappa_2^2]^2 \sin^4\theta + 4\kappa_2^2\kappa_3^2\cos^2\theta \\ &= [\kappa_L\kappa_R - \kappa_1\kappa_3]^2\sin^4\theta + 4\kappa_2^2\kappa_3^2\cos^2\theta \end{aligned} \tag{2.62}$$

This is the dispersion relation for the medium. The expression for the refractive index can be written explicitly:

$$n^2 = \frac{-B \pm D}{2A} \tag{2.63}$$

This can be manipulated to give

$$n^2 = 1 - \frac{A+B+C}{A + \frac{1}{2}(B \pm D)} \tag{2.64}$$

which is a form that is often used.

The dispersion relation or, equivalently the refractive index formula, contains all the information about the variation of the phase of a plane electromagnetic wave in the medium. The study of plane waves in a plasma is largely the study of the properties of the refractive index. Later we shall devote considerable space to classifying these properties. It is useful, however, to delay this study until we have described how plane waves can be superimposed to synthesize arbitrary linear wave structures.

Electron Plasma

For an electron plasma in which the effect of positive ions may be neglected we can use the approximate expressions (2.38) and (2.39) to find simplified expressions for the coefficients of the quadratic. If each is multiplied by $1 - Y^2$ these are:

$$A = 1 - X - Y^2 + XY^2\cos^2\theta \tag{2.65}$$

$$B = -2[(1-X)^2 - Y^2] - XY^2(1+\cos^2\theta) \tag{2.66}$$

$$C = (1-X)[(1-X)^2 - Y^2] \tag{2.67}$$

$$D = X\sqrt{Y^4\sin^4\theta + 4Y^2(1-X)^2\cos^2\theta} \tag{2.68}$$

The expression for the refractive index (2.64) then becomes

$$n^2 = 1 - \frac{X}{1 - \dfrac{Y^2 \sin^2\theta}{2(1-X)} \pm \sqrt{\dfrac{Y^4 \sin^4\theta}{4(1-X)^2} + Y^2 \cos^2\theta}} \tag{2.69}$$

This is the well-known Appleton-Hartree formula which was derived independently by Appleton [5] and Hartree [48]. An equivalent formula was developed earlier by Lassen [74] and some authors now call it the Appleton-Lassen formula.

Multi-Ion Plasmas

At lower frequencies the effects of positive ions become important. Terms of order m_e/m_i can be neglected in the expressions for the refractive index. Define the fractional abundance of the ith species of ion as $\eta_i \equiv N_i/N$. We can write $X_i = X\eta_i Y_i/Y$. We choose to write the expressions for the coefficients in a form which is valid for all frequencies. The expressions for the coefficients A, B and C in (2.58) are thus written as the sum of the expressions for an electron plasma with the addition of ion terms. It is again convenient to multiply them by $1-Y^2$. The ion terms are only important when $Y \gg 1$ and this approximation is used only in the ion terms. Thus at high frequency the electron terms dominate in the form previously derived while at low frequencies the approximate expressions for the ion terms become important.

Let

$$\alpha = \sum_i \frac{\eta_i Y_i}{1-Y_i^2}; \quad \beta = \sum_i \frac{\eta_i Y_i^2}{1-Y_i^2}; \quad \gamma = \sum_i \sum_j \frac{\eta_i \eta_j Y_i Y_j}{(1-Y_i)(1+Y_j)}$$

Then

$$A = 1 - X - Y^2 + XY^2 \cos^2\theta + \alpha XY \sin^2\theta \tag{2.70}$$

$$\begin{aligned} B = {} & -2\left[(1-X)^2 - Y^2\right] - XY^2(1+\cos^2\theta) + 2\beta X(1-X)\sin^2\theta \\ & -\alpha XY\left[3 - \cos^2\theta - X(1+\cos^2\theta)\right] + \gamma X^2 \sin^2\theta \end{aligned} \tag{2.71}$$

$$C = (1-X)\left[(1-X)^2 - Y^2 + 2\alpha XY - 2\beta X(1-X) - \gamma X^2\right] \tag{2.72}$$

$$\begin{aligned} D^2 = {} & \left[-XY^2 + \alpha XY(1+Y) - 2\beta X(1-X) - \gamma X^2\right]^2 \sin^4\theta \\ & +4X^2Y^2(1-X)^2\cos^2\theta(1+\beta)^2 \end{aligned} \tag{2.73}$$

The case of a plasma with one species of ion is obtained by setting $\eta_1 = 1$, $\eta_i = 0$, $i \neq 1$. These expressions can be useful but it is often just as convenient to use the expressions (2.59), (2.60), (2.61) and (2.62)

Low Frequency Approximations

In most parts of the magnetosphere the electron plasma frequency is much greater than the electron gyrofrequency. If, in addition, the wave frequency is of the order of the gyrofrequency ($\omega_{pe}^2 \gg \Omega_e^2 \sim \omega^2$) then we can neglect the effect of heavy ions and

$$X \gg Y^2 \sim 1$$

Then

$$\begin{aligned} A &\simeq -X(1 - Y^2\cos^2\theta) &(2.74)\\ B &\simeq -2X^2 &(2.75)\\ C &\simeq -X^3 &(2.76)\\ D &\simeq 2X^2Y\cos\theta &(2.77)\end{aligned}$$

In this case

$$n^2 \simeq -\frac{X}{1 \pm Y\cos\theta} \qquad (2.78)$$

If, on the other hand the wave frequency is much less than the gyrofrequency ($\omega_{pe}^2 \gg \Omega_e^2 \gg \omega^2$) the effect of ions must be included. In this case $X \gg Y^2 \gg 1$. Possible approximations are

$$\begin{aligned} A &\simeq -X(1 - Y^2\cos^2\theta) + \alpha XY\sin^2\theta] &(2.79)\\ B &\simeq -2X^2 - \alpha X^2Y(1+\cos^2\theta) &(2.80)\\ C &\simeq -X^3(1-\gamma) &(2.81)\\ D &\simeq 2X^2Y\left[(1+\beta)^2\cos^2\theta + \frac{1}{4}\alpha^2\sin^4\theta\right]^{\frac{1}{2}} &(2.82)\end{aligned}$$

There are some difficulties in taking limiting values as $\omega \to 0$ and the complete expressions (2.59), (2.60), (2.61) and (2.62) may be more useful for computation.

Hydromagnetic Limit

The extreme low frequency case is the hydromagnetic limit in which $\Omega_i^2 \gg \omega^2$ or

$$Y_1 \gg 1$$

In this case

$$\begin{aligned} A &= -\cos^2\theta\\ B &= (1/U_A^2)(1+\cos^2\theta)\\ C &= -1/U_A^4\end{aligned}$$

and the quadratic equation factorizes, giving two values for the refractive index:

$$\begin{aligned} n^2 &= 1/U_A^2 \\ n^2\cos^2\theta &= 1/U_A^2 \end{aligned}$$

Dependence of Frequency on Wave Number

If desired the dispersion relation can, of course, be written as a relationship between ω and k. In this case it is easiest to express k as a function of ω, noting that $n = ck/\omega$. Then

$$A'k^4 + B'k^2 + C' = 0 \tag{2.83}$$

where

$$A' = A;\ B' = (\omega/c)^2 B;\ C' = (\omega/c)^4 C$$

2.5.3 Propagated, Evanescent and Inhomogeneous Waves

The square of the refractive index is the root of a quadratic equation. Inspection of its properties shows that, in the loss-free medium considered here, $D \equiv B^2 - 4AC$ is always positive or zero. Thus n^2 is real and n is either real or purely imaginary. The wave varies in space as $\exp(i\mathbf{k}.\mathbf{r}) = \exp[i\omega(n_x x + n_y y + n_z z)/c]$. First consider a wave propagated in the z-direction for which $n_x = n_y = 0$, $n_z = n$. If n is real then planes of constant phase are given by $\omega[(nz/c) - t] =$ constant. These planes of constant phase move in the z-direction with the phase speed c/n. The wave is a *propagated wave.* On the other hand if n is purely imaginary there is no phase variation in space and the signal is purely oscillatory in time; the wave is not propagated and the phase is constant everywhere. Instead of a periodic disturbance in space there is a factor $\exp[\pm\omega|n|z/c]$ which represents a disturbance which grows or decays exponentially with z. This disturbance cannot exist throughout all space or it would become infinite at $z = \pm\infty$. The signal must be determined on some boundary and decay exponentially with distance from this boundary. Such a disturbance is called an *evanescent wave.*

The situation can arise where boundary conditions on a plane ($z = 0$, say) fix the values of n_x and n_y. Since $n_z^2 = n^2 - n_y^2 - n_x^2$ it is possible that n_z is purely imaginary even though the other quantities are real. In this case planes of constant phase are given by

$$n_x x + n_y y - ct = \text{constant}$$

while the signal decays with z exponentially as $\exp(-\omega|n_z|/c)$, so that planes of constant phase are propagated perpendicular to the boundary while planes of constant amplitude are parallel to it. This is illustrated in Figure 2.1. In the lower part of the diagram an incident propagated wave

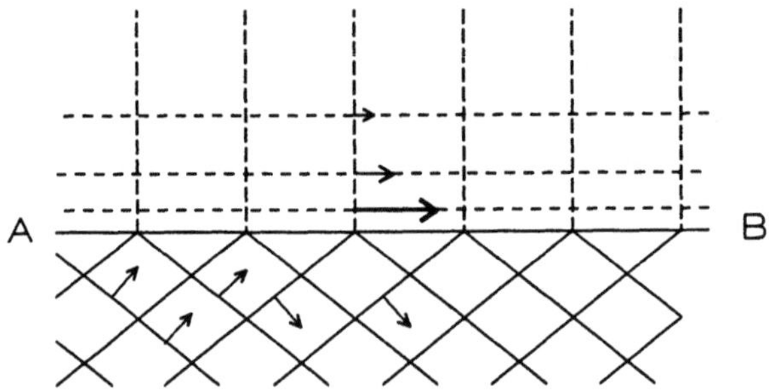

Figure 2.1: Planes of constant phase (dashed lines) and planes of constant amplitude (dotted lines) in an inhomogeneous wave at a boundary AB between two media.

is totally internally reflected with parallel planes of constant amplitude and phase coincident. In the upper part of the diagram n_z is imaginary and the wave is not propagated. While such a disturbance is sometimes called an evanescent wave it is really more complicated and it is better to refer to it as an *inhomogeneous wave.*

2.6 Polarization

2.6.1 Polarization of Electromagnetic Waves

Once the refractive index has been determined it can be substituted in (2.57) in order to find the ratio $E_x : E_y : E_z$. This is a definition of the polarization of the wave relative to a coordinate system which is referred to the magnetic field direction and in which the wave normal may be in any direction. It is, however, conventional in radio propagation theory to define the polarization of a wave in terms of the field components transverse to the wave. It is therefore more convenient to express these ratios in a different coordinate system which is related to the wave normal direction rather than to the magnetic field direction. Both these uses of the term *polarization* occur in the literature. When, in addition, the same word is used for **P**, the polarization of the medium, there is room for confusion. In this book the context will make it clear what is intended. In wave propagation problems we use "polarization" of a wave in the radio propagation sense. Later (§2.7.3) we shall introduce a labeling of different modes in terms of polarization referred to the magnetic field direction. In this case we shall refer to the "polarization label".

2.6.2 Definition of Polarization

Let the z'-axis be parallel to $\mathbf{k}$ and let $\mathbf{B}$ lie in the x'-z' plane. Define polarization ρ as

$$\rho = E_{y'}/E_{x'} \tag{2.84}$$

Maxwell's equations then show that $B_{x'}/B_{y'} = -\rho$ and thus in free space, or in an isotropic medium, ρ gives complete information about the relationship between the field components. In an anisotropic medium, however, there is also a component of the electric field parallel to the wave normal. It can be written in terms of a ratio

$$\eta = E_{z'}/E_{x'} \tag{2.85}$$

2.6.3 Relation Between Field Components

In the x'-y'-z' coordinate system $l_x = -\sin\theta$, $l_y = 0$, $l_z = \cos\theta$ and $k_z = k$, $k_x = k_y = 0$. The minus sign in the expression for l_x arises because θ is measured from $\mathbf{B}$ to $\mathbf{k}$. We can substitute (2.37) in (2.56) to give

$$\begin{pmatrix} \kappa_1\cos^2\theta + \kappa_3\sin^2\theta - n^2 & i\kappa_2\cos\theta & (\kappa_1-\kappa_3)\sin\theta\cos\theta \\ -i\kappa_2\cos\theta & \kappa_1 - n^2 & -i\kappa_2\sin\theta \\ (\kappa_1-\kappa_3)\sin\theta\cos\theta & i\kappa_2\sin\theta & \kappa_1\sin^2\theta + \kappa_3\cos^2\theta \end{pmatrix} \begin{pmatrix} E_{x'} \\ E_{y'} \\ E_{z'} \end{pmatrix} = 0 \tag{2.86}$$

It follows from these self-consistent equations that

$$\begin{aligned} E_{x'} : E_{y'} : E_{z'} &= 1 : \rho : \eta \\ &= \kappa_2\kappa_3\cos\theta : -i[An^2 - \kappa_1\kappa_3] : \kappa_2(\kappa_3 - n^2)\sin\theta \end{aligned} \tag{2.87}$$

An alternative way of expressing this is

$$\begin{aligned} E_{x'} : E_{y'} : E_{z'} = 1 : \rho : \eta = {} & An^2 + B + \kappa_1\kappa_3 : \\ & -i\kappa_2\kappa_3\cos\theta : -[(\kappa_1-\kappa_3)n^2 - \kappa_L\kappa_R + \kappa_1\kappa_3]\sin\theta\cos\theta \end{aligned} \tag{2.88}$$

This implies a useful relationship

$$(An^2 - \kappa_1\kappa_3)(An^2 + B + \kappa_1\kappa_3) = \kappa_2^2\kappa_3^2\cos^2\theta \tag{2.89}$$

In general ρ and η are complex numbers. The argument of ρ is the phase difference between $E_{x'}$ and $E_{y'}$. At a particular value of z' the projection of $\mathbf{E}$ on the x'–y' plane traces out an ellipse and rotates with period $2\pi/\omega$. The wave is *elliptically polarized.* The projection of $\mathbf{E}$ on the x'–z' and y'–z' planes can be treated similarly by considering the phase of η.

The magnetic field components can be found from Faraday's law of induction (1.26):

$$B_{x'} = -nE_{y'}/c; \; B_{y'} = nE_{x'}/c; \; B_{z'} = 0 \tag{2.90}$$

It is useful to give an explicit equation for finding the polarization ρ without first finding the refractive index. In (2.87) we note that $2An^2 = -B \pm D$. Then

$$2i\rho\kappa_2\kappa_3 \cos\theta + 2\kappa_1\kappa_3 = -B \pm D \tag{2.91}$$

Then we can use the expressions for B and D, (2.60) and (2.62), and square both sides to get a quadratic equation for ρ

$$\rho^2 - 2iG\frac{\sin^2\theta}{\cos\theta}\rho + 1 = 0 \tag{2.92}$$

where

$$G = \frac{\kappa_1\kappa_3 - \kappa_L\kappa_R}{2\kappa_2\kappa_3} \tag{2.93}$$

This can be solved for ρ and the corresponding value of n found from

$$n^2 = \frac{(\kappa_1 + i\rho\kappa_2\cos\theta)\kappa_3}{A} \tag{2.94}$$

2.6.4 Special Cases of the Polarization

In the coordinate system chosen here there is considerable simplification. We see (2.87) that ρ is always purely imaginary and η is always real. Let $\rho = i\mathcal{R} = \mathcal{R}e^{i\pi/2}$. Suppose that

$$E_{x'} = \Re\{Ae^{i\alpha}\exp[i(kz - \omega t)]\}$$

where A is real and $e^{i\alpha}$ is a phase factor. Then

$$\begin{aligned} E_{x'} &= A\cos(kz - \omega t + \alpha) \\ E_{y'} &= A\mathcal{R}\cos(kz - \omega t + \alpha + \tfrac{1}{2}\pi) \\ &= -A\mathcal{R}\sin(kz - \omega t + \alpha) \\ E_{z'} &= A\eta\cos(kz - \omega t + \alpha) \end{aligned}$$

It follows that

$$\begin{aligned} E_{x'}^2 + \frac{E_{y'}^2}{\mathcal{R}^2} &= A^2 \\ \eta E_{x'} - E_{z'} &= 0 \\ \frac{E_{y'}^2}{\mathcal{R}^2} + \frac{E_{z'}^2}{\eta^2} &= A^2 \end{aligned}$$

Thus the projection of the electric field on the x'–y' plane is an ellipse with its axes aligned with the coordinate axes. If $\mathcal{R} > 1$ the major axis is aligned with the x'-axis; if $\mathcal{R} < 1$ the major axis is aligned with the y'-axis. The wave is elliptically polarized. If $\mathcal{R}$ is positive the electric vector rotates in a clockwise sense when viewed in the direction of the wave normal; the wave is *right-hand polarized.* If $\mathcal{R}$ is negative it rotates in the opposite sense and is *left-hand polarized.* Note that this *radio convention* for polarization is opposite from the convention used in optics where the polarization is defined by viewing the rotation of the electric vector while facing towards the oncoming wave.

The projection of the electric field on the y'–z' plane is also an ellipse; the sense of rotation depends on the signs of $\mathcal{R}$ and η. The projection on the x'–z' plane is a straight line making an angle $\arctan \eta$ with the x'-axis.

Circular Polarization

When $\rho = i$ and $\eta = 0$ the wave is right-hand circularly polarized and when $\rho = -i$ and $\eta = 0$ it is left hand circularly polarized. In Figure 2.2a we show a "snapshot" of the electric vector in a left-hand circularly polarized wave at positions along the z-axis. It should be noted that the locus of the tip of $\mathbf{E}$ for a *left*-hand polarized wave is a *right*-hand spiral in space. If we imagine that this spiral rotates anti-clockwise about the z-direction we can see that the wave will progress in the positive z-direction. Alternatively if we imagine the spiral advancing bodily along the z-axis we see that $\mathbf{E}$ at a fixed value of z will rotate anti-clockwise. The corresponding situation for right-hand circular polarization is shown in Figure 2.2b.

Linear Polarization Along the Axes

When $\rho = 0$ the wave is linearly polarized along the x'-axis. When $\rho = \infty$ it is linearly polarized along the y-axis. These are special cases of elliptical polarization with the minor axis of the ellipse becoming zero.

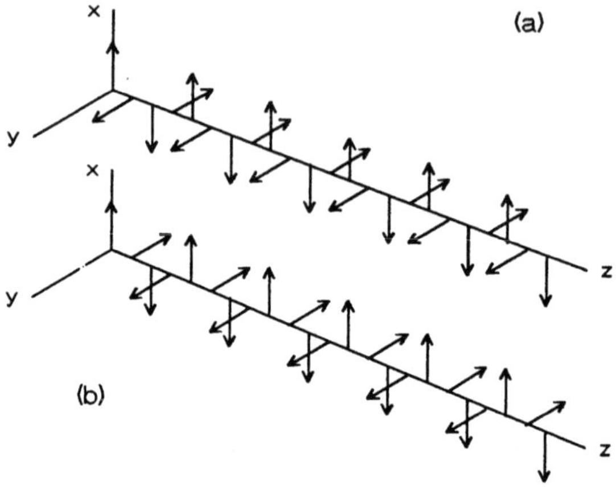

Figure 2.2: Circular polarization. (a) Left-hand polarization. (b) Right-hand polarization

Electron Plasma

When the effect of positive ions can be neglected we can use the approximate expressions (2.38) and (2.39) to yield the ratio

$$1 : \rho : \eta = XY\cos\theta : i\left(1 - \frac{n^2}{1-X}\right)(1 - X - Y^2) : XY\sin\theta\left(1 - \frac{n^2}{1-X}\right) \tag{2.95}$$

The general statements about circular and linear polarization above also apply to this special case.

2.7 Properties of the Refractive Index

The refractive index contains a great deal of information about the nature of waves in a plasma. It is a function of frequency, electron density, ion species density, magnetic field strength and direction. It can show a wide variety of types of behavior depending on the values of these parameters. In this section we derive some results which illustrate some of these types of behavior. At the end of Chapter 3 a comprehensive set of diagrams showing the dependence of the refractive index on these parameters is presented.

2.7.1 Propagation Parallel to the Magnetic Field

When $\theta = 0$ the wave normal is parallel to the magnetic field. In this case there is a substantial simplification of the expressions for the refractive index and polarization of the wave. The quadratic equation for n^2 (2.58) now has coefficients

$$A = 1; \quad B = -2\kappa_1 = -2(\kappa_L + \kappa_R): \quad C = \kappa_L \kappa_R \tag{2.96}$$

where a common factor κ_3 has been cancelled and we have used (2.29). The quadratic equation factorizes and the two roots are

$$n^2 = \kappa_L; \quad n^2 = \kappa_R \tag{2.97}$$

The polarization corresponding to these two cases can be found from (2.87). There is no component of the electric field parallel to the wave normal ($\eta = 0$) and the polarizations for the two values of n respectively are

$$\rho = -i; \quad \rho = +i \tag{2.98}$$

Thus one wave is propagated with refractive index $\sqrt{\kappa_L}$ and is left-hand circularly polarized while the other wave is propagated with refractive index $\sqrt{\kappa_R}$ and is right-hand circularly polarized. They are called the L- and R-waves respectively.

2.7.2 Propagation Perpendicular to the Magnetic Field

When $\theta = \pi/2$ the wave normal is perpendicular to the magnetic field and again there is a substantial simplification of the expressions for the refractive index and polarization of the wave. The quadratic equation for n^2 (2.58) now has coefficients

$$A = \kappa_1; \quad B = \kappa_2^2 - \kappa_1^2 + \kappa_1\kappa_3; \quad C = \kappa_3(\kappa_1^2 - \kappa_2^2) \tag{2.99}$$

Again the quadratic factorizes and we get

$$n^2 = \kappa_3 \quad \text{or} \quad \frac{\kappa_1^2 - \kappa_2^2}{\kappa_1} \tag{2.100}$$

In the first case it can be seen from (2.88) that $\rho = 0$ and $\eta = 0$. Thus this wave is linearly polarized with the electric vector parallel to the magnetic field. Since κ_3 is independent of $\mathbf{B}_0$ the refractive index is exactly the same as it would be if there were no applied magnetic field. This is easy to understand as the plasma particles oscillate parallel to the magnetic field and thus experience no magnetic force. This wave is therefore called the *ordinary* wave or O-wave.

In the second case, from (2.87), the ratios of the field components are

$$0 \ : \ -i(\kappa_1^2 - \kappa_2^2 - \kappa_1\kappa_3) \ : \ \kappa_2\left(\kappa_3 - \frac{\kappa_1^2 - \kappa_2^2}{\kappa_1}\right)$$

Thus the transverse components of the wave are linearly polarized perpendicular to the magnetic field but there is a component of the electric field parallel to the wave normal which is in phase quadrature to the transverse component. The electric vector therefore oscillates in an ellipse whose plane is perpendicular to the magnetic field but which contains the wave normal. The ratio of the field components is given by

$$E_z/E_y = -i\kappa_2/\kappa_1$$

This wave is called the *extraordinary* wave or X-wave.

2.7.3 Propagation at an Intermediate Angle – The Refractive Index Surface

If we keep the frequency, ion and electron densities, and magnetic field strength constant and vary the direction of propagation, θ, then for each of the two possible waves n^2 and the polarization vary continuously. In general the polarization is elliptical with a component in the direction of the wave normal.

We can define a *refractive index surface* in refractive index space. This is the locus of the tip of the refractive index vector. Because n is a function only of θ it is a surface of revolution about the direction of the magnetic field. The properties of waves in the medium depend on the properties of the refractive index surface. It is a useful means of classifying the types of waves which may be propagated.

We may apply labels to the surfaces in several ways. For transverse propagation the wave will be either ordinary, O, or extraordinary, X. We apply the labels O and X to the surface for all angles of propagation. Alternatively for propagation parallel to the magnetic field the wave will be either left-handed, L, or right-handed, R. We apply the polarization labels L and R to the surfaces for all angles of propagation. This *polarization label* can lead to confusion. In general the wave is elliptically polarized. When $\theta = \pi/2$ it is linearly polarized. When it is propagated in the *opposite* direction to the magnetic field the L-wave is right-handed and the R-wave is left-handed. The polarization label identifies the polarization of the wave when the wave normal direction is rotated to be parallel with the magnetic field.

There are other sources of confusion if the other parameters vary. A wave may change its polarization label if X or Y is varied, without changing its O- or X-labeling. For a full discussion see Stix [100].

A further possible labeling arises from the fact that the wave with the larger value of n^2 has a smaller phase speed. This leads to the labeling *fast* wave or F-wave and *slow* wave or S-wave.

2.7.4 Cut-off, Resonance and Cross-over Frequencies

There are important critical values of the refractive index which occur when n^2 changes sign. This occurs for $n = 0$ and for $n = \infty$. The condition $n = 0$ is called a *cut-off* of n and the condition $n = \infty$ is called a *resonance*. Other important transitions occur when the two roots of the quadratic equation for n^2 become equal. The condition for this to occur is called a *cross-over*. The reason for these names will become apparent.

Conditions for Cut-off

The condition for a cut-off is $C = 0$. From (2.61) this implies that

$$\kappa_L = 0 \text{ or } \kappa_R = 0 \text{ or } \kappa_3 = 0 \tag{2.101}$$

Note that each of these conditions depends on the frequency, the electron and ion densities, and the magnetic field strength but they do not depend on the angle between $\mathbf{B}_0$ and k. The reason for the name "cut-off" is that if any of these quantities is varied so that the cut-off condition is satisfied then on one side of the cut-off n is real and the wave is propagated, while on the other side n is imaginary and the wave is evanescent or "cut off". For example, in the ionosphere, the electron density is a function of height. If a wave is propagated upwards it may approach a cut-off. It is evanescent on the other side of the cut-off and cannot be propagated there. A detailed analysis (Budden [19]) shows that the wave is, in fact reflected at this level.

2.7.5 Conditions for Resonance

Normally the condition for a resonance is $A = 0$. From (2.59) the condition $A = 0$ occurs for propagation at an angle θ to the magnetic field such that

$$\tan^2\theta = -\kappa_3/\kappa_1 \tag{2.102}$$

This condition can only be satisfied if κ_1 and κ_3 are of opposite sign. Suppose a parameter such as frequency, ion or electron density, or magnetic field strength is varied. If either κ_1 or κ_3 changes sign we pass from a region where there is an angle θ for which resonance can occur to a region where it cannot. The conditions for this to be so are

$$\kappa_1 = 0 \text{ or } \infty; \;\; \kappa_3 = 0 \tag{2.103}$$

A glance at the expression for κ_3 shows that it cannot be infinite.

When $\kappa_1 = 0$ waves in the direction $\theta = \pi/2$ have a resonance; when κ_L or κ_R are infinite then C/A is infinite for $\theta = 0$ and waves in this direction have a resonance. These are called *principal resonances.*

The reason for the term "resonance" becomes apparent if we consider some examples.

Longitudinal Resonance – Gyroresonance: When $\theta = 0$ there is a principal resonance where κ_L or κ_R is infinite. In order to understand the nature of the resonance consider the expression for the current density. When $\theta = 0$ the condition for resonance of the right-hand wave corresponding to $n^2 = \kappa_R$ is that κ_R should be infinite. The constitutive relation for a right-hand circularly polarized wave is given by (2.27)

$$J_R = -i\omega\epsilon_0(\kappa_R - 1)E_R$$

Clearly for finite E_R the current density is infinite, hence the term *resonance.* Inspection of the explicit expression for κ_R shows that this occurs at $Y = 1$, that is at a wave frequency equal to the electron gyrofrequency. It is easy to understand the physical mechanism. The electric vector rotates with the same frequency as electrons. An electron gyrating about the magnetic field with the right phase experiences an electric field which is always in the same direction as its velocity. The wave therefore steadily increases the electron energy until the assumption of linearity breaks down.

Exactly the same situation obtains in the case that κ_L is infinite except that this happens for left-hand waves at the ion gyrofrequencies.

Transverse Resonance – Hybrid Resonances: When $\theta = \pi/2$ there is a resonance where $\kappa_1 = 0$. This is a principal resonance in the X-wave whose refractive index is given by $n^2 = \kappa_L\kappa_R/\kappa_1$. The ratio of the field components is $E_{z'}/E_{y'} = -i\kappa_2/\kappa_1$ and so the electric field parallel to the wave normal becomes infinite. This corresponds to an oscillation of the electric field along the wave normal at the frequencies determined by the condition for resonance. Using the explicit expression (2.17) for κ_1 this gives

$$1 - \frac{\omega_{p,e}^2}{\omega^2 - \Omega_e^2} - \sum_i \frac{\omega_{p,i}^2}{\omega^2 - \Omega_i^2} = 0 \tag{2.104}$$

If we regard this as an equation for ω^2 we see that it has $n + 1$ roots where n is the number of ion species. These are the *hybrid resonance frequencies.*

In a single-ion plasma, if we make the approximation $m_e/m_i \ll 1$, it can easily be shown [100] that there is an *upper hybrid resonance frequency* given by

$$\omega_{UH}^2 = \omega_{p,e}^2 + \Omega_e^2 \tag{2.105}$$

and a *lower hybrid resonance frequency* given by

$$\frac{1}{\omega_{LH}^2} = \frac{1}{\omega_{p,i}^2 + \Omega_i^2} + \frac{1}{\Omega_e \Omega_i} \tag{2.106}$$

For multi-ion plasmas these expressions remain approximately correct and there are additional hybrid resonances between each pair of ion gyrofrequencies.

Resonance at Intermediate Angles – Resonance Cone: In general when the coefficient A becomes very small there is an approximate factorization of the quadratic for n^2,

$$(An^2 + B)(n^2 + C/B) = 0$$

The root which becomes infinite corresponds to the first factor and An^2 remains finite. The polarization relation (2.87) shows that the component $E_{z'}$ in the direction of the wave normal becomes very large while the other components are negligible. The wavelength is inversely proportional to n and therefore becomes small. Faraday's law requires that $\mathbf{B} = \mathbf{k} \times \mathbf{E}/\omega$ and the magnetic field is thus negligible. The wave is therefore often described as an *electrostatic* wave.

The directions of propagation for which this resonance takes place lie on a cone whose axis coincides with the magnetic field direction and whose apex angle is $2\theta_{\text{res}}$.

Plasma Resonance When $\theta = 0$ the coefficients A, B and C have a common factor κ_3. Thus the dispersion relation is satisfied by $\kappa_3 = 0$. This is simply the condition for the wave frequency to be equal to the plasma frequency and the result is a *plasma resonance* in which the plasma oscillates at the plasma frequency in the direction of the magnetic field. The resonance does not involve n and there is thus no spatial dependence of the phase. The resonance is a pure oscillation. The condition $\kappa_3 = 0$ therefore corresponds both to a cut-off and a resonance. One of the characteristic waves has a cut-off at this point and in addition there is a plasma resonance.

2.7.6 Conditions for Cross-over

A quadratic equation has equal roots when the discriminant, $D^2 \equiv B^2 - 4AC$, is zero. Examination of (2.62) shows that this is never negative. It can, however, be zero. There are two cases where this is so.

$\theta = \pi/2$: When the wave normal is at right angles to the magnetic field then $D = 0$ when $\kappa_L\kappa_R = \kappa_1\kappa_3$. The O-wave and X-wave then have the same speed and a wave with arbitrary polarization is propagated unchanged. Waves which are not at right angles to the field do not have this property. On one side of the cross-over the O label applies to the R-wave and the X to the L. On the other side of the cross-over O corresponds to L and X to R.

$\theta = 0$: When the wave normal is parallel to the magnetic field then $D = 0$ when $\kappa_3 = 0$ or when $\kappa_2 = 0$. The first of these is also the condition for plasma resonance. The second never occurs for a one-ion plasma. For a multi-ion plasma it does occur for frequencies between the ion gyrofrequencies. If we use the expression for a two-ion plasma we obtain the condition

$$\eta_2 Y_1^2 + \eta_1 Y_2^2 = 1 \tag{2.107}$$

or

$$\omega^2 = \eta_1 \Omega_2^2 + \eta_2 \Omega_1^2 \tag{2.108}$$

This condition is independent of the plasma frequency. For propagation along the magnetic field the L- and R-waves have the same speed and a wave with arbitrary polarization is propagated unchanged. Again the correspondence between O, X and R, L changes on either side of the cross-over.

2.7.7 The Plasma Frequency

In the above we have seen that the condition $\kappa_3 = 0$ or, to a very good approximation $X = 1$, may represent a cut-off, a resonance or a cross-over frequency. Which of these it is in a particular case depends on the mode of wave propagation. We shall draw attention to cases when they arise.

2.8 Energy Flux in a Plane Wave

We can use the time averaged Poynting vector (1.31) to find the average energy flux in a plane wave. Together with (2.90) this gives

$$\begin{aligned}
\Pi_{x'} &= -\frac{1}{2\mu_0 c}\Re(n^* E_{x'}^* E_{z'}) \\
\Pi_{y'} &= -\frac{1}{2\mu_0 c}\Re(n^* E_{y'}^* E_{z'}) \\
\Pi_{z'} &= \frac{1}{2\mu_0 c}\Re[n^*(|E_{x'}|^2 + |E_{y'}^2|)]
\end{aligned} \tag{2.109}$$

If n^2 is negative the refractive index is imaginary $\Pi_{x'} = \Pi_{z'} = 0$. However $\Pi_{y'}$ is not in general zero and there can be energy propagation perpendicular to the wave normal. This is because in the anisotropic medium the wave is inhomogeneous rather than purely evanescent. If n^2 is positive the refractive index is real and

$$\Pi_{x'} = -\frac{n}{2\mu_0 c} E^*_{x'} E_{z'};\ \Pi_{y'} = 0;\ \Pi_{z'} = \frac{n}{2\mu_0 c} \Re[|E_{x'}|^2 + |E^2_{y'}] \qquad (2.110)$$

It follows that the time averaged energy flux vector has a component in the direction of the wave normal; the directions of phase propagation and energy propagation are not the same. Let the angle between the direction of $\mathbf{\Pi}$ and $\mathbf{k}$ be α. Then

$$\tan\alpha = \Pi_{x'}/\Pi_{y'} = -\eta/(1 + |\rho|^2)$$

We use (2.88) and (2.89) to show

$$\tan\alpha = \frac{[(\kappa_1 - \kappa_3)n^2 - \kappa_L\kappa_R + \kappa_1\kappa_3]\sin\theta\cos\theta}{2An^2 + B} \qquad (2.111)$$

Obviously in the symmetric cases of waves propagated perpendicular or parallel to $\mathbf{B}$ the directions of phase and energy propagation are the same. This is not generally true.

2.9 Summary

- Wave propagation in a uniform cold plasma with a steady magnetic field superimposed can be treated by assuming that each charged particle species behaves as a separate interpenetrating fluid. In such a fluid the properties of waves are modified by natural frequencies of oscillation. The plasma frequencies are associated with the relative displacements of ions and electrons. The gyrofrequencies are the frequencies with which ions and electrons gyrate about the background magnetic field.

- In the presence of a wave with small amplitude the equation of motion of each species can be linearized. A conductivity relationship between current density and electric field can then be derived. Because of the anisotropy due to the background field this is a linear tensor relationship. It is convenient to express the conductivity in a matrix form. Associated expressions for the susceptibility and dielectric tensors can be found.

- If the conductivity relation is substituted in Maxwell's equations and plane waves with exponential space and time variation assumed, a set of equations can be found relating the various components of the wave fields. The condition for self-consistency of this set of homogeneous equations is the dispersion relation, an equation relating wavelength and frequency. We choose to write it as a relationship between refractive index and frequency. It is a quadratic in the square of the refractive index. This implies that, for a particular direction of propagation, there are two characteristic waves in such a plasma, each with its own phase speed and polarization. Approximations can be found for the refractive index and polarization in a number of different frequency regimes.

- Waves can be classified as ordinary or extraordinary according to the polarization for propagation at right angles to the field. They can be classified as right-handed or left-handed according to their polarization for propagation along the field.

- There are a number of general properties of the refractive index which appear when parameters such as the frequency, plasma frequency, gyrofrequency and wave normal direction are varied. For some values of the parameters the refractive index becomes zero. This is known as a cut-off. The refractive index may become infinite, corresponding to a resonance. At a cross-over frequency the polarization labels change from ordinary to extraordinary or from right- to left-handed.

- Because of the anisotropy of the medium the direction of the energy flux in a characteristic wave is not the same as the direction of the wave normal.

Chapter 3

Waves in a Uniform Cold Magnetoplasma – 2. Rays and Wave Packets

3.1 Introduction

In this chapter we go on to consider more general disturbances than the infinite plane waves so far considered. This can be done because of the possibility of superposing a number of infinite plane waves to synthesize an arbitrary signal. This is the *principle of superposition* which applies to linear waves. The disturbances which we have considered have been assumed to be small enough for the equations to be linearized. The principle of superposition is simply an expression of the fact that an arbitrary small disturbance can be represented as a Fourier synthesis of such plane waves. Plane waves of different frequencies and directions having the appropriate amplitude and phase are added together to give the resultant wave.

We shall then use these techniques to study the nature of signals which are spatially and temporally limited to form *wave packets*. We shall show how the direction and speed of propagation of a wave packet is related to the direction and speed of energy propagation. This will allow us to consider the nature of waves emanating from a point source. Finally this will lead to a scheme for classification of the types of wave which can exist in a magnetoplasma.

3.2 Wave Packets and Rays

3.2.1 Superposition of Plane Waves

As an example of the principle of superposition consider a disturbance consisting of a superposition of plane waves, all propagated in the same direction. At time $t = 0$ suppose that a particular field component can be represented by the function $f(z)$. The medium has a dispersion relation $\omega = \omega(k)$. We suppose that the disturbance occurs over a limited range of values of z. We can perform a spatial Fourier analysis such that

$$f(z) = \int_{-\infty}^{\infty} \mathcal{F}(k) \exp ikz \, dk \tag{3.1}$$

This represents the disturbance as a synthesis of periodic spatial oscillations with wavelength $2\pi/k$ and amplitude $\mathcal{F}(k)dk$. The dispersion relation requires that it oscillate with angular frequency $\pm\omega$ given by the dispersion relation. Each component plane wave in the spectrum travels with its own phase speed, ω/k. If the disturbance is moving in the positive z-direction we choose the minus sign for ω. At time t each plane wave will have advanced a different distance and the disturbance will be a superposition of the waves as they each advance with their own phase speed:

$$f(z,t) = \int_{-\infty}^{\infty} \mathcal{F}(k) \exp\left(ikz - \omega(k)t\right)dk \tag{3.2}$$

This gives the disturbance as a function of z and t. In general, as time progresses, the disturbance, or *wave packet* will move bodily and also change shape. In certain circumstances the shape remains unchanged while the wave packet advances with a speed which is different from the phase speed. Only in the case of a non-dispersive medium, where all the plane waves have the same phase speed, is the speed of the wave packet the same as the phase speed.

In this section we discuss the concept of a wave packet and show how it moves with the velocity of energy propagation. This allows us to gain a better understanding of the propagation of waves which are not plane.

3.2.2 Method of Stationary Phase

In this section we outline an approximate method for evaluating the integral which represents a Fourier synthesis of a spectrum of plane waves. There are two such important methods: the method of steepest descents and the method of stationary phase. We shall confine ourselves to a heuristic treatment of the latter which can easily be interpreted physically. A more rigorous treatment is given by Budden [19].

The method of stationary phase is a method of finding approximate values of integrals of the form

$$\int \mathcal{F}(k) \exp[i\phi(k)]dk$$

The integrand contains an amplitude factor, $\mathcal{F}(k)$, and a phase $\phi(k)$. It represents a component wave in the spectrum. In our case the phase is of the form $\phi = kz - \omega(k)t$. Assume that the amplitude factor $\mathcal{F}$ is sharply peaked at $k = k_0$. Where $\mathcal{F}$ is appreciable we can expand ϕ in a Taylor series about k_0:

$$\phi = \phi_0 + \left(\frac{d\phi}{dk}\right)_0 (k - k_0) + O[(k - k_0)^2]$$

In general, provided that $(d\phi/dk)_0$ is not too small, the integrand oscillates, there is cancellation and a small contribution to the integral. The component waves in the spectrum interfere destructively. If the phase does not depend on k there is no oscillation and no cancellation. The interference of the component waves is constructive. This occurs where the phase variation is stationary with respect to k, that is where

$$\frac{\partial \phi}{\partial k} = 0 \tag{3.3}$$

Only for values of k where this condition is satisfied are there significant contributions to the integral. Thus the value of the integral is a weighted mean of the value of the integrand near the values of k for stationary phase. Budden [19] carries out a more general analysis and gives an explicit expression for the integral but we will not need it here.

3.2.3 One-Dimensional Waves Spatially Limited in the Direction of Propagation

Consider a disturbance which is a function only of z' and t. We first assume a pulselike disturbance strongly localized near $z' = 0$. Then $\mathcal{F}$ is real and varies slowly with k. In the extreme case of a δ-function pulse $\mathcal{F}$ is constant. The maximum contribution to the integral occurs at those points where the phase is stationary, that is where $\partial(kz')/\partial k = z' = 0$. At time t each component plane wave has phase $kz' - \omega(k)t$ where ω is related to k through the dispersion relation. The condition for stationary phase is now

$$\frac{\partial}{\partial k}[kz' - \omega(k)t] = 0$$

that is where

$$z' = \frac{\partial \omega}{\partial k} t \tag{3.4}$$

The phase is stationary and therefore the integral is maximum at a position z' which changes with t. Thus the disturbance travels in the z'-direction with speed

$$V_L = \frac{\partial \omega}{\partial k} \tag{3.5}$$

This speed is often called the group velocity. This is misleading. We have seen that in anisotropic media the velocity of energy propagation is not in the same direction as the wave normal. Because we have chosen our disturbance to be independent of x and y we cannot detect any motion perpendicular to the wave normal. V_L is actually the component of the group velocity in the direction of the wave normal.

An alternative derivation assumes a disturbance with a well defined wavelength, λ_0. This implies that it cannot be strongly localized like a pulse. Instead $\mathcal{F}$ must be strongly peaked at $k = k_0$, the wave number corresponding to λ_0. In this case we need only consider contributions to the integral from a small range of values of k near k_0. We write

$$k = k_0 + \delta k$$

Then, by Taylor's theorem,

$$\omega = \omega(k_0) + \left(\frac{\partial \omega}{\partial k}\right)_{k=k_0} \delta k$$

We may write

$$f(z', 0) \simeq \langle \mathcal{F}(k_0) \rangle \exp[i(k_0 z' - \omega_0 t)] \int_{k_0 - \Delta k}^{k_0 + \Delta k} \exp\left\{ i \left[z' - \left(\frac{\partial \omega}{\partial k}\right)_0 t \right] \delta k \right\} dk$$

where $\langle \mathcal{F}(k_0) \rangle$ is the mean value of $\mathcal{F}$ over the range $k_0 - \Delta k \ll k \ll k_0 + \Delta k$ within which $\mathcal{F}$ is appreciable. The quantity outside the integral sign then represents an infinite plane wave of frequency ω_0 and wavelength $2\pi/k_0$, advancing with the phase speed, ω/k. The integral is an envelope function of z' which has an unvarying shape but moves with speed $V_L = -(\partial \omega/\partial k)_0$.

In Figure 3.1 we illustrate this intuitively. We use lines to represent the crests of three component plane waves. In the upper part of the diagram we show them superimposed in such a way that the crests marked AA' coincide. At this point the disturbance has a maximum. After a time interval t each wave has advanced a distance $V_P t$ at its own phase speed and so different crests marked BB' now coincide. The maximum of the disturbance has thus advanced a different distance $V_L t$.

The phase and group speeds can easily be found graphically from a plot of the dispersion relation in the form ω *versus* k as shown in Figure 3.2. Clearly the phase speed for a particular value of k is the slope of the line OP drawn from the origin O to point P which corresponds to this value of k. The group speed is the slope of AB, the tangent to the curve at P.

3.2.4 Waves Spatially Limited in a Direction Perpendicular to the Wave Normal – Rays

We now consider a different synthesis of plane waves. Instead of synthesizing a disturbance spatially limited parallel to the wave normal, by adding waves of different wavelengths propagated in the same direction, we synthesize a disturbance limited in space, by adding waves of the same frequency propagated in different directions. In an anisotropic medium these waves have different wavelengths.

Figure 3.1: Illustration of group velocity by the superposition of three plane waves. Each line represents a wave crest. See text for explanation.

Suppose that a monochromatic disturbance $f(x, y, 0)$ is defined on the plane $z = 0$. The coordinate system chosen is now general and we do not assume that the magnetic field is in the z-direction. A factor $\exp(-i\omega t)$ is assumed. A two-dimensional Fourier transform allows us to represent it by

$$f(x, y, 0) = \int_{-\infty}^{\infty} \int_{-\infty}^{\infty} \mathcal{F}(k_x, k_y) \exp[i(k_x x + k_y y)] dk_x \, dk_y \tag{3.6}$$

A plane wave has the form $\exp[i(k_x x + k_y y + k_z z - \omega t)]$ and the dispersion relation demands that k_z is a function of k_x, k_y and ω. The extension of the disturbance throughout space is then

$$f(x, y, z) = \int_{-\infty}^{\infty} \int_{-\infty}^{\infty} \mathcal{F}(k_x, k_y) \exp\{i[k_x x + k_y y + k_z(k_x, k_y, \omega) z]\} dk_x \, dk_y \tag{3.7}$$

This represents f as an *angular spectrum of plane waves*. It is convenient to adopt a set of spherical polar coordinates in **k**-space, with polar axis parallel to z. Recall that the refractive index is cylindrically symmetrical about the direction of $\mathbf{B}_0$. We choose the z-axis parallel to the magnetic field. Then k is a function of ω and θ. The Jacobian of the transformation

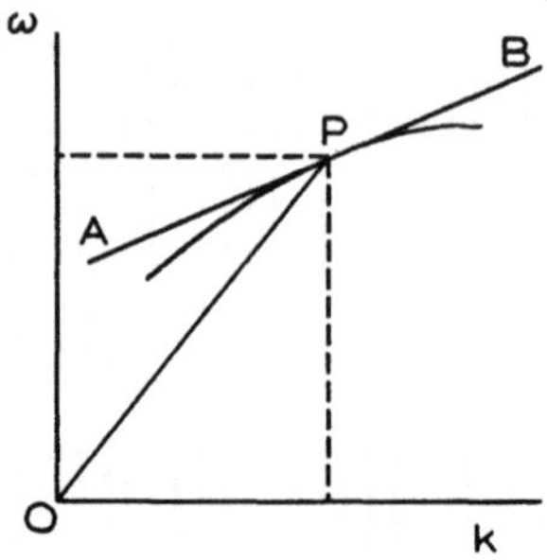

Figure 3.2: Phase speed and group speed at P. The slope of OP represents the phase speed and the slope of AB represents the group speed.

is $\mathcal{J} = k \sin\theta[k\cos\theta + \sin\theta(\partial k/\partial\theta)]$. In these coordinates the integral becomes

$$f(x,y,z) = \int_0^{2\pi}\int_0^{\pi} \mathcal{G}(\theta,\phi)\exp\{i(\omega n/c)[(x\cos\phi + y\sin\phi)\sin\theta + z\cos\theta]\}d\theta\, d\phi \tag{3.8}$$

where $\mathcal{G} = \mathcal{J}\mathcal{F}$ and we have set $k = \omega n/c$ where n is the refractive index.

The method of stationary phase can be applied to this integral for both the ϕ and θ coordinates. Because of the cylindrical symmetry we are only concerned with θ. Differentiate the phase with respect to θ and equate the result to zero. There is no loss of generality in setting $\phi = 0$ in the result which becomes

$$\frac{\partial n}{\partial\theta}(x\sin\theta + z\cos\theta) + n(x\cos\theta - z\sin\theta) = 0 \tag{3.9}$$

For a given value of θ this implies that there is a linear relationship between x and z; the only strong contribution to the integral comes from points lying on a straight line. This line is called the *ray path* . As shown in Figure 3.3a we define β to be the angle measured from the z-axis to this line so that $\tan\beta = x/z$. The stationary phase condition may then be written

$$\frac{\partial n}{\partial\theta}(\sin\theta\sin\beta + \cos\theta\cos\beta) + n(\sin\beta\cos\theta - \cos\beta\sin\theta)$$
$$\equiv \frac{\partial n}{\partial\theta}\cos(\theta-\beta) - n\sin(\theta-\beta) = 0 \tag{3.10}$$

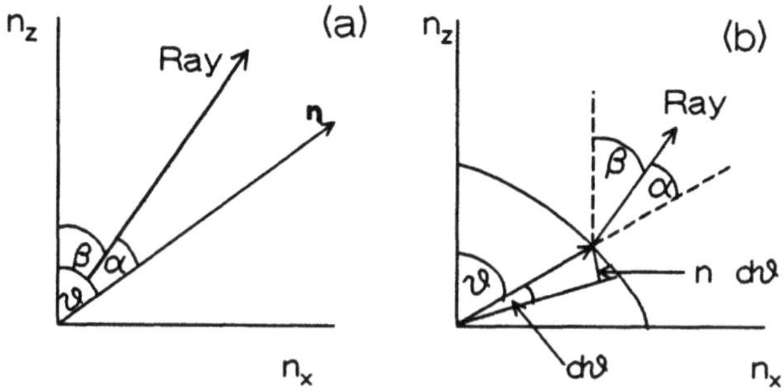

Figure 3.3: Relationships between wave normal and ray directions. See text for explanation.

But, as shown in Figure 3.3a, $\alpha = \theta - \beta$ is the angle measured from the ray to the wave normal. Thus

$$\tan\alpha = -\frac{1}{n}\frac{\partial n}{\partial\theta} \tag{3.11}$$

In order to compute this explicitly for a magnetoplasma we differentiate (2.58) to give

$$\begin{aligned}\tan\alpha &= -\frac{\frac{\partial A}{\partial\theta}n^2 + \frac{\partial B}{\partial\theta}}{4An^2 + 2B} \\ &= \frac{[(\kappa_1 - \kappa_3)n^2 - \kappa_L\kappa_R + \kappa_1\kappa_3]\sin\theta\cos\theta}{2An^2 + B}\end{aligned} \tag{3.12}$$

This is identical with (2.111) which proves that the ray path coincides with the direction of energy propagation. Further it can be seen from the geometry of Figure 3.3b that this implies the very important result that *the ray is normal to the refractive index surface.*

An intuitive picture of the superposition of plane waves whose wavelength depends on angle is given in Figure 3.4. This shows quite clearly how the path along which constructive interference takes place differs considerably from the direction of the wave normals of the component waves.

Once the signal has been confined to a narrow ray such as this it can be seen that the wave crests move along the direction of the ray with a speed $V_R = V_P/\cos\alpha$. The velocity which is directed along the ray and which has this magnitude is the *ray velocity*, $\mathbf{V}_R$.

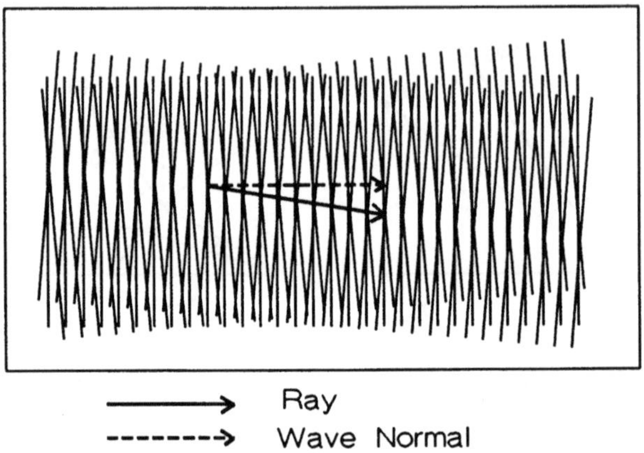

Figure 3.4: Superposition of plane waves with wavelength which depends on angle. The path along which constructive interference occurs is the ray path.

3.2.5 Waves Limited in Time and Space – Wave Packets

We can synthesize a disturbance which is limited both in time and space by superimposing plane waves of different wavelengths and directions. This implies a representation of the form

$$f(x,y,z,t) = \iiint \mathcal{F}(k_x,k_y,k_z)\exp\{i[k_x x + k_y y + k_z z - \omega(k_x,k_y,k_z)t]\}\,dk_x\,dk_y\,dk_z \tag{3.13}$$

It is not difficult to apply the method of stationary phase in the same way as before. The result is a disturbance which is spatially limited in all directions. It is called a *wave packet.* By requiring the phase to be stationary with respect to variations in k_x, k_y and k_z we can show that the disturbance moves with a velocity whose components are $\partial\omega/\partial k_x$, $\partial\omega/\partial k_y$, $\partial\omega/\partial k_z$. This velocity, $\mathbf{V}_G$, is called the *group velocity*

3.2.6 Velocity Relationships

The relationships between all the different velocities are summarized in Figure 3.5. In the following we refer to the various panels of the diagram.

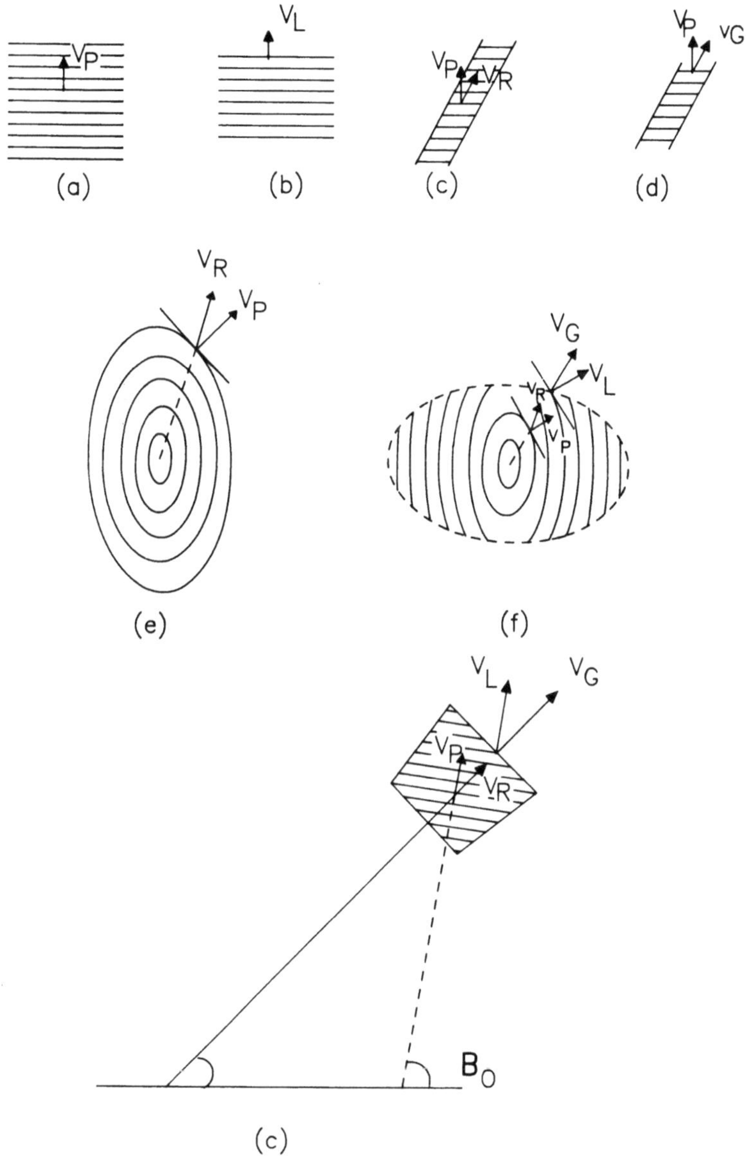

Figure 3.5: Summary of properties of various velocities. For explanation see text. (After Walker [115])

(a) A train of infinite plane wave fronts advances with the phase velocity, $\mathbf{V}_P$, perpendicular to the wave fronts. The direction of $\mathbf{V}_P$ defines the direction of the wave normal. The magnitude of the phase velocity is c/n.

(b) The envelope of a short train of infinite plane wave fronts advances with a velocity $\mathbf{V}_L$ in the direction of the wave normal. The wave fronts move through this envelope with velocity $\mathbf{V}_P$.

(c) A train of waves is confined spatially to a narrow pencil by superposing infinite trains of plane waves propagated in slightly different directions, θ. An infinite train of waves thus spatially confined travels along the ray with the ray velocity, $\mathbf{V}_R$. Suppose an observer is located at some distance from the source and observes the wave over a region of space that is small compared with the lateral extent of the ray. He will see approximately plane waves moving in the direction of the phase velocity. He would be wrong to project the wave normal backwards to find the location of the source. The position of the source is on a line projected back in the ray direction and if a time t has elapsed since a particular wave front left the source it will have traveled a distance V_Rt.

(d) The envelope of a short train of waves which is spatially confined to a narrow ray travels along the ray with the group velocity. The component of $\mathbf{V}_G$ in the direction of the wave normal is $\mathbf{V}_L$. Thus $V_G = V_L/\cos\alpha$.

(e) If a monochromatic point source radiates continuously in all directions it emits a succession of *wave fronts* which are surfaces of constant phase. For example the crest of a wave is a wave front. The wave fronts move out radially with the ray velocity. At time t a particular wave front has moved a distance V_Rt, as described in (c). Since the magnitude of the ray velocity is a function of θ each portion of the wave front will have traveled a different distance. The locus of the tip of the ray velocity vector as θ varies is called the *ray velocity surface*. It thus has the shape of a wave front originating from a point source. This refutes the common misconception that the locus of the tip of the phase velocity vector or *phase velocity surface* has the shape of a wave front (e.g. Stix [100], Chen [25]).

(f) If a point source is switched on and thereafter radiates continuously, the envelope of the disturbance travels out radially with the group velocity. This envelope is called the *signal front*. As before, we can define a *group velocity surface* which is the locus of the tip of the group velocity vector and has the same shape as a signal front. Within the signal front the wave fronts move out radially with the ray velocity. Of course a precise full-wave calculation would show that the signal front does not remain sharply defined. Well outside the signal front there is complete destructive interference of the component waves in the spectrum; well inside it the signal is essentially a monochromatic wave as described in (e). The arrival of the signal front at a point must be regarded as signalling the first sharp rise in amplitude of the signal before it settles down to its continuous wave behavior.

(g) All these relationships are summarized through the illustration of a short spatially confined wave packet. The phase velocity makes an angle θ with the magnetic field. The ray and group velocities make an angle β with the magnetic field and thus an angle α with the wave normal so that $\beta = \alpha - \theta$ as shown.

3.3 Classification of Waves in a Cold Plasma

3.3.1 Topology of the Refractive Index Surface

The refractive index surface is a surface of revolution in refractive index space about the direction of the magnetic field. The nature of the surface depends on the frequency, the electron and ion densities and the magnetic field strength. These can all be related to the parameters X and Y if we note that $X_1 = (m_e/m_1)X$ and $Y_1 = (m_e/m_1)Y$. There are always two surfaces for a given X and Y corresponding to the two values of refractive index. Either or both of these may have no real values of n if the wave is cut off. A surface may have one of three basic topologies. If there is no resonance cone the surface is a distorted sphere examples of which are shown in Figure 3.6 (a), (b), and (c). If there is a resonance cone with resonance angle θ_{crit} then the surface is asymptotic to the cone for $\theta \to \theta_{\text{crit}}$ and either n is real for $\theta \gg \theta_{\text{crit}}$ or for $\theta \ll \theta_{\text{crit}}$. Examples of the first case are shown in Figure 3.6 (d) and (e) and of the second in (f) and (g).

In addition to the basic topological features the surface may have points of inflection. Stott [102] has made a systematic study of the occurrence of these points of inflection in general media. Figure 3.6 shows all the types of interest in a cold plasma. While these points of inflection do not change the nature of the refractive index surface they have a dramatic effect on the topology of the ray velocity surface. To see these effects consider Figure 3.6a. There are two points of inflection on the refractive index surface in this case. The direction of the ray velocity is normal to the refractive index surface. Consider what happens to the direction of the ray velocity, ϕ, as the direction of the wave normal, θ, is increased from zero. At first ϕ increases with increasing θ until a point

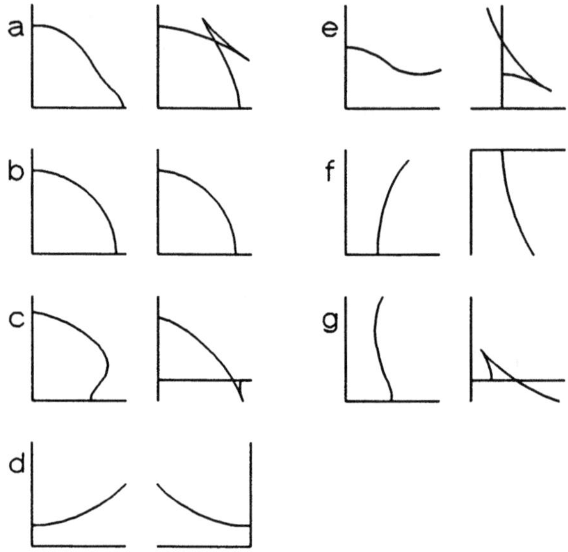

Figure 3.6: Topology of the refractive index surface and ray velocity surfaces. The magnetic field is in the x-direction. In each panel the refractive index surface in one quadrant is shown on the left and the ray velocity surface in the corresponding quadrant is shown on the right.

of inflection θ_S is reached. At this point $\phi = \phi_S$. As θ increases further ϕ *decreases* so that a cusp is formed. The process is repeated at the second point of inflection forming a second cusp. This cusp is very important if we remember that the ray velocity surface represents the shape of a wave front originating from a point source. As an example of its importance consider Figure 3.7. An observer located at O will see waves with three different wavelengths propagated in three different directions! These are not different modes but are all part of the same wave front. The angle ϕ_S defines a *Storey cone* [101]. It should not be confused with a resonance cone. There is no resonance at this angle but the wave shows a caustic behavior. Discussion of the details is beyond the scope of this book but see for example Walker and McKenzie [117] for an analogous case.

For given values of X and Y the ray velocity surface may or may not have a cusp or Storey cone. A Storey cone develops when two inflection points of the refractive index surface coalesce. The conditions for which this happens can be derived at the cost of a great deal of tedious algebra. It was done for a plasma for which the effect of heavy positive

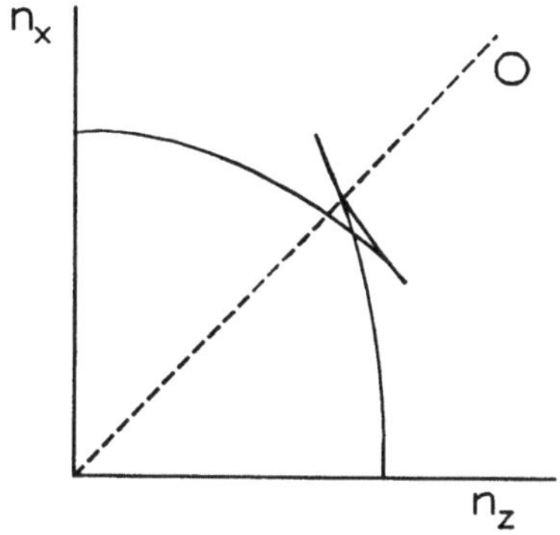

Figure 3.7: Wave fronts with a cusp spreading from a point source.

ions is neglected by Clemmow and Mullaly [28] and for a plasma with one species of positive ion by Walker [116]. These conditions represent a relationship between X and Y which can be plotted as a boundary in an X-Y parameter space. On one side of the boundary the ray surface has singular behavior in the form of a cusp or Storey cone; on the other it does not. In the language of catastrophe theory (e. g. [12]) the ray surface is structurally unstable on such a boundary and in moving across the boundary there is a *smooth unfolding* of the singularity. For this reason Walker [116] has called such a boundary a *ray surface unfolding* or briefly an *unfolding*.

3.3.2 The CMA Diagram

Consider an X–Y parameter space. In addition to the unfolding boundaries the conditions for cut-off or resonance also form boundaries in this space. On one side of such a cut-off boundary one of the refractive index surfaces disappears while it exists on the other side. On one side of a resonance boundary one of the surfaces has no resonance cone. At the boundary there is a resonance at $\theta = 0$ or $\pi/2$. On the other side the surface has a resonance cone. Thus the refractive index surface changes its topology whenever there is a cut-off or a principal resonance. The ray surface also changes its topology at an unfolding boundary. These conditions are all summarized in Table 3.1 for a one-ion plasma. The additional boundaries which occur between the ion gyrofrequencies are summarized in Table 3.2

All these boundaries are shown on plots in the X–Y space in Figure 3.8 for an electron plasma, a one-ion plasma and a two-ion plasma. In the upper left panel X and Y are of order unity and the effect of

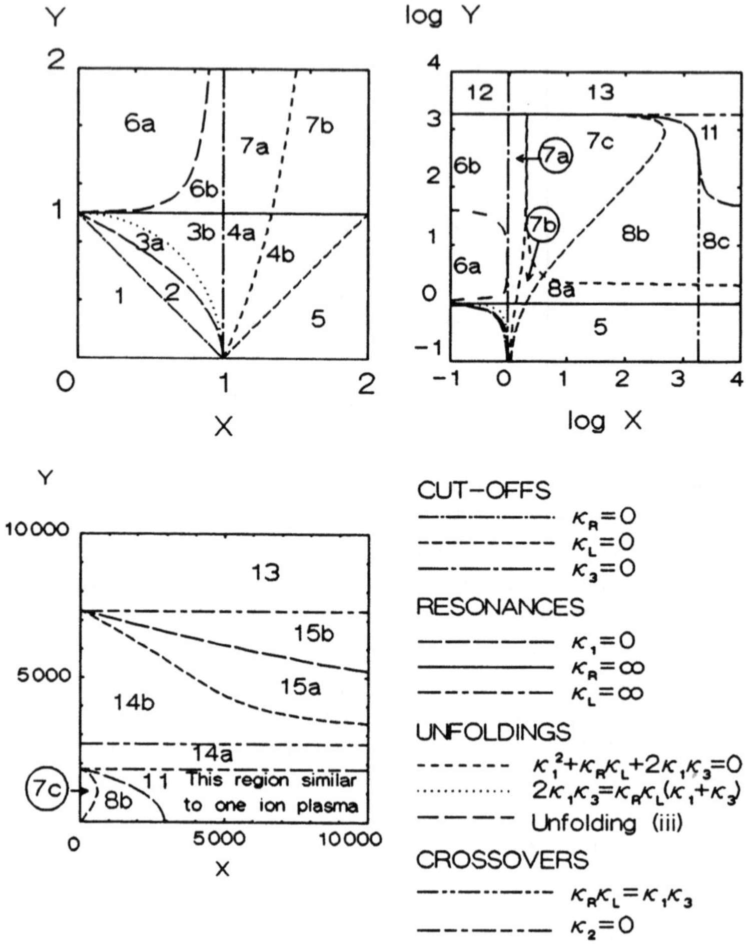

Figure 3.8: CMA diagrams. Upper left panel: electron magnetoplasma. Upper right panel: H^+ magnetoplasma. Lower left panel: magnetoplasma with equal number densities of H^+ and He^+ ions. Only the region between the ion gyrofrequencies is shown.

TYPE		*CONDITION*	*FORMULA FOR ONE-ION PLASMA*
Cut-off	i	$\kappa_R = 0$	$X \simeq 1 - Y$ (all positive X)
	ii	$\kappa_L = 0$	$X = (1 + Y)(1 - Y_1)$
	iii	$\kappa_3 = 0$	$X \simeq 1$
Resonance	i	$\kappa_1 = 0$	$X = (1 - Y^2)(1 - Y_1^2)(1 - YY_1)^{-1}$
	ii	$\kappa_R = \infty$	$Y = 1$
	iii	$\kappa_L = \infty$	$Y_1 = 1$
	iv	$\kappa_3 = 0$	$X = 1$
Unfolding	i	$\kappa_1^2 + \kappa_R\kappa_L$ $+ 2\kappa_1\kappa_3 = 0$	$X = 2[2 - Y^2(1 - Y_1^2)$ $\pm Y(1 - Y_1^2)^{1/2}]$ $\times[4 - Y^2(1 - Y_1^2)]^{-1}$
	ii	$2\kappa_1\kappa_3 =$ $\kappa_R\kappa_L(\kappa_1 + \kappa_3)$	$X \simeq \sqrt{1 - Y^2}$ (all positive X
	iii	No simple general expression	$F(Y^2 - G)$ $= 2Y(1 - F\sqrt{G})$ where $F \simeq 1 - Y_1^2$ and $G \simeq FY^2(1 - X - Y_1^2)^{-1}$
Cross-over	i	$\kappa_R\kappa_L = \kappa_1\kappa_3$	
	ii	$\kappa_3 = 0$	$X = 1$

Table 3.1: Conditions for boundaries in the CMA diagram.

the heavy positive ions is negligible. This diagram therefore applies to any magnetoplasma, whatever the ion species. In the upper right panel $m_e/m_1 = 0.000544$, thus the diagram represents an H^+ plasma. Other single ion plasmas have qualitatively similar behavior. Multi-ion plasmas have additional boundaries corresponding to the various ion cyclotron frequencies. The lower left panel shows the region between these boundaries for a two-ion plasma consisting of 50% H^+ and 50% He^+. As far as possible the numbering of the regions follows the scheme of Stix [100]. Regions 9 and 10 cannot be resolved and occur for $X \ll 1$ just below the $\kappa_L = \infty$ boundary. The subdivision of the regions by unfolding boundaries follows the scheme of Walker [116]. The labels on the unfolding boundaries correspond to those in Table 3.1

These are examples of *CMA diagrams* named after Clemmow and Mullaly [28] who introduced such a diagram for the case where heavy ions are ignored but the unfoldings are included, and Allis [3] who included the effect of ions in a more general formulation but considered only cut-offs and resonances. The diagrams are a means of classifying the waves

TYPE	CONDITION	FORMULA FOR TWO-ION PLASMA
Cut-off	$\kappa_L = 0$	$X = \frac{Y(1-Y_1)(1-Y_2)}{1-\eta_2 Y_1 - \eta_1 Y_2}$
	$\kappa_3 = 0$	$X \simeq 1$
Cross-over	$\kappa_2 = 0$	$\eta_2 Y_1^2 + \eta_1 Y_2^2 = 1$
Resonance	$\kappa_1 = 0$	$X = \frac{Y(1-Y_1^2)(1-Y_2^2)}{\eta_1 Y_1(1-Y_2^2)+\eta_2 Y_2(1-Y_1^2)}$

Table 3.2: Conditions for boundaries in the CMA diagram for a two ion plasma. $m_1 = 1$ amu, $m_2 = 4$ amu. The two ion species have equal number densities.

in a magnetoplasma. Within each region the topology of the refractive index and ray velocity surfaces remains unchanged. As a boundary is crossed either one of the refractive index surfaces or one of the ray surfaces changes its topology.

Also shown in the upper right panel of Figure 3.8 is the cross-over boundary $\kappa_R\kappa_L = \kappa_1\kappa_3$. On this boundary, when $\theta = \pi/2$, n^2 is the same for both ordinary and extraordinary waves and the refractive index surfaces touch. The labeling O or X switches from one surface to the other at this point although there is no change in topology.

The details of how the unfolding boundaries behave near the ion gyrofrequency ($Y_1 = 1$) are discussed by Walker [116]. They cannot normally be resolved from the ion gyroresonance.

The boundary $\kappa_3 = 0$ or $X \simeq 1$ requires further discussion. It has already been noted that it can represent a cut-off, a resonance or a cross-over frequency.

Consider Figure 3.9. This shows the behavior of the refractive index surfaces in regions 3b and 4a, on either side of the boundary $X = 1$. The left hand panel represents conditions in region 3b. There are two modes, *FOL* and *SXR* (the so-called Z-mode). The right hand panel represents conditions in region 4a. There is only one mode, the *XL* mode. The central panel represents the degenerate surfaces at $X = 1$. They consist of a spherical surface of unit radius and a straight line in the direction of the magnetic field. The OL-mode has a cut-off at $X = 1$. If we approach the boundary from region 4a it degenerates to the portion OP of the straight line, representing a plasma resonance rather than a wave.

The other mode has both a principal resonance and a cross-over. If we approach the boundary from region 3b the XR-surface degenerates to the sphere plus the portion PQ of the straight line. As we approach the boundary from region 4a the XL-surface degenerates to the sphere plus the portion of the straight line to the right of Q. The XL-surface thus has a principal resonance in that the angle of the resonance cone increases to 90^0 and the surface becomes closed. It also switches its polarization label to XR as we cross the boundary.

Consider now Figure 3.10. This shows the behavior of the refractive index surfaces in regions 6b and 7a. The format is the same as that of Figure 3.9 except that the most significant parts of the diagram are reproduced on a larger scale in the lower set of panels. The left hand panels, representing conditions in region 6b, also show two modes, FOL and SXR. The right hand panels represent conditions in region 7a. There are two modes, SOR (the so-called whistler mode) and FXL. In this case the OL-mode in region 6b is not connected to the OR-mode in region 7a. The OL-mode has a cut-off, degenerating to the portion OP of the straight line. The OR-mode has a principal resonance with the angle of the resonance cone closing to zero. The XR-mode in region 6b has a cross-over and changes its labeling to XL in region 7a.

In Tables 3.3 and 3.4 the types of surface shown in Figure 3.6 are related to the regions of the diagram. Column 1 gives the region and subregion of the CMA diagram. Column 2 labels the wave as fast, F (slow, S), ordinary, O (extraordinary, X), and right-handed, R (left-handed, L). Column 3 gives the shape of the refractive index surface according to Figure 3.6. Column 4 gives some common names used to describe these modes. In addition the labeling of the surfaces F-S, R-L and O-X is given. Only those regions which are shown on the diagrams are shown. The regions very near the ion gyrofrequency which cannot be resolved are neglected.

An alternative CMA plot has been given by Budden [19] for an electron plasma. Instead of plotting X *vs.* Y he plotted ω/Ω_e *vs.* ω_{pe}/Ω_e. In Figure 3.11 we show such a plot generalized for an H^+ plasma. Again in (a) the effect of the heavy positive ions is negligible and the diagram therefore applies to any magnetoplasma, whatever the ion species. In (b) $m_e/m_1 = 0.00054$, thus the diagram represents an H^+ plasma. The numbering of the regions follows the scheme of Figure 3.8. Regions 9 and 10 cannot be resolved and occur for very small ω_{pe} near the $\kappa_L = \infty$ boundary. This version of the CMA diagram is useful in two cases. If we wish to see how refractive index varies with frequency for a particular plasma we simply follow a horizontal line on the diagram. On the other hand, in many problems the magnetic field varies slowly while the electron density varies rapidly. In such a case we can study the regions encountered by a fixed frequency wave by following a vertical line on the diagram.

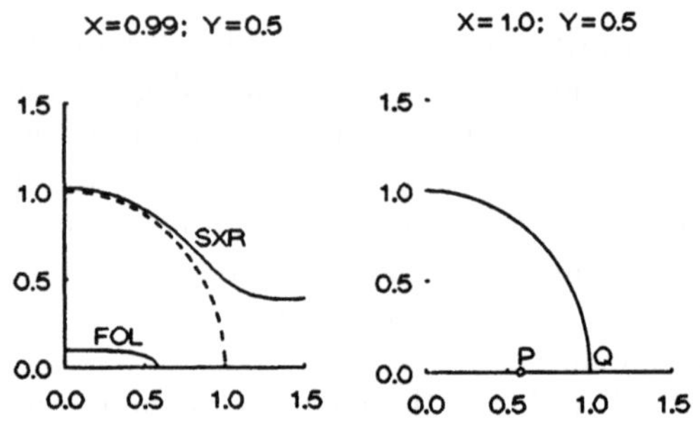

Figure 3.9: Behavior of the refractive index surface on either side of the $X = 1$ boundary for $Y < 1$.

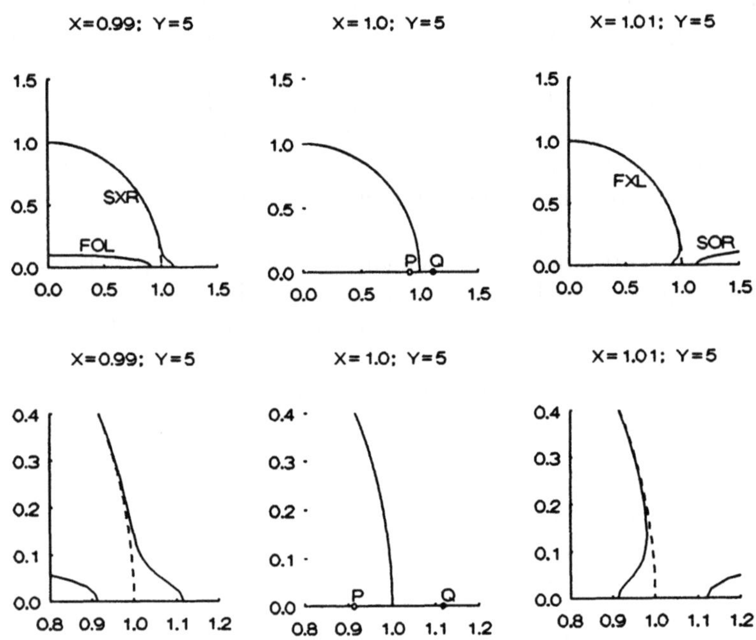

Figure 3.10: Behavior of the refractive index surface on either side of the $X = 1$ boundary for $Y > 1$.

REGION	*TYPE*	*SHAPE*	*NAME*
1	SOL	b	Free Space Ordinary mode
	FXR	b	Free Space Extraordinary mode
2	OL	b	Free Space Ordinary mode
3a	FOL	b	Free Space Ordinary mode
	SXR	d	Z-mode
3b	FOL	b	Free Space Ordinary mode
	SXR	e	Z-mode
4a	XL	c	Z-mode
4b	XL	a	Z-mode
5	—	—	Forbidden Region
6a	FOL	b	Ordinary mode
	SXR	b	Z-mode
6b	FOL	b	Ordinary mode
	SXR	a	Z-mode
7a	SOR	f	Whistler mode
	FXL	c	Z-mode
7b	SOR	f	Whistler mode
	FXL	b	Z-mode
7c	SOR	g	Whistler mode
	FXL	b	Z-mode
8a	OR	f	Whistler mode
8b	OR	g	Whistler mode
8c	XR	g	Whistler mode
9,10	Unresolved from ion gyrofrequency		
11	XR	b	Whistler mode
12	FOR	b	Ordinary mode
	SXL	b	Extraordinary mode
13	FXR	b	Whistler mode
	SOL	g	Ion cyclotron whistler

Table 3.3: Types of refractive index surface for a one ion plasma.

REGION	*TYPE*	*SHAPE*	*NAME*
14a	FXR	b	Whistler mode
	SOL	g	Ion cyclotron whistler
14b	SOR	b	Whistler mode
	FXL	g	Ion cyclotron whistler
15a	OR	g	Whistler mode
15b	OR	b	Whistler mode

Table 3.4: Types of refractive index surface between the ion gyrofrequencies for a two ion plasma.

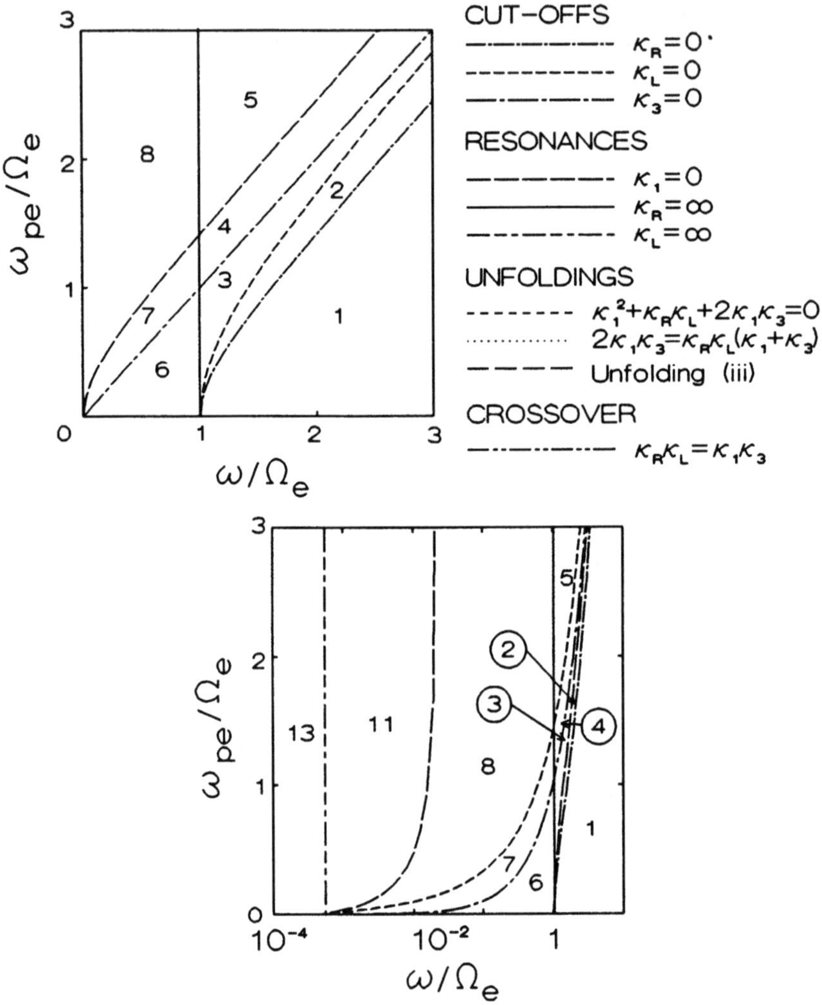

Figure 3.11: CMA diagram as modified by Budden [19] for a H^+ magnetoplasma. The abscissa is on a linear scale in the upper diagram and on a logarithmic scale in the lower to emphasize different regions.

3.3.3 Examples of the Refractive Index, Ray Velocity and Group Velocity Surfaces

Some examples of the refractive index, ray velocity and group velocity surfaces are shown in Figures 3.12 to 3.17. Walker [115] has performed more extensive calculations of the shapes of the various surfaces for an electron plasma in all the regions of the CMA diagram. Other examples are given elsewhere in this book when specific problems are discussed.

Each diagram is divided into three panels. In each of these the magnetic field is horizontal. The surfaces are surfaces of rotation about this axis. Each set of panels corresponds to a single region of the CMA diagram. The three panels in each subsection represent

(a) One quadrant of the refractive index surface.

(b) The corresponding quadrant of the ray velocity surface normalized to c, the speed of light.

(c) The corresponding quadrant of the group velocity surface normalized to c.

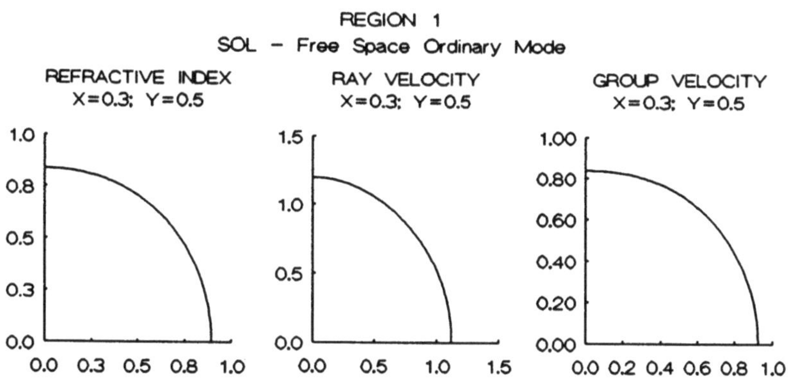

Figure 3.12: Refractive index surface, ray velocity surface and group velocity surface – 1.

These diagrams can help visualization of propagation in anisotropic media. Time devoted to studying them is well spent and will improve one's understanding of propagation in a magnetoplasma.

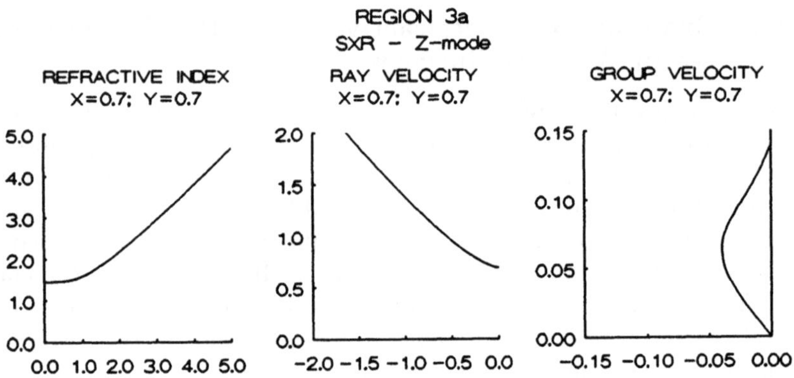

Figure 3.13: Refractive index surface, ray velocity surface and group velocity surface – 2.

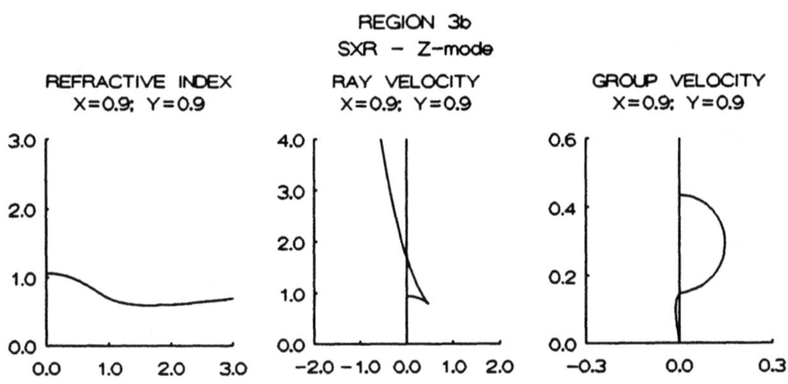

Figure 3.14: Refractive index surface, ray velocity surface and group velocity surface – 3.

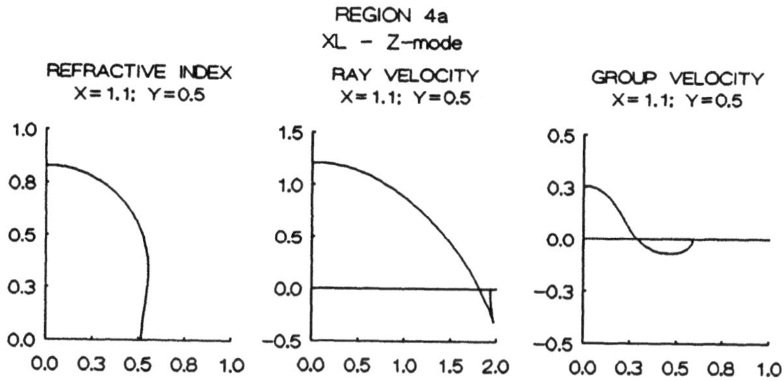

Figure 3.15: Refractive index surface, ray velocity surface and group velocity surface – 4.

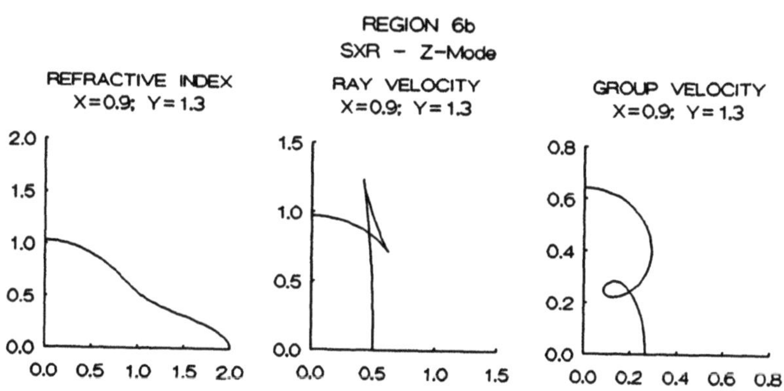

Figure 3.16: Refractive index surface, ray velocity surface and group velocity surface – 5.

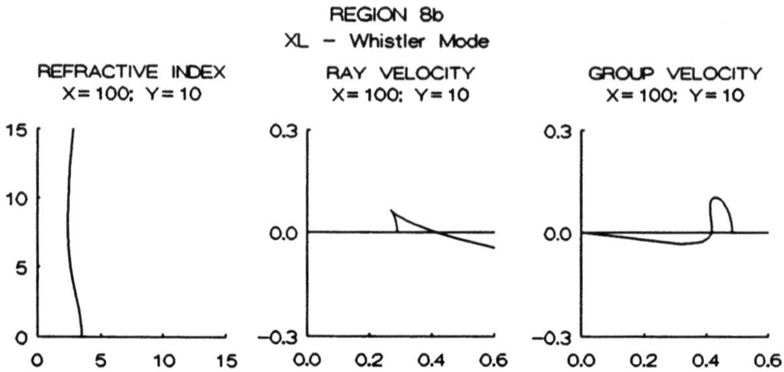

Figure 3.17: Refractive index surface, ray velocity surface and group velocity surface – 6.

3.4 Refractive Index and Dispersion Relation

We are now in a position to make systematic plots of the behavior of the dispersion relation and the refractive index. First let us consider how the refractive index depends on frequency. If we examine Figure 3.11, and neglect regions 9 and 10 which are too small to be resolved, we see that there are essentially three different cases: (i) $\omega_{pe} < \Omega_e$, in which case, as frequency increases, we encounter, in order, regions 13, 11, 8, 7, 6, 3, 2, 1; (ii) $\Omega_e < \omega_{pe} < \sqrt{2}\Omega_e$ in which case we encounter regions 13, 11, 8, 7, 4, 3, 2, 1; and (iii) $\omega_{pe} > \sqrt{2}\Omega_e$ in which case we encounter regions 13, 11, 8, 5, 4, 3, 2, 1. We could plot refractive index *vs.* normalized frequency for each of these cases but instead we choose to multiply n by normalized frequency getting $n\omega/\Omega_e \equiv ck/\Omega_e$ and plot the resulting normalized wave number against normalized frequency. The reason for this is that a line drawn from the origin to a point on the graph has slope $ck/\omega \equiv c/V_P \equiv n$ while the tangent to the graph at this point has slope $c\,dk/d\omega \equiv c/V_L$ where V_L is the component of the group velocity in the direction of the wave normal. These diagrams therefore allow us to deduce both the phase speed the speed of energy propagation of a plane wave. A set of such diagrams is shown in Figures 3.18, 3.19 and 3.20. The different scales of each diagram show various frequency ranges of interest. Labels L and R correspond to waves propagating parallel to $\mathbf{B}_0$ with left- and right-handed polarization respectively. Labels O and X correspond to ordinary and extraordinary waves propagating perpendicular to $\mathbf{B}_0$ respectively. Intermediate curves are plotted for $\theta = 45^0$. The shaded regions show the region within which the curves must lie for values of θ between 0^0 and 90^0.

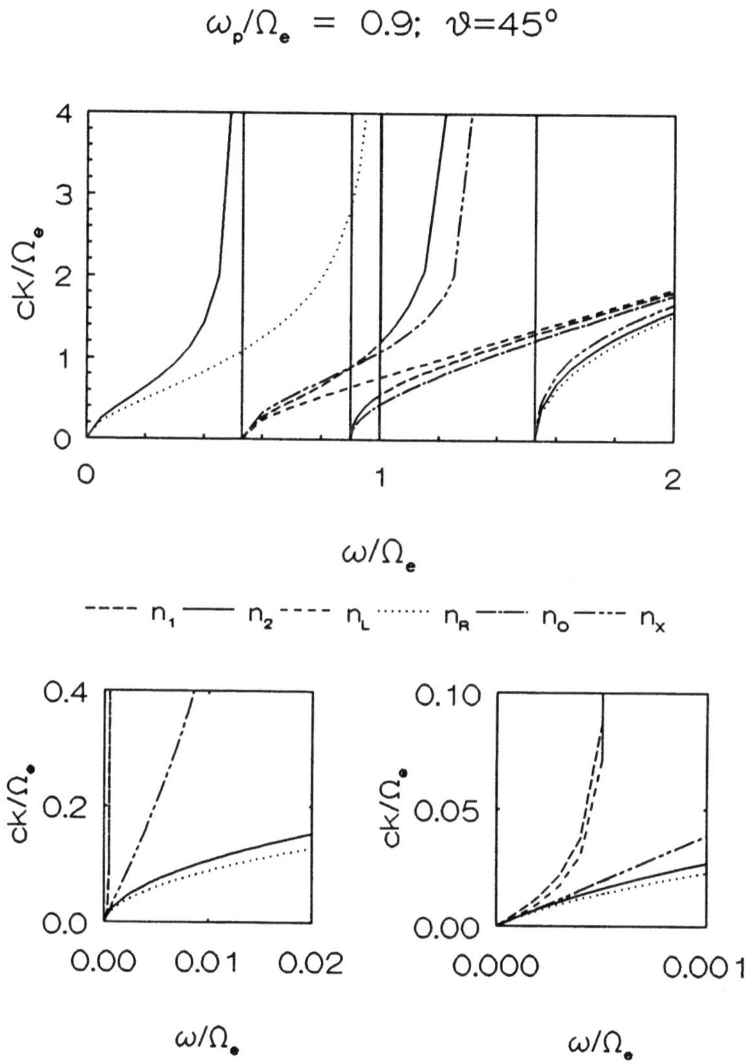

Figure 3.18: Dispersion relation for $\omega = 0.9\,\Omega_e$.

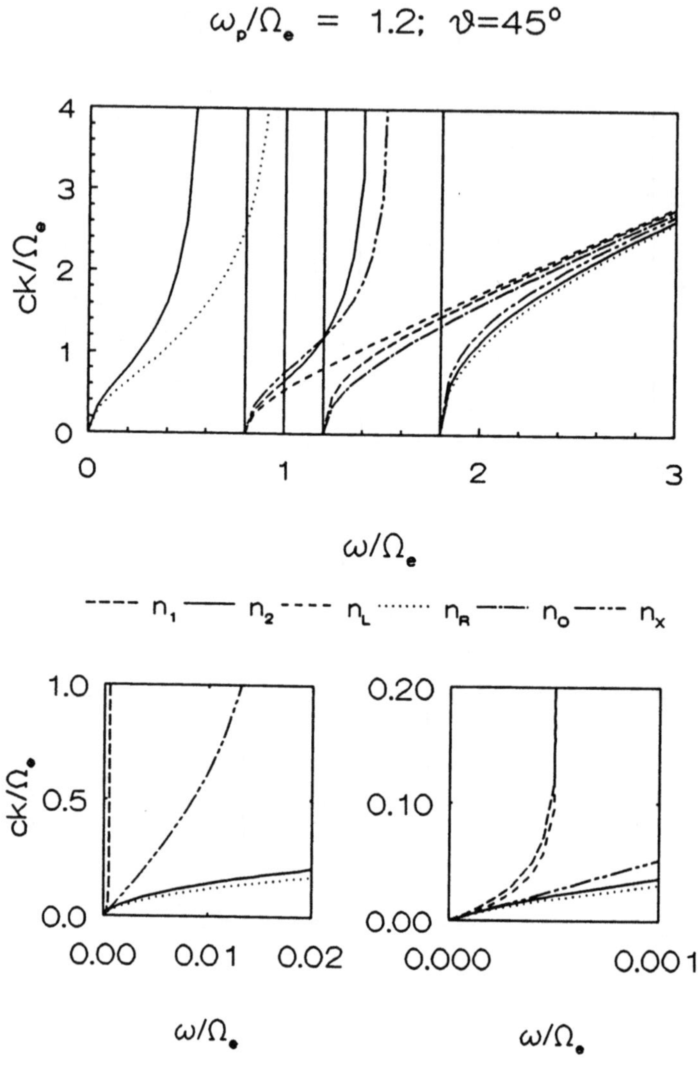

Figure 3.19: Dispersion relation for $\omega = 1.2\,\Omega_e$.

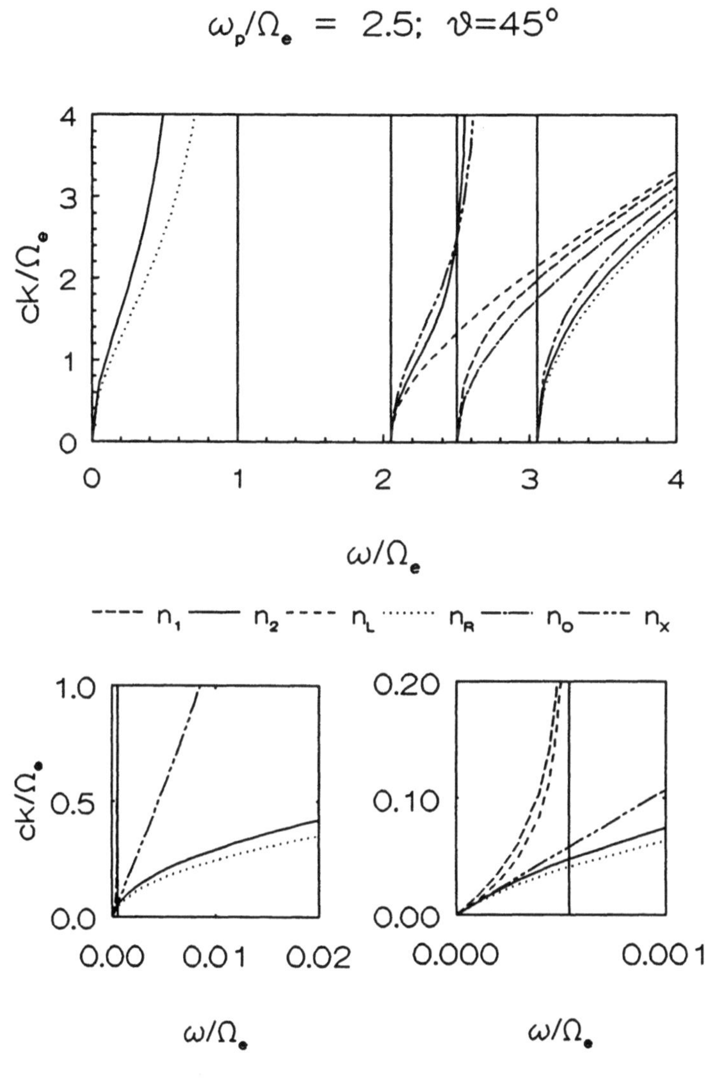

Figure 3.20: Dispersion relation for $\omega = 2.5\,\Omega_e$.

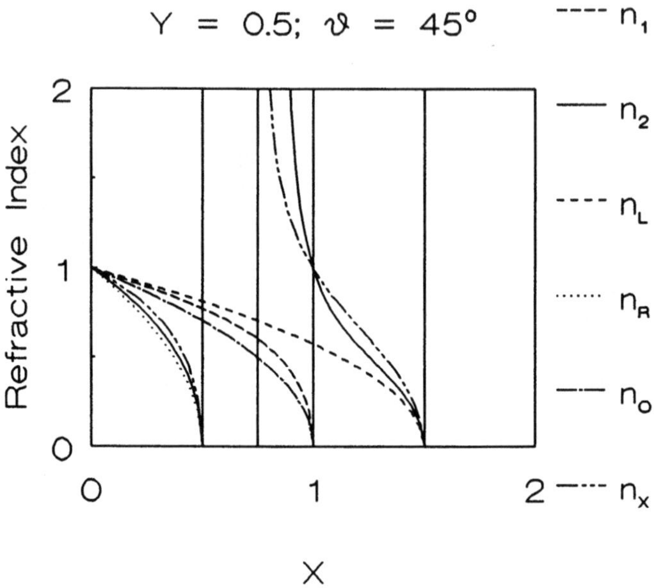

Figure 3.21: n *versus* X for $Y = 0.5$.

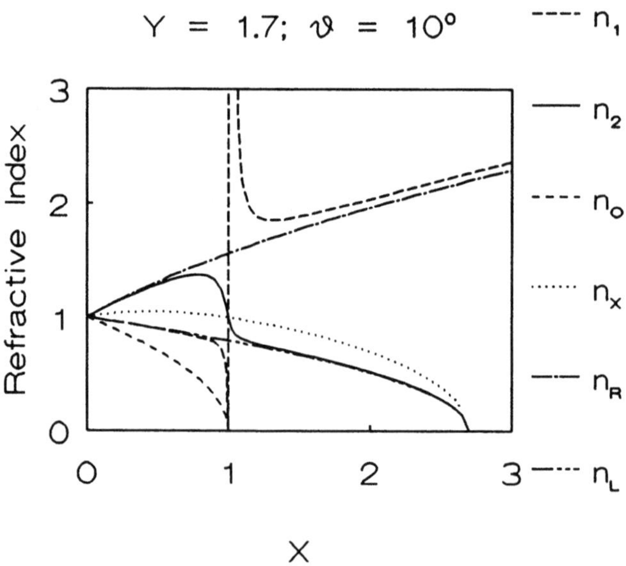

Figure 3.22: n *versus* X for $Y = 1.7$.

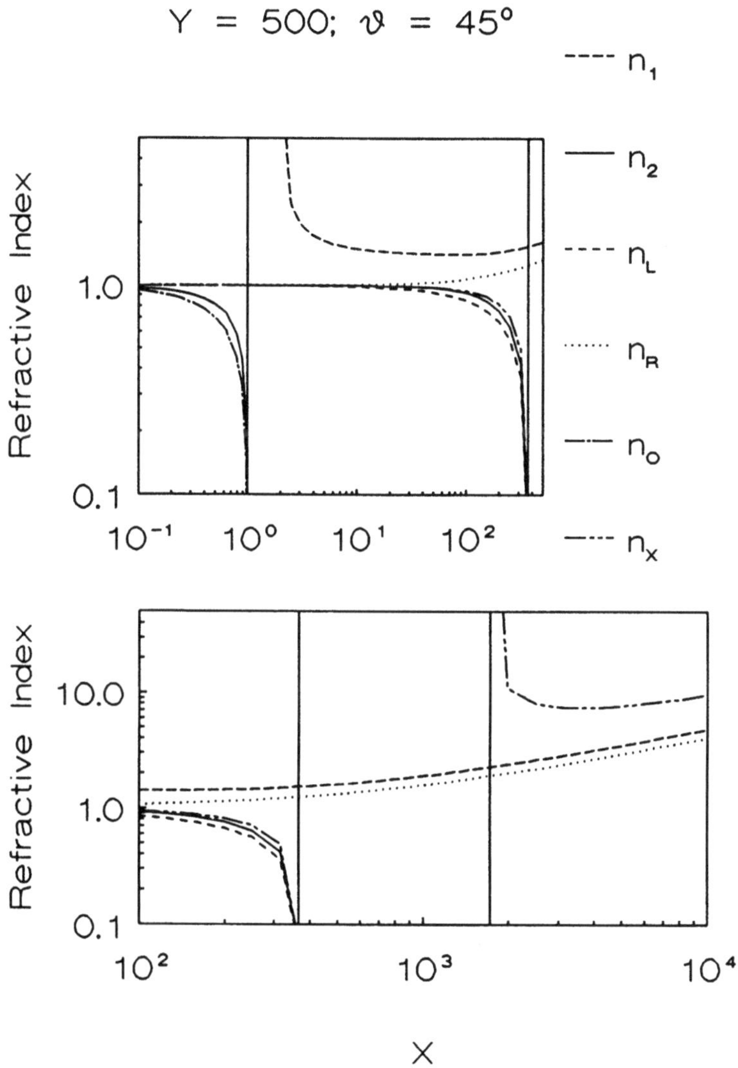

Figure 3.23: n *versus* X for $Y = 500$.

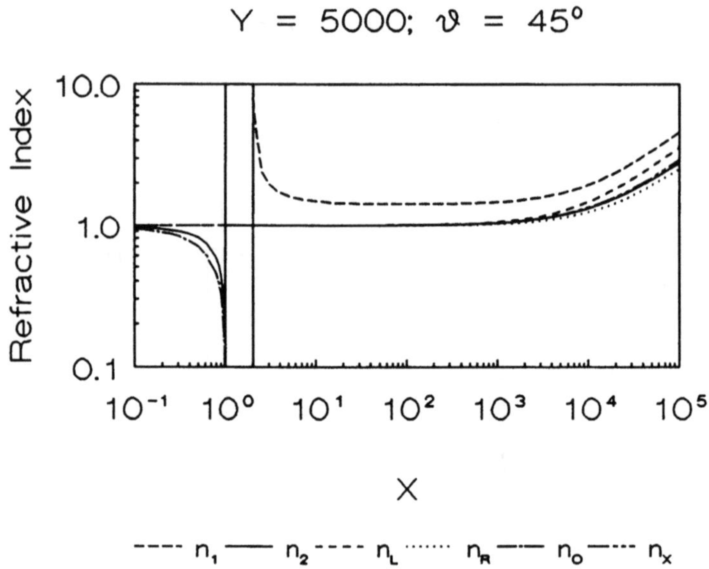

Figure 3.24: n *versus* X for $Y = 5000$.

The next plot of interest holds frequency and magnetic field constant and allows electron density to vary. This implies that Y is constant and X varies. Such a plot is useful in a region where the electron density varies much more rapidly that the magnetic field. A study of the upper right panel of Figure 3.8 shows that there are essentially four cases: (i) $Y < 1$ in which case as X increases the wave encounters regions 1, 2, 3, 4 and 5, (ii) $1 < Y < 1/Y_1$ in which case it encounters regions 6, 7 and 8, (iii) $1/Y_1 < Y < m_1/m_e$ in which case it encounters region 11 in addition, and (iv) $Y > m_1/m_e$ in which case it encounters regions 12 and 13. Again we have omitted cases involving regions 9 and 10 which are not resolved. In Figures 3.21, 3.22, 3.23 and 3.24 we show plots for these cases. In these figures labels L and R correspond to waves propagating parallel to $\mathbf{B}_0$ with left- and right-handed polarization respectively. Labels O and X correspond to ordinary and extraordinary waves propagating perpendicular to $\mathbf{B}_0$ respectively. Intermediate curves are plotted for $\theta = 45^0$. The shaded regions show the region within which the curves must lie for values of θ between 0^0 and 90^0.

3.5 Summary

- Arbitrary wave disturbances in a uniform cold plasma can be synthesized as superpositions of plane waves. A wave packet is a disturbance in a limited region of space. It travels with the group velocity in the ray direction. Plane waves move through it with the phase velocity. Phase and group velocities have different magnitudes and directions. The speed at which phase advances along a ray is the ray velocity. The ray direction is the direction of energy propagation.

- The refractive index surface for a given frequency is the locus of the tip of the refractive index vector for all possible wave normal directions. The ray direction for a plane wave is normal to the refractive index surface at the point where the refractive index vector for the wave intersects it.

- The ray surface for a given frequency is the locus of the tip of the ray velocity vector for all possible ray directions. It has the shape of the wave front which would be emitted by an isotropic point source in the medium.

- As the plasma frequency and gyrofrequency change the refractive index surface changes shape. The CMA diagram is a diagram representing boundaries in a parameter space X–Y where X and Y are normalized plasma frequency and gyrofrequency parameters. The major boundaries are the resonances and cut-offs. The topology of the refractive index surface changes across these boundaries. The ray surface undergoes additional changes in topology across subsidiary boundaries called unfoldings. The CMA diagram is an important means of classifying the waves which can occur in a given medium. Related diagrams can be defined using different parameter spaces.

- In this chapter many examples of refractive index surface, ray surface, group velocity surface and dispersion relations are plotted for different regions of the CMA diagram. Contemplation of these is a powerful means of gaining insights into wave behavior.

Chapter 4

Propagation of Electromagnetic Waves in a Non-Uniform Cold Magnetoplasma

4.1 Introduction

Until now we have considered waves in a uniform magnetoplasma. In the magnetosphere the ion and electron densities and the magnitude and direction of the magnetic field are functions of position. The general problem of solving the wave equation in such a medium is very difficult and is often intractable analytically so that numerical techniques must be used. If, however, the scale on which the plasma parameters vary is large compared with the wavelength then a variety of approximate methods can be used. In this book we shall confine ourselves to waves where this is possible. Methods of solving the more general problem numerically in a plane stratified medium are discussed in detail by Budden [17, 19] with particular application to radio waves in the ionosphere. Except in the ionosphere and within field-aligned irregularities in the magnetosphere the plasma parameters vary on a length scale comparable with the earth's radius and it is only for such wavelengths that our treatment breaks down.

4.2 Plane Stratified Media

If the magnitude and direction of the magnetic field and the ion and electron densities all vary in a single direction only the medium is said to be plane stratified. This is only a reasonable model in relatively few parts of the magnetosphere. Nevertheless it is instructive to consider the nature of wave propagation in such a medium.

4.2.1 Laws of Reflection and Transmission – Snell's Law

The simplest situation is a sharp boundary between two uniform magnetoplasmas. Suppose this boundary is the plane $z = 0$. Suppose that a characteristically polarized wave, either ordinary or extraordinary, is incident on such a boundary, with its wave normal making an angle θ with the z-axis and its plane of incidence (the plane containing the wave normal and the normal to the boundary) making an angle ϕ with the x-axis. The phase of the incident wave varies with x and y as $\exp[ik_0(n_x x + n_y y)] = \exp[ik_0 n \sin\theta(x\cos\phi + y\sin\phi)]$. There will in general be two characteristically polarized waves reflected and two transmitted. These waves must obey appropriate boundary conditions which must hold for all x and y. This is only possible if all the field components in the waves on either side of the boundary have identical phase variation in x and y. This requires that (1) ϕ is the same for the transmitted and reflected waves; the incident, reflected and transmitted waves all lie in the same plane, (2) the reflected wave makes the same angle θ with the z-axis; the angle of incidence is equal to the angle of reflection and (3) the angle θ' made by the transmitted wave with the z-axis is given by $n \sin\theta = n' \sin\theta'$, where n' is the refractive index in the second medium. This last condition is *Snell's Law* and holds even when the refractive index is itself a function of θ as is the case in a magnetoplasma. It can be expressed in the form $n_y = constant$ so that the spatial variation $\exp[ik_0(n_x x + n_y y)]$ is the same on either side of the boundary.

It is straightforward, but algebraically tedious, to define and find appropriate reflection and transmission coefficients. Methods are given by Budden [19]. We do not need to go into such detail. It is intuitively fairly obvious that, if the two media differ in their properties by only a small amount, the wave will be transmitted only slightly changed; the amplitude of the reflected wave will be very small and the transmitted wave will chiefly consist of a wave of the same characteristic polarization as the incident wave. This is important when considering the transition to continuously varying media.

4.2.2 Continuously Varying Medium

As a first approximation to a continuous medium consider a plane stratified medium which consists of a number of homogeneous slabs. Denote quantities associated with the ith slab by subscripts i. Suppose that a characteristic wave is incident on slab 1 at $z = 0$. Suppose also that the properties of adjacent slabs differ so slightly that reflection and coupling with the other characteristic polarization can be ignored. Then the characteristic wave is transmitted from slab to slab, obeying Snell's law at each boundary. The phase change in the z-direction in the ith slab is $k_0 n_i \Delta z_i$ where Δz_i is the thickness of the slab. The phase behavior at the top of the nth slab is then

$$\exp[ik_0(n_x x + n_y y + \sum_{i=1}^{n} n_i \Delta z_i)]$$

In the limit where $\Delta z_i \to 0$ this becomes

$$\exp[ik_0(n_x x + n_y y + \int_0^z n(z,\theta)\, dz)]$$

provided that coupling and reflection can be ignored when the limit is taken. This is permissible if the difference in the properties of adjacent slabs tends to zero appropriately as the slab thickness tends to zero. Otherwise cumulative reflection and coupling at the boundaries may invalidate the assumption that these processes can be ignored. A proper treatment such as that of Budden [19] shows that the requirement is that the properties of the medium must vary with z on a scale that is long compared with the wavelength.

This formulation is called the *phase integral* solution of the problem because of the integral $\int n\,dz$ which appears in the expression for the phase. A better treatment would also allow for amplitude changes resulting from the requirement that energy must be conserved along a ray. Such amplitude changes as well as the phase integral are included in a WKB solution of the problem [19] but we do not need to consider them.

4.2.3 The Booker Quartic Equation

In the phase integral solution n_x and n_y are fixed by the boundary conditions. The refractive index $n = (n_x^2 + n_y^2 + n_z^2)^{1/2}$ must still obey (2.58). In the expressions for the coefficients (2.59) (2.60) note that

$$\cos\theta = \frac{l_x n_x + l_y n_y + l_z n_z}{n}$$

where $l_x = B_x/B$, $l_y = B_y/B$, $l_z = B_z/B$ and are the x, y and z direction cosines of $\mathbf{B}$ respectively. If we substitute this in the expressions for the coefficients A, B, C and collect powers of n_z the result is a quartic equation:

$$\alpha n_z^4 + \beta n_z^3 + \gamma n_z^2 + \delta n_z + \epsilon \tag{4.1}$$

where

$$\begin{aligned}
\alpha &= \kappa_1 + (\kappa_3 - \kappa_1) l_z^2 && (4.2)\\
\beta &= 2(l_x n_x + l_y n_y) l_z (\kappa_3 - \kappa_1) && (4.3)\\
\gamma &= -\kappa_L \kappa_R (1 - l_z^2) - \kappa_1 \kappa_3 (1 + l_z^2) + 2\kappa_1 (n_x^2 + n_y^2) \\
&\quad -(\kappa_1 - \kappa_3)[l_z^2 (n_x^2 + n_y^2) + (l_x n_x + l_y n_y)^2] && (4.4)\\
\delta &= 2(l_x n_x + l_y n_y) l_z [\kappa_L \kappa_R - \kappa_1 \kappa_3 + (n_x^2 + n_y^2)(\kappa_3 - \kappa_1)] && (4.5)\\
\epsilon &= [\kappa_L \kappa_R - \kappa_1 (n_x^2 + n_y^2)][\kappa_3 - (n_x^2 + n_y^2)] \\
&\quad +(l_x n_x + l_y n_y)^2 [\kappa_L \kappa_R - \kappa_1 \kappa_3 + (n_x^2 + n_y^2)(\kappa_3 - \kappa_1)] && (4.6)
\end{aligned}$$

This is called the Booker quartic equation. It was first derived for an electron plasma by Booker [13]. It is just another form of the dispersion relation. However, the x- and y-components of the refractive index are fixed by the boundary conditions and n_z is the variable which is found from the dispersion relation.

For an electron plasma the coefficients take the form

$$\alpha = 1 - X - Y^2 + XY^2 l_z^2 \tag{4.7}$$

$$\beta = 2(l_x n_x + l_y n_y) l_z XY^2 \tag{4.8}$$

$$\begin{aligned}\gamma = & -2(1 - X - Y^2)(1 - n_x^2 - n_y^2 - X) \\ & + XY^2[(1 - l_z^2) + (1 - l_y^2)n_x^2 + (1 - l_x^2)n_y^2]\end{aligned} \tag{4.9}$$

$$\delta = -2(1 - l_x^2 - l_y^2)(l_x n_x + l_y n_y) l_z XY^2 \tag{4.10}$$

$$\begin{aligned}\epsilon = & (1 - n_x^2 - n_y^2 - X)[(1 - n_x^2 - n_y^2)(1 - X - Y^2) - X(1 - X)] \\ & -(1 - n_x^2 - n_y^2)(l_x n_x + l_y n_y)^2 XY^2\end{aligned} \tag{4.11}$$

Consider now a plane stratified medium in which the spatial behavior of the waves is given by a phase factor containing a phase integral

$$\exp\{-ik_0[n_x x + n_y y + \int^z n_z \, dz]\}$$

If we apply the method of stationary phase the only contribution to the integral will come from points for which the phase is stationary with respect to variations in n_x and n_y. If we differentiate with respect to each of these we get the two conditions

$$x = -\int^z \frac{\partial n_z}{\partial n_x} dz \quad \text{i. e.} \quad \frac{dx}{dz} = -\frac{\partial n_z}{\partial n_x}$$

$$y = -\int^z \frac{\partial n_z}{\partial n_y} dz \quad \text{i. e.} \quad \frac{dy}{dz} = -\frac{\partial n_z}{\partial n_y}$$

and these are the equations determining a ray path in a plane stratified medium. The quartic equation can be differentiated to give the derivatives of n_z. The following pair of equations for the ray path is the result:

$$\frac{dx}{dz} = \frac{\frac{\partial \beta}{\partial n_x} n_z^3 + \frac{\partial \gamma}{\partial n_x} n_z^2 + \frac{\partial \delta}{\partial n_x} n_z + \frac{\partial \epsilon}{\partial n_x}}{4\alpha n_z^3 + 3\beta n_z^2 + 2\gamma n_z + \delta} \tag{4.12}$$

$$\frac{dy}{dz} = \frac{\frac{\partial \beta}{\partial n_y} n_z^3 + \frac{\partial \gamma}{\partial n_y} n_z^2 + \frac{\partial \delta}{\partial n_y} n_z + \frac{\partial \epsilon}{\partial n_y}}{4\alpha n_z^3 + 3\beta n_z^2 + 2\gamma n_z + \delta} \tag{4.13}$$

The Booker quartic is just another form of the equation for the refractive index surface. It is of the form $F(n_x, n_y, n_z) = 0$. The normal to the refractive index surface is simply the gradient of F in refractive index space which has components $\partial F/\partial n_x$, $\partial F/\partial n_y$, $\partial F/\partial n_z$. This normal is in the ray direction. The components of the element $d\mathbf{r}$ of a ray thus have the ratio:

$$dx : dy : dz = \partial F/\partial n_x : \partial F/\partial n_y : \partial F/\partial n_z$$

We thus see that the expression for the ray path is another form of the condition that the ray is normal to the refractive index surface.

4.2.4 The Poeverlein Construction

Consider a signal which is propagated through a plane stratified medium. In order to find the path of the ray Poeverlein [88] devised a graphical construction. This is illustrated in Figure 4.1a. This represents a set of refractive index surfaces, A, B, C, D, all plotted with the same origin in refractive index space. Each surface corresponds to a different level $z = z_i$ in the stratified medium. The coordinate system has been chosen

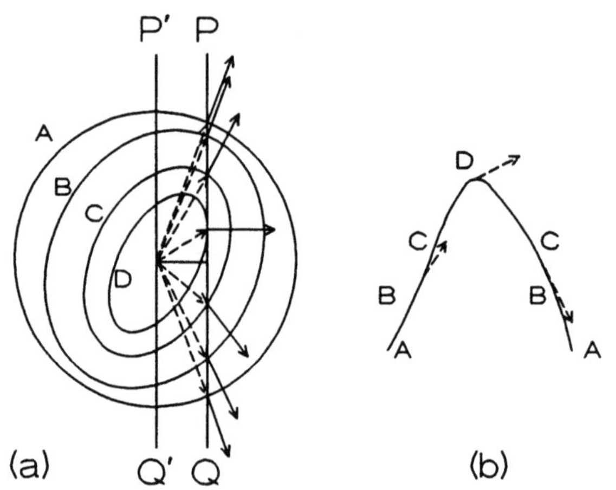

Figure 4.1: The Poeverlein construction for a wave in the magnetic meridian.

so that $n_y = 0$ and Snell's law requires that n_x is constant along the ray. The line PP' corresponding to $n_x = S$ is plotted on the diagram. The point where this line intersects the refractive index surface corresponding to $z = z_i$ represents the tip of the refractive index vector at this level. The direction of the ray is normal to the surface at this point and is shown as an arrow on the diagram at all the successive levels. By this means we can deduce the direction of the ray at each level and sketch it as shown in Figure 4.1b. This is the graphical equivalent of tracing the ray using the formula (4.12) with $n_y = 0$. It is a projection of the ray path on the plane containing the wave normal. In general the ray does not lie in this plane unless the wave normal is in the magnetic meridian. A number of interesting examples of this technique are given by Budden [19].

4.3 Ray Tracing in General Media

In the plasma surrounding the earth only the lowest regions, between about 100 and 500 km (the ionosphere), can reasonably be treated as horizontally stratified. In the magnetosphere the plasma density and the magnitude and direction of the magnetic field vary from point to point. The scale of this variation is generally large compared with wavelengths of interest and in these circumstances general ray tracing techiques are used. In this section we introduce the basic equations for numerical tracing of rays in a non-uniform medium, following closely the approach of Budden [19]. Such techniques are based on the phase memory concept. We assume that in these circumstances the phase of the wave varies in space as

$$\exp i\Phi(\mathbf{r})$$

where Φ represents the phase of the wave and is a single-valued function of position. When the medium is sufficiently slowly varying then this phase can be approximated by the phase integral

$$\Phi(x, y, z) \simeq \int^{\mathbf{r}} \mathbf{n}.d\mathbf{r}$$

Details of this approximation are given by Budden [19, §14.2]. Then clearly the phase Φ acts as a potential function for the refractive index vector $\mathbf{n}$ and

$$\mathbf{n} = \nabla\Phi$$

and thus

$$\nabla \times \mathbf{n} = 0 \tag{4.14}$$

The dispersion relation is a relationship between $\mathbf{n}$ and $\mathbf{r}$ which may be written

$$F(\mathbf{r}, \mathbf{n}) = 0 \tag{4.15}$$

At a fixed position $\mathbf{r}$ this relationship is the equation of the refractive index surface. As before the ray is normal to this surface. Suppose that in time dt a wavefront advances $d\mathbf{r}$ along the direction of the ray. Then $d\mathbf{r}/dt$ is the ray velocity, V_R, and is normal to the refractive index surface. Thus

$$\frac{d\mathbf{r}}{dt} = A\nabla_n F \tag{4.16}$$

where A is a constant of proportionality. In Cartesian coordinates,

$$\frac{dx}{dt} = A\frac{\partial F}{\partial n_x}, \quad \frac{dy}{dt} = A\frac{\partial F}{\partial n_y}, \quad \frac{dz}{dt} = A\frac{\partial F}{\partial n_z} \tag{4.17}$$

The dispersion relation can be regarded as giving **n** as a function of **r**:

$$F[\mathbf{r}, \mathbf{n}(\mathbf{r})] = 0 \tag{4.18}$$

Since it holds at all points in space

$$\frac{\delta F}{\delta x} = \frac{\partial F}{\partial x} + \frac{\partial F}{\partial n_x}\frac{\partial n_x}{\partial x} + \frac{\partial F}{\partial n_y}\frac{\partial n_y}{\partial x} + \frac{\partial F}{\partial n_z}\frac{\partial n_z}{\partial x} = 0$$

or, from (4.14) and (4.17),

$$\frac{\partial F}{\partial x} + A^{-1}\left[\frac{dx}{dt}\frac{\partial n_x}{\partial x} + \frac{dy}{dt}\frac{\partial n_x}{\partial y} + \frac{dz}{dt}\frac{\partial n_x}{\partial z}\right] = 0$$

The last three terms are simply $A^{-1}dn_x/dt$. We can get similar equations for dn_y/dt and dn_z/dt. Thus

$$\frac{d\mathbf{n}}{dt} = -A\nabla F \tag{4.19}$$

or, in Cartesian coordinates,

$$\frac{dn_x}{dt} = -A\frac{\partial F}{\partial x}, \quad \frac{dn_y}{dt} = -A\frac{\partial F}{\partial y}, \quad \frac{dn_z}{dt} = -A\frac{\partial F}{\partial z} \tag{4.20}$$

The constant A depends on the units in which the independent variable t is expressed. It is possible, for example, to write $A = 1$. In this case, however, t is not linearly related to the time elapsed. If we wish t to represent the time then since

$$\mathbf{V}_R.\mathbf{n} = c$$

we see that

$$A = c\left(n_x\frac{\partial F}{\partial n_x} + n_y\frac{\partial F}{\partial n_y} + n_z\frac{\partial F}{\partial n_z}\right)^{-1} \tag{4.21}$$

There are a number of possible choices for the function F. Any form of the dispersion relation may be used. For example the Booker quartic equation is a suitable form of the equation $F = 0$. There is, however, a choice which simplifies the form of A. This is the form used by Haselgrove [50] in her pioneering studies of ray tracing in plasmas. In this formulation it is noted that the formula for the refractive index gives n as a function of position $\mathbf{r}$ and the angle θ between the wave normal and the magnetic field direction. This angle is given by

$$\cos\theta = (l_x n_x + l_y n_y + l_z n_z)/\sqrt{(n_x^2 + n_y^2 + n_z^2)}$$

Thus we may write n as a function of $\mathbf{r}$ and $\mathbf{n}$

$$n = \mathcal{N}(\mathbf{r}, \mathbf{n})$$

The dispersion relation can then be written

$$F \equiv \frac{\mathcal{N}(\mathbf{r}, \mathbf{n})}{\sqrt{(n_x^2 + n_y^2 + n_z^2)}} - 1 = 0 \qquad (4.22)$$

In these circumstances Budden [19] shows that $A = c$ so that the ray tracing equations become

$$\frac{dx}{dt} = c\frac{\partial F}{\partial n_x}, \; \frac{dy}{dt} = c\frac{\partial F}{\partial n_y}, \; \frac{dz}{dt} = c\frac{\partial F}{\partial n_z} \qquad (4.23)$$

and

$$\frac{dn_x}{dt} = -c\frac{\partial F}{\partial x}, \; \frac{dn_y}{dt} = -c\frac{\partial F}{\partial y}, \; \frac{dn_z}{dt} = -c\frac{\partial F}{\partial z} \qquad (4.24)$$

The set of equations (4.17) and (4.20) are a set of first order differential equations which determine the ray. They can be integrated numerically step by step to find the ray path. They can be expressed in an appropriate coordinate system to suit the problem under investigation. A number of authors have presented forms of the equations in Cartesian and spherical polar coordinates. These are summarised by Budden [19]. In cylindrically symmetric or plane stratified models substantial simplification can be obtained by restricting attention to the case where the wave normal is in the magnetic meridian. In the most general case the advantages of expressing the model in an appropriate coordinate system may be outweighed by the complexity of the ray tracing equations; there are advantages to using Cartesian coordinates in these circumstances.

Ray tracing is a specialist subject and a detailed study is beyond the scope of this book. In Part II, when studying applications of the theory to actual magnetospheric phenomena we present several examples of rays which have been traced numerically.

4.4 Summary

- When the plasma is non-uniform it is not, in general, possible to obtain analytical solutions for waves in the medium.

- In the case of a sharp plane boundary separating two uniform media the boundary conditions require that Snell's law holds. In the opposite extreme, that of a plane stratified medium which varies with position on a distance scale that is long compared with the wavelength, a phase integral approach may be used. The refractive index is regarded as a known function of position and the cumulative change in phase as the wave progresses is found by integrating it with respect to position.

- The Booker quartic equation relates the Cartesian components of the refractive index. Combination of this with Snell's law and the phase integral expression leads to an expression for the ray path along which energy is propagated. The ray path can be visualized by using a graphical construction due to Poeverlein. This makes use of the fact that the ray is normal to the refractive index surface and follows the direction of this normal for different surfaces corresponding to successive positions along the path.

- If the medium is not plane stratified but still varies slowly with position a general set of first order ray tracing equations can be developed which may be integrated numerically in a given problem. The process is effectively a numerical implementation of the Poeverlein construction. It is equivalent to carrying out Huyghen's construction where the shape of the elementary wavelets is the shape of the ray surface.

Chapter 5

Waves in a Uniform Warm Magnetoplasma

5.1 Introduction

So far we have ignored the effect of the plasma pressure on wave propagation. This was equivalent to assuming that the wave speed was much greater than the thermal speed of the particles. In the magnetosphere there is a wide variety of populations of energetic particles, often far from thermal equilibrium. These particles can give rise to new forms of wave as well as modifying those already studied. If they are far from thermal equilibrium they can provide a source of free energy for wave generation or amplification. It is therefore seldom appropriate to treat the particles as if they constituted a gas in thermal equilibrium, characterized by a scalar pressure, a uniform temperature and a simple equation of state. Nevertheless this is what we shall do in this chapter. Our excuse is that this gives us a physical intuition for the kind of effects that can be introduced; when we use a more complicated kinetic or quasi-linear approach we shall have a better appreciation of the results we obtain. The simple picture may guide us to look for appropriate solutions to the more complicated problems.

In order to find the appropriate constitutive relation for the plasma we shall treat the electrons and each positive ion species as a fluid which obeys the fluid equations (1.9) (1.12) (1.13). We shall find that this introduces new features to the constitutive relation. Previously the conductivity and dielectric tensors were point relations depending only on the frequency and the values of the particle densities and magnetic field. Because there are spatial gradients of the pressure involved in the equation of motion the constitutive relation now depends on the spatial scale, (that is the wavelength) as well as the frequency. The refractive index thus appears explicitly in the constitutive relation and this increases the algebraic complexity of the problem.

This approach omits a very important effect. Particles which have almost the same velocity as the wave remain in phase with the wave fields over times that are long compared with the period. If a significant number of particles have this velocity then there can be substantial energy exchange between particles and waves. For plasmas in thermal equilibrium this energy is transferred from wave to particles resulting in the phenomenon of *Landau damping.* This effect is omitted in the fluid approach; many of the waves which arise in the fluid approach are severely damped. A rough rule of thumb is that if the wave speed is much larger than the root mean square speed of the particles the damping is small because few particles remain in phase with the wave. The discussion of Landau damping requires a kinetic approach.

It will be convenient in this chapter to change the notation slightly. We shall only deal with a single ion plasma and we shall denote quantities associated with electron and ion fluids by subscripts e and i respectively.

5.2 Characteristic Speeds

In a neutral gas sound waves are propagated with speed $V_S = \sqrt{\gamma P/\rho_m}$. For an ideal gas $P = NKT$ and, since $\rho_m = Nm$,

$$V_S^2 = \frac{\gamma KT}{m}$$

Similar expressions occur in our treatment of warm plasmas. It is convenient to define various sound speeds and to normalize them with respect to the speed of light. Define the *normalized electron and ion sound speeds*, U_e, U_i by

$$U_{e,i}^2 = \frac{\gamma_{e,i} KT_{e,i}}{m_{e,i}c^2} \tag{5.1}$$

It is also convenient to define the *normalized hybrid sound speed*, U, by

$$U^2 = \frac{K(\gamma_e T_e + \gamma_i T_i)}{(m_e + m_i)c^2} \simeq \frac{K(\gamma_e T_e + \gamma_i T_i)}{m_i c^2} \tag{5.2}$$

Note the following useful relations:

$$X_e U_i^2 + X_i U_e^2 = X_e U^2 \tag{5.3}$$

$$Y_e U_i^2 + Y_i U_e^2 = Y_e U^2 \tag{5.4}$$

These normalized quantities are of the order of the ratio of the kinetic energy of the particles to the rest mass energy. We shall only consider non-relativistic particles and thus

$$U_e^2,\ U_i^2,\ U^2 \ll 1 \tag{5.5}$$

5.3 The Constitutive Relation

5.3.1 Linearization of the Fluid Equations

We write the pressure and density as a constant zero order quantity plus a varying first order perturbation, $P_0 + p$, $\rho_0 + \rho = Nm + \rho$. The particle velocity $\mathbf{v}$ is of first order and since the differential operators operate only on first order quantities we can drop the operator $\mathbf{v}.\nabla$. We assume space and time behavior $\exp\{i[\mathbf{k}.\mathbf{r} - \omega t\}$ so that $\partial/\partial t \equiv -i\omega$, $\nabla \equiv i\mathbf{k}$. Then, to first order the fluid equations (1.9) (1.12) (1.13) become

$$\begin{aligned} -i\omega N m\mathbf{v} &= Nq\mathbf{E} + \mathbf{J} \times \mathbf{B}_0 - i\mathbf{k}p \\ \rho &= \rho_0 \mathbf{k}.\mathbf{v}/\omega \\ p &= (\gamma P_0/\rho_0)\rho \end{aligned} \qquad (5.6)$$

We note that $\mathbf{k} = \omega\mathbf{n}/c$, $\mathbf{J} = Nq\mathbf{v}$ and write $V_s^2 = \gamma P_0/\rho_0$ where V_s is the sound speed for the particle species. Then these can be combined to give a linearized equation of motion of a fluid element

$$\mathbf{J} + i\frac{\mathbf{J} \times \mathbf{\Omega}}{\omega} - \frac{V_s^2}{c^2}\mathbf{n}(\mathbf{n}.\mathbf{J}) = i\epsilon_0 \frac{\omega_p^2}{\omega}\mathbf{E} \qquad (5.7)$$

In Cartesian tensor notation this is

$$\{\delta_{jk} - U^2 n_j n_k + \frac{i}{\omega}\epsilon_{jkl}\Omega_l\} J_k = i\epsilon_0 \frac{\omega_p^2}{\omega} E_j \qquad (5.8)$$

where, in this case, U is the normalized sound speed for the particle species and not the hybrid sound speed. We can write this out in the coordinate system in which $\mathbf{B}_0$ is in the z-direction and $\mathbf{k}$ is in the x–z plane

$$\frac{i\epsilon_0\omega_p^2}{\omega}\begin{pmatrix} E_x \\ E_y \\ E_z \end{pmatrix} =$$

$$\begin{pmatrix} 1 - U^2n^2\sin^2\theta & \mp i\Omega/\omega & -U^2n^2\sin\theta\cos\theta \\ \pm i\Omega/\omega & 1 & 0 \\ -U^2n^2\sin\theta\cos\theta & 0 & 1 - U^2n^2\cos^2\theta \end{pmatrix}\begin{pmatrix} J_x \\ J_y \\ J_z \end{pmatrix} \qquad (5.9)$$

The upper sign corresponds to positive ions and the lower to electrons. This equation must be inverted to give the contribution of each particle species to the constitutive relation for the medium.

5.3.2 The Dielectric Tensor

We choose to give the constitutive relation in the form of a dielectric tensor, κ_{jk}, which is related to the conductivity tensor by (2.35). We solve (5.9) for each species and sum over the particles, getting for the elements of κ_{jk}

$$\kappa_{11} = 1 - \frac{X_e(1 - n^2U_e^2\cos^2\theta)}{1 - Y_e^2 - U_e^2n^2(1 - Y_e^2\cos^2\theta)}$$

$$
\begin{aligned}
& \quad -\sum_i \frac{X_i(1-n^2U_i^2\cos^2\theta)}{1-Y_i^2-U_i^2n^2(1-Y_i^2\cos^2\theta)} \\
\kappa_{22} &= 1-\frac{X_e(1-n^2U_e^2)}{1-Y_e^2-U_e^2n^2(1-Y_e^2\cos^2\theta)} \\
& \quad -\sum_i \frac{X_i(1-n^2U_i^2)}{1-Y_i^2-U_i^2n^2(1-Y_i^2\cos^2\theta)} \\
\kappa_{33} &= 1-\frac{X_e(1-Y_e^2-n^2U_e^2\sin^2\theta)}{1-Y_e^2-U_e^2n^2(1-Y_e^2\cos^2\theta)} \\
& \quad -\sum_i \frac{X_i(1-Y_i^2-n^2U_i^2\sin^2\theta)}{1-Y_i^2-U_i^2n^2(1-Y_i^2\cos^2\theta)} \\
\kappa_{12}=-\kappa_{21} &= \frac{iX_eY_e(1-U_e^2n^2\cos^2\theta)}{1-Y_e^2-U_e^2n^2(1-Y_e^2\cos^2\theta)} \\
& \quad -\sum_i \frac{iX_iY_i(1-U_i^2n^2\cos^2\theta)}{1-Y_i^2-U_i^2n^2(1-Y_i^2\cos^2\theta)} \\
\kappa_{13}=\kappa_{31} &= -\frac{X_eU_e^2n^2\sin\theta\cos\theta}{1-Y_e^2-U_e^2n^2(1-Y_e^2\cos^2\theta)} \\
& \quad -\sum_i \frac{X_iU_i^2n^2\sin\theta\cos\theta}{1-Y_i^2-U_i^2n^2(1-Y_i^2\cos^2\theta)} \\
\kappa_{23}=-\kappa_{32} &= -\frac{iX_eY_eU_e^2n^2\sin\theta\cos\theta}{1-Y_e^2-U_e^2n^2(1-Y_e^2\cos^2\theta)} \\
& \quad +\sum_i \frac{iX_iY_iU_i^2n^2\sin\theta\cos\theta}{1-Y_i^2-U_i^2n^2(1-Y_i^2\cos^2\theta)} \qquad (5.10)
\end{aligned}
$$

5.4 Dispersion Relations and Refractive Index

When the elements of the dielectric tensor are substituted into the wave equation (2.56) the result is

$$
\begin{pmatrix} \kappa_{11}-n^2\cos^2\theta & \kappa_{12} & n^2\sin\theta\cos\theta+\kappa_{13} \\ -\kappa_{12} & \kappa_{22}-n^2 & \kappa_{23} \\ n^2\sin\theta\cos\theta+\kappa_{13} & -\kappa_{23} & \kappa_{33}-n^2\sin^2\theta \end{pmatrix} \begin{pmatrix} E_x \\ E_y \\ E_z \end{pmatrix} = 0 \qquad (5.11)
$$

Alternatively we can write this in the x', y', z' coordinate system in which the z'-axis is parallel to the wave normal and $\mathbf{B}_0$ is in the x'–z' plane:

$$
\begin{pmatrix} \kappa'_{11}-n^2 & \kappa'_{12} & \kappa'_{13} \\ -\kappa'_{12} & \kappa'_{22}-n^2 & \kappa'_{23} \\ \kappa'_{13} & -\kappa'_{23} & \kappa'_{33} \end{pmatrix} \begin{pmatrix} E_{x'} \\ E_{y'} \\ E_{z'} \end{pmatrix} = 0 \qquad (5.12)
$$

where

$$\begin{aligned}
\kappa'_{11} &= \kappa_{11}\cos^2\theta + \kappa_{33}\sin^2\theta - 2\kappa_{13}\sin\theta\cos\theta \\
\kappa'_{22} &= \kappa_{22} \\
\kappa'_{33} &= \kappa_{11}\sin^2\theta + \kappa_{33}\cos^2\theta + 2\kappa_{13}\sin\theta\cos\theta \\
\kappa'_{12} &= \kappa_{12}\cos\theta + \kappa_{23}\sin\theta \\
\kappa'_{13} &= (\kappa_{11} - \kappa_{33})\sin\theta\cos\theta + \kappa_{13}(\cos^2\theta - \sin^2\theta) \\
\kappa'_{23} &= \kappa_{23}\cos\theta - \kappa_{12}\sin\theta
\end{aligned} \tag{5.13}$$

As usual, for this set of homogeneous equations to have a non-trivial solution the determinant of the matrix must be zero. This condition may be written

$$\begin{aligned}
& n^4\{\kappa_{33}\cos^2\theta + \kappa_{11}\sin^2\theta + 2\kappa_{13}\sin\theta\cos\theta\} \\
&+ \; n^2\{\kappa_{13}^2 - \kappa_{11}\kappa_{33} + 2[\kappa_{12}\kappa_{23} - \kappa_{13}\kappa_{22}]\sin\theta\cos\theta \\
&\qquad -[\kappa_{22}\kappa_{33} + \kappa_{23}^2]\cos^2\theta - [\kappa_{11}\kappa_{22} + \kappa_{12}^2]\sin^2\theta\} \\
&+ \; \kappa_{11}\kappa_{22}\kappa_{33} + 2\kappa_{12}\kappa_{23}\kappa_{13} + \kappa_{11}\kappa_{23}^2 - \kappa_{22}\kappa_{13}^2 + \kappa_{33}\kappa_{12}^2 \\
&= \; 0
\end{aligned} \tag{5.14}$$

While this appears to be a quadratic in n^2 it should not be forgotten that the elements of κ_{ij} also depend on n^2. These expressions are far more complicated than those for a cold plasma. The conductivity relation is not a point relation; it depends on spatial gradients of the field vectors so that the refractive index appears explicitly in the expressions for the elements of the dielectric tensor. It is often possible to make valid approximations at this stage rather than to press ahead with the general expression. We shall do so before committing ourselves to the labour of finding a more general dispersion relation.

5.4.1 The Electrostatic Approximation

The consequence of considering the effect of plasma pressure is to modify the cold plasma dispersion relation and introduce additional waves. The parameters which determine the size of these effects are the normalized rms speeds of the ions and electrons, U_e and U_i. We have already seen that these are very small for non-relativistic particles. The additional waves which are introduced can be expected to have refractive indices which are comparable with $U_{e,i}^{-1}$ which is very large. The refractive indices found from the cold plasma approach are generally of the order of $\sqrt{|\kappa_1|}$, $\sqrt{|\kappa_2|}$ and $\sqrt{|\kappa_3|}$. These are usually much less than $U_{e,i}^{-1}$ except at very low frequencies in the hydromagnetic limit. Another exception occurs near a resonance cone where n^2 for a cold plasma tends to infinity.

We therefore make an approximation in which we assume that n^2 is very large. To zero order (5.12) can be written

$$\begin{pmatrix} -n^2 & 0 & 0 \\ 0 & -n^2 & 0 \\ 0 & 0 & \kappa'_{33} \end{pmatrix} \begin{pmatrix} E_{x'} \\ E_{y'} \\ E_{z'} \end{pmatrix} = 0 \tag{5.15}$$

The first two equations then yield $E_{x'}$, $E_{y'} = 0$. The third equation is

$$\kappa'_{33} E_z = 0$$

which has a non-trivial solution if

$$\kappa'_{33}(n^2) = 0 \tag{5.16}$$

Provided that this equation yields a value for n^2 which satisfies the assumption that it is large, this describes a wave with electric field approximately directed along the direction of the wave normal. A second order approximation then shows that

$$E_{x'} \simeq \frac{\kappa'_{13}}{n^2} E_{z'}; \quad E_{y'} \simeq \frac{\kappa'_{23}}{n^2} E_{z'}$$

For this wave $\mathbf{k} \times \mathbf{E} = 0$ so that the magnetic field is approximately zero. For this reason it is often called an *electrostatic wave* . We shall see that the electrostatic approximation in a warm plasma can be derived as a special case of a more general approach.

We could have derived the dispersion relation of the electrostatic wave by replacing Maxwell's equations with Poisson's equation for the electric potential. Sometimes this is the only possible technique when dealing with situations which are too complicated to permit the warm plasma approach.

5.4.2 General Expression for the Refractive Index

For a plasma with one species of ion the expression for n^2 is a polynomial of fourth order. After a great deal of tedious algebra it may be written

$$\begin{aligned} &(n^2-1)^2 \\ &\quad + \frac{X_e}{\Delta_e}\Big\{-n^4(1-Y_e^2\cos^2\theta+2U_e^2)+n^2\Big[(2+U_e^2)(2-X_e) \\ &\quad -Y_e^2(1+\cos^2\theta)\Big]-(1-Y_e^2)-(2-X_e)(1-X_e)\Big\} \\ &\quad + \frac{X_i}{\Delta_i}\Big\{-n^4(1-Y_i^2\cos^2\theta+2U_i^2)+n^2\Big[(2+U_i^2)(2-X_i) \end{aligned}$$

$$
\begin{aligned}
& -Y_i^2(1+\cos^2\theta)\Big] - (1-Y_i^2) - (2-X_i)(1-X_i)\Big\} \\
+ & \frac{X_e X_i}{\Delta_e \Delta_i}\Big\{2n^4\Big[U_e^2(1-Y_i^2\cos^2\theta) \\
& + U_i^2(1-Y_e^2\cos^2\theta) + U_e^2U_i^2(1+Y_eY_i\cos^2\theta)\Big] \\
+ & n^2\Big[-4+(Y_e^2+Y_i^2)(1+\cos^2\theta) - 2Y_eY_i\sin^2\theta \\
& -U_e^2[(1-X_e)(1-Y_i^2\cos^2\theta)+2(1-X_i)(1+Y_eY_i\cos^2\theta) \\
& +(1-Y_i^2)] \\
& -U_i^2[(1-X_i)(1-Y_e^2\cos^2\theta)+2(1-X_e)(1+Y_iY_e\cos^2\theta) \\
& +(1-Y_e^2)]\Big] \\
+ & (1-Y_e^2)(2-X_i)+(1-Yi^2)(2-X_e) \\
& + 2(1+Y_eY_i)(1-X_e-X_i)\} \\
= & \ 0
\end{aligned}
\qquad (5.17)
$$

where

$$\Delta_{e,i} = 1 - Y_{e,i}^2 - n^2U_{e,i}^2(1-Y_{e,i}^2\cos^2\theta) \qquad (5.18)$$

We now multiply this equation by $\Delta_e\Delta_i$. The result is a fourth order equation in the square of the refractive index:

$$A_8n^8 + A_6n^6 + A_4n^4 + A_2n^2 + A_0 = 0 \qquad (5.19)$$

The coefficients are

$$
\begin{aligned}
A_8 &= U_e^2U_i^2(1-Y_e^2\cos^2\theta)(1-Y_i^2\cos^2\theta) && (5.20)\\
A_6 &= -U_e^2(1-Y_e^2\cos^2\theta)(1-Y_i^2) - U_i^2(1-Y_i^2\cos^2\theta)(1-Y_e^2) \\
&\quad -2U_e^2U_i^2(1-Y_e^2\cos^2\theta)(1-Y_i^2\cos^2\theta) \\
&\quad +X_eU_i^2(1-Y_i^2\cos^2\theta)(1-Y_e^2\cos^2\theta+2U_e^2) \\
&\quad +X_iU_e^2(1-Y_e^2\cos^2\theta)(1-Y_i^2\cos^2\theta+2U_i^2) && (5.21)\\
A_4 &= A + 2U_e^2(1-Y_e^2\cos^2\theta)(1-Y_i^2) + 2U_i^2(1-Y_i^2\cos^2\theta)(1-Y_e^2) \\
&\quad +U_e^2U_i^2(1-Y_e^2\cos^2\theta)(1-Y_i^2\cos^2\theta) \\
&\quad -X_eU_i^2(1-Y_i^2\cos^2\theta)[(2+U_e^2)(2-X_e)-Y_e^2(1+\cos^2\theta)] \\
&\quad -X_iU_e^2(1-Y_e^2\cos^2\theta)[(2+U_i^2)(2-X_i)-Y_i^2(1+\cos^2\theta)] \\
&\quad -2X_eU_e^2(1-Y_i^2) - 2X_iU_i^2(1-Y_e^2) \\
&\quad +2X_eX_i[U_i^2(1-Y_e^2\cos^2\theta)+U_e^2(1-Y_i^2\cos^2\theta) \\
&\quad +U_e^2U_i^2(1+Y_eY_i\cos^2\theta)] && (5.22)\\
A_2 &= B - U_e^2(1-Y_e^2\cos^2\theta)(1-Y_i^2) - U_i^2(1-Y_i^2\cos^2\theta)(1-Y_e^2) \\
&\quad +X_e\{U_i^2(1-Y_i^2\cos^2\theta)[1-Y_e^2+(2-X_e)(1-X_e)] \\
&\quad +U_e^2(1-Y_i^2)(2-X_e)\}
\end{aligned}
$$

$$
\begin{aligned}
& +X_i\{U_e^2(1-Y_e^2\cos^2\theta)[1-Y_i^2+(2-X_i)(1-X_i)] \\
& +U_i^2(1-Y_e^2)(2-X_i)\} \\
& -X_eX_i\{U_e^2[(1-X_e)(1-Y_i^2\cos^2\theta)+2(1-X_i)(1+Y_eY_i\cos^2\theta) \\
& +1-Y_i^2]+ \\
& U_i^2[(1-X_i)(1-Y_e^2\cos^2\theta)+2(1-X_e)(1+Y_eY_i\cos^2\theta) \\
& +1-Y_e^2]\} \qquad (5.23) \\
A_0 &= C \qquad (5.24)
\end{aligned}
$$

where A, B and C are the coefficients for a one ion plasma given by (2.70), (2.71) and (2.72). These expressions are far more complicated than for a cold plasma. Much simplification is possible in many realistic situations. We shall consider only a limited number of cases which can be of importance in the magnetosphere.

5.4.3 Neglect of Heavy Ion Effects – "Electron Plasma Waves"

At frequencies very much greater than the gm gyrofrequency, $\sqrt{\Omega_e\Omega_i}$, both X_i and Y_i are negligible. If, in addition, the electron and ion temperatures are comparable, $U_e^2 \gg U_i^2$ and all terms with subscript i can be dropped. The expression for the refractive index is now a cubic in n^2:

$$A_6n^6 + A_4n^4 + A_2n^2 + A_0 = 0 \qquad (5.25)$$

where

$$
\begin{aligned}
A_6 &= -U_e^2(1-Y_e^2\cos^2\theta) \qquad (5.26) \\
A_4 &= A + 2U_e^2(1-X_e-Y_e^2\cos^2\theta) \qquad (5.27) \\
A_2 &= B - U_e^2[(1-X_e)^2-Y_e^2\cos^2\theta] \qquad (5.28) \\
A_0 &= C \qquad (5.29)
\end{aligned}
$$

and A, B, C are the coefficients for a cold electron plasma (2.65), (2.66), (2.67).

The square of the normalized electron sound speed is extremely small for all temperatures encountered in the magnetosphere. Thus, in most circumstances, the coefficient A_6 is very small compared with the other coefficients, A_4, A_2 and A_0. In these circumstances the cubic can be approximately factorized

$$(A_6n^2 + A)[n^4 + (B/A)n^2 + C/A] \simeq 0 \qquad (5.30)$$

We have set $A_4 \simeq A$, $A_2 \simeq B$. The second factor simply represents the same electromagnetic waves that we found for a cold plasma. The other factor gives

$$n^2 \simeq \frac{1 - Y_e^2 - X_e(1 - Y_e^2\cos^2\theta)}{U_e^2(1 - Y_e^2\cos^2\theta)} \tag{5.31}$$

and represents a new kind of wave which only occurs for finite plasma pressure. This wave is usually called an *electron plasma wave* but, as we shall see, this nomenclature can be misleading. Of importance is the case of the longitudinal electron plasma wave. Here $\theta = 0$ and

$$n^2 \simeq (1 - X_e)/U_e^2$$

This is a true electrostatic wave with electric field oscillations parallel to the wave normal. Near the plasma frequency, when X is close to unity, the speed of these waves is much greater than the r. m. s. particle speed and Landau damping is expected to be weak. For frequencies far from the plasma frequency it may be significant.

The approximations we have used break down in the neighbourhood of a resonance cone. Here the approximate factorization of the cubic fails because the first two terms in the cubic for n^2 are of the same order of magnitude. In this case there is a different approximate factorization:

$$(A_6 n^4 + An^2 + B)(n^2 + C/B) = 0 \tag{5.32}$$

The quadratic factor represents a coupling between the so-called electron plasma wave and that electromagnetic wave which, when the pressure is neglected, has a resonance cone.

As an example of this kind of behavior we show the effect of including the plasma pressure on the Z-mode in Figures 5.1 and 5.2. Consider Figure 5.1. The dashed curve is the refractive index for the cold plasma Z-mode in region 3a of the CMA diagram. There is a resonance at angle θ_{res} given by

$$\cos^2\theta_{res} = -(1 - X_e - Y_e^2)/X_e Y_e^2$$

The full curve shows how the surface is modified by warm plasma effects. The choice $U_e = 0.05$ corresponds to electrons with thermal energies of about 0.75 keV. There is no longer a resonance cone; propagation is now allowed for all wave normal directions. For propagation along the magnetic field the refractive index is large and given by $n^2 \simeq (1 - X_e)/U_e^2$. This is simply the longitudinal electron plasma wave. As θ increases from zero n decreases. If coupling with electromagnetic waves is ignored the plasma wave has a zero at the resonance cone. When the coupling is allowed for the wave couples to the Z-mode at the resonance cone and deforms continuously into the Z-mode as θ increases. This shows very clearly why it is misleading to regard electrostatic and electromagnetic waves, as in some sense, distinct. It also indicates why it is misleading

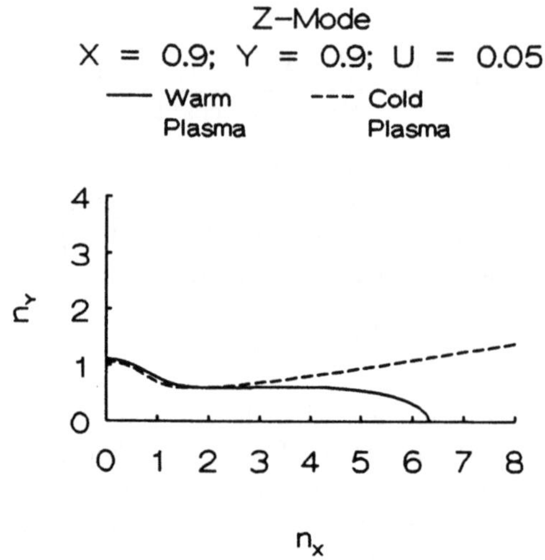

Figure 5.1: Modification of Z-mode in region 3a of the CMA diagram when plasma pressure effects are included. The dashed line shows the unmodified mode.

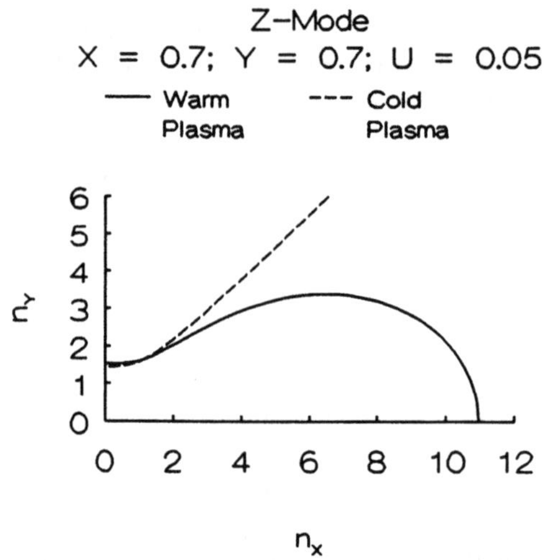

Figure 5.2: Modification of Z-mode in region 3b of the CMA diagram when plasma pressure effects are included. The dashed line shows the unmodified mode.

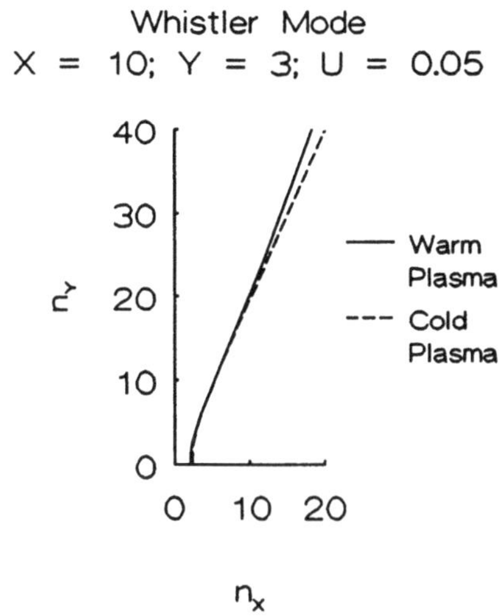

Figure 5.3: Modification of whistler mode in region 8b of the CMA diagram when plasma pressure effects are included. The dashed line shows the unmodified mode.

to speak of electron plasma waves as a different type of wave from an electromagnetic wave.

A different type of behavior is shown in Figure 5.3. Here the uncoupled electron plasma wave has a resonance where $A_6 = 0$ ($\cos^2\theta = 1/Y_e^2$) and a zero at the resonance cone ($\cos^2\theta_{res} = -(1 - X_e - Y_e^2)/X_eY_e^2$). The coupling to the whistler mode occurs at the resonance cone and has the effect of creating a single refractive index surface with a resonance cone at a somewhat larger angle. Landau damping can be expected to become important for these waves as the direction of the wave normal approaches the resonance cone and the wave speed becomes comparable with the r. m. s. particle speed.

5.4.4 Low Frequency Approximation for the Refractive Index

We adopt the low frequency approximation of section 2.5.2 and note that U_e^2, $U_i^2 \ll 1$. In each of the coefficients A_j only the highest power of X_e is retained and products $U_i^2U_e^2$ are neglected. The expressions (2.79), (2.80) and (2.81) are used for A, B and C. The result is

$$A_8 \simeq U_e^2U_i^2(1 - Y_e^2\cos^2\theta)(1 - Y_i^2\cos^2\theta) \simeq 0 \qquad (5.33)$$

$$A_6 \simeq X_e U^2(1 - Y_e^2 \cos^2\theta)(1 - Y_i^2 \cos^2\theta) \tag{5.34}$$

$$A_4 \simeq X_e[Y_e^2 \cos^2\theta(1 - Y_i^2) - (1 - Y_e Y_i) + 2X_e U^2(1 - Y_i Y_e \cos^2\theta)] \tag{5.35}$$

$$A_2 \simeq X_e^2[Y_e Y_i(1 + \cos^2\theta) - 2 + X_e U^2] \tag{5.36}$$

$$A_0 \simeq -X_e^3 \tag{5.37}$$

The setting of $A_8 = 0$ can be justified if X_e is sufficiently large and U sufficiently small. In this case the fourth order equation has an approximate factor $(n^2 + X_e U^2/U_e^2 U_i^2)$. Equating this factor equal to zero gives a wave which is always evanescent and not of physical interest. Equating the other factor to zero gives

$$A_6 n^6 + A_4 n^4 + A_2 n^2 + A_0 = 0$$

The expression for the refractive index is thus a cubic equation in n^2. This approximation is equivalent to that made by Stringer [105]. His results were obtained by a different method. He develops a wide variety of approximate formulae for use in different situations and presents representative plots of the dispersion relation in the form ω *versus* k. We shall only give examples of a few special cases in the following subsections.

5.4.5 Ion Acoustic Waves

We can obtain a further simplification if $U^2 \ll X_e^{-1}$. In this case the terms in U^2 can be neglected in the expressions for A_4 and A_2. There is a common factor X_e which can be divided out. Then, except where θ is almost exactly 90^0, the coefficients become

$$A_6 \simeq -U^2 Y_e^2 \cos^2\theta(1 - Y_i^2 \cos^2\theta) \tag{5.38}$$

$$A_4 \simeq Y_e^2 \cos^2\theta(1 - Y_i^2) \tag{5.39}$$

$$A_2 \simeq X_e[Y_e Y_i(1 + \cos^2\theta) - 2] \tag{5.40}$$

$$A_0 \simeq -X_e^2 \tag{5.41}$$

There are two approximate factors:

$$[(A_6/A_4)n^2 + 1][A_4 n^4 + A_2 n^2 + A_0] = 0 \tag{5.42}$$

The second factor represents waves in a cold plasma, unaffected by the plasma pressure. The first factor yields

$$n^2 = -\frac{A_4}{A_6} = \frac{1 - Y_i^2}{(1 - Y_i^2 \cos^2\theta)U^2} \tag{5.43}$$

or

$$n^2 = \left\{\frac{m_i c^2}{K(\gamma_i T_i + \gamma_e T_e)}\right\}\left\{\frac{\omega^2 - \Omega_i^2}{\omega^2 - \Omega_i^2 \cos^2\theta}\right\} \tag{5.44}$$

These are called *ion acoustic waves.* When the wave normal is aligned with the magnetic field they travel with the hybrid sound speed cU. It should be noted that, at these very low frequencies the electrons respond immediately to the wave field and the electron fluid is thus isothermal with $\gamma_e = 1$. The ion fluid responds adiabatically.

In Figures 5.4 and 5.5 examples are shown of the refractive index and

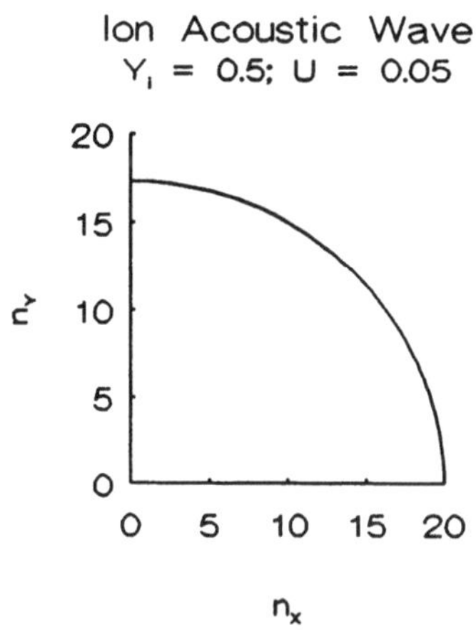

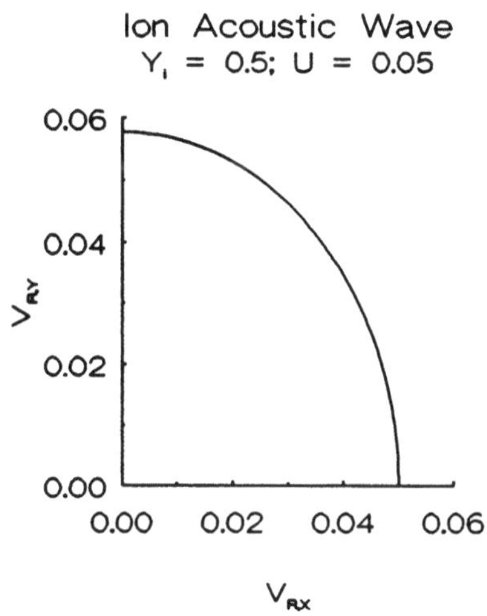

Figure 5.4: Refractive index surface and ray surface for ion acoustic wave for which $\omega > \Omega_i$

ray surfaces for ion acoustic waves. It will be noted that, if $\omega > \Omega_i$, the wave propagates in all directions while, if $\omega < \Omega_i$, there is a resonance cone where $\cos\theta = 1/Y_i$. Because the wave speed is comparable with the thermal speed, Landau damping may be significant but may only become overwhelming near the resonance cone.

5.4.6 Hydromagnetic Limit

In the hydromagnetic limit we assume $Y_i^2 \gg 1$ but do not assume that $U^2 \ll X_e^{-1}$. The coefficients then become

$$A_6 \simeq U^2 Y_e^2 Y_i^2 \cos^4\theta \tag{5.45}$$
$$A_4 \simeq -Y_e^2 Y_i^2 \cos^2\theta - 2X_e Y_e Y_i U^2 \cos^2\theta \tag{5.46}$$
$$A_2 \simeq X_e[Y_e Y_i(1+\cos^2\theta) + X_e U^2] \tag{5.47}$$
$$A_0 \simeq -X_e^2 \tag{5.48}$$

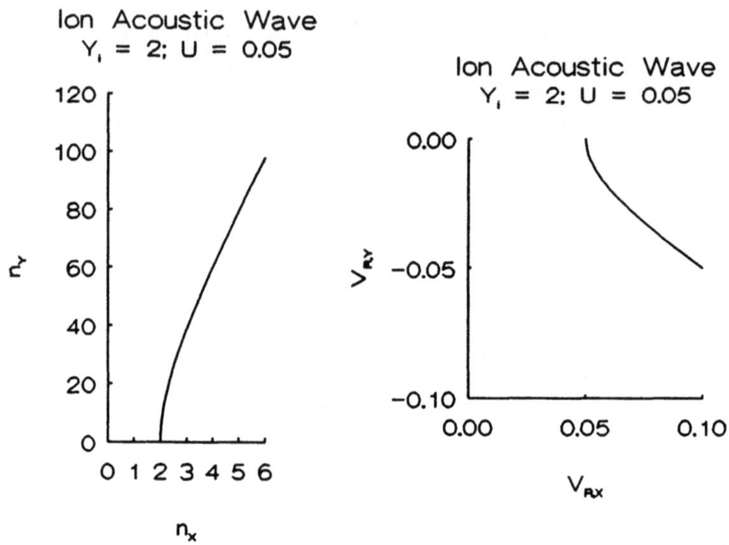

Figure 5.5: Refractive index surface and ray surface for ion acoustic wave. $\omega < \Omega_i$

We can express these in terms of the normalized Alfvén speed $U_A = (Y_e Y_i / X_e)^{-1/2}$. Divide the coefficients by X_e^2. The cubic then factorizes and can be written

$$(n^2 U_A^2 \cos^2\theta - 1)[n^4 U_A^2 U^2 \cos^2\theta - n^2(U_A^2 + U^2) + 1] = 0 \qquad (5.49)$$

The first factor represents the slow Alfvén wave of equation (2.83). The second represents two waves, one a modified form of the fast Alfvén wave of equation (2.83) and the other a modified sound wave, the *magnetosonic wave*. Examples of the refractive index and ray surfaces for these waves are shown in Figures 5.6 and 5.7 Each of the refractive index diagrams on the left shows three curves. In each case the refractive index of the slow Alfvén wave is represented by a vertical line; the refractive index surface is a plane perpendicular to the magnetic field. The direction of energy propagation is normal to this surface and thus is always exactly along the magnetic field. There are two other surfaces corresponding to magnetosonic waves. One is closed (type *b* in Figure 3.6) and can be called the fast magnetosonic wave. The other is open but is different in shape from any shown in Figure 3.6. It can be called the slow magnetosonic wave. The surface for the slow Alfvén wave is tangent to one of these surfaces. This means that for propagation exactly along the magnetic field there is a degeneracy and one of the magnetosonic waves propagates with the Alfvén speed. Whether it is the slow or the fast magnetosonic

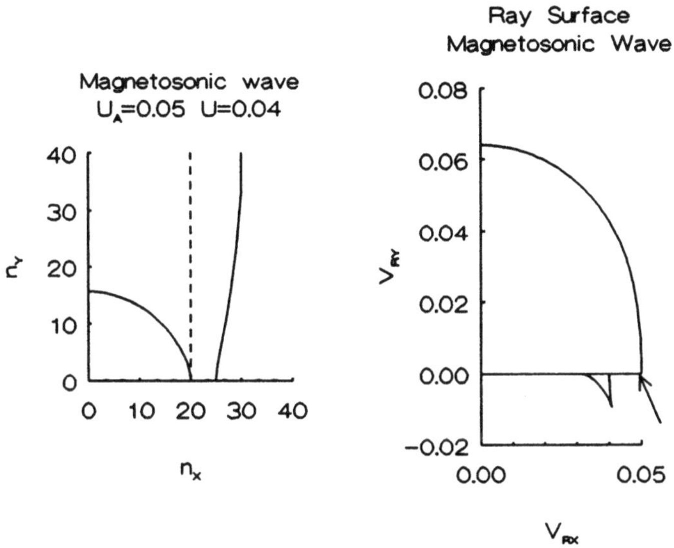

Figure 5.6: Refractive index surface and ray surface for hydromagnetic wave. $U_A > U$.

wave depends on whether U_A is less than or greater than U. In either case the other magnetosonic wave is propagated parallel to the magnetic field with the sound speed U. The fast wave is propagated perpendicular to the field with a hybrid speed $\sqrt{U_A^2 + U^2}$. The other two waves cannot be propagated perpendicular to the field.

The ray surfaces of the magnetosonic waves are shown in the right hand panels of Figures 5.6 and 5.7. They can be a different shape from any of those shown in Figure 3.6. While the ray surface corresponding to the closed refractive index surface is the same as type *b* in Figure 3.6 that for the open surface is quite different from any shown there. This is because of the different arrangement of the points of inflection.

The slow Alfvén wave has a ray direction exactly parallel to the field. The ray velocity always has a magnitude exactly equal to the Alfvén speed and thus the ray surface for this wave is degenerate and consists of a single point indicated by the arrow on the ray diagram.

5.5 Polarization

It is possible to find a general expression for the polarization from (5.12). As in the case of a cold plasma, we define $\rho = E_{y'}/E_{x'}$ and $\eta = E_{z'}/E_{x'}$.

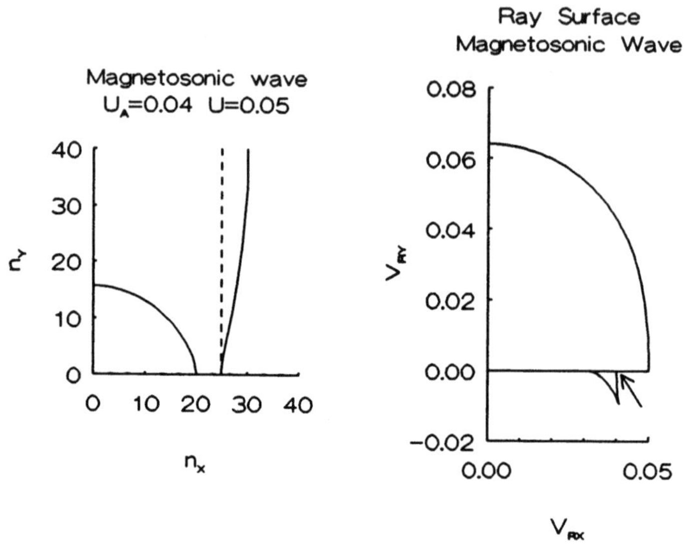

Figure 5.7: Refractive index surface and ray surface for hydromagnetic wave. $U_A < U$.

Then it is easy to show that

$$\begin{aligned} 1:\rho:\eta &= \kappa'_{13}(n^2-\kappa'_{22})+\kappa'_{12}\kappa'_{23} \\ &: \kappa'_{23}(n^2-\kappa'_{11})-\kappa'_{13}\kappa'_{12} \\ &: (n^2-\kappa'_{22})(n^2-\kappa'_{11})+{\kappa'_{12}}^2 \end{aligned} \tag{5.50}$$

This expression allows us to calculate the polarization in the most general case. We shall, however only consider some special cases. For a very wide range of conditions we can make an excellent approximation by noting that one of the roots of the cubic for n^2 is very large compared with the others and with the elements of κ'_{jk}. Then the polarization relation becomes

$$1:\rho:\eta \simeq \kappa'_{13}:\kappa'_{23}:n^2 \tag{5.51}$$

Thus $|\eta|$ is very large and the electric field is approximately in the direction of the wave normal. The polarization of the other two waves is then given accurately by the expression for a cold plasma (2.87). These two waves generally have a substantial transverse field except near a resonance cone when one of the other values of n^2 becomes large. It is this approximation that allows the separation into waves that can be thought of as "electrostatic" and "electromagnetic". The separation is not always possible and, as described above, an electrostatic wave can change continuously into an electromagnetic wave as the angle between the wave normal and the magnetic field is changed.

5.6 Summary

- The cold plasma approximation breaks down when the wave speed is comparable with the particle speed. In these circumstances the warm plasma approximation can sometimes be used. In this approximation the effects of the finite particle temperature are taken into account by treating the various particle species as fluids with a finite pressure. This treatment ignores important kinetic effects arising from the resonance of particles with the wave field.

- The thermal state of the plasma can be characterized by characteristic electron and ion sound speeds. These speeds are assumed to be small compared to the speed of light. If they also turn out to be small compared with wave speeds calculated from the theory then the warm plasma approximation is a good one.

- The dispersion relation for a cold plasma is modified by the introduction of an extra term into the fluid equation from which it is derived. This term is a force due to the gradient of the pressure. The scale length of this gradient is, of course, the wave length. The result is a dielectric tensor whose elements depend not only on the frequency but also on the wave number. Such a medium is said to exhibit spatial dispersion because of this dependence.

- A useful approximation is the electrostatic approximation. This assumes that the refractive index is very large and decouples the additional plasma waves that arise from the introduction of spatial dispersion from the nearly transverse cold plasma waves. A simplified expression for the refractive index can then be found.

- The electrostatic approximation fails near resonances of the cold plasma waves where the cold plasma phase speeds may match those of the warm plasma waves. A more complete theory leads to a much more complicated expression for the refractive index which is a quartic equation in n^2. Because of the assumption that the characteristic sound speeds are small compared with the speed of light this can be reduced to a cubic in n^2.

- When heavy ion effects can be neglected electron plasma waves occur near the electron plasma frequency. These may couple with the Z-mode or the whistler mode near the resonance cone.

- When a low frequency approximation is made another new set of waves appears. Important among these are ion acoustic waves.
- In the hydromagnetic limit the slow Alfvén wave of cold plasma theory is unaffected. The fast mode splits into two magnetosonic waves.
- The expression for the wave polarization shows that, when n is large the wave is approximately linearly polarized along the wave normal. This justifies the electrostatic approximation.

Chapter 6

Waves in a Hot Plasma – 1. General Features

6.1 Introduction

In the previous chapter we have treated a plasma as a mixture of ion and electron fluids each characterized by an isotropic pressure. Each element of the fluid moves locally under the influence of the local field of the wave. In so doing we have ignored an important effect. If a particle of the fluid has a speed comparable with the wave speed then during one period of the oscillation the particle moves distances comparable with the wavelength and therefore experiences completely different forces from those experienced by a particle with a different energy. These particles may remain in phase with the wave over many wave periods and this leads to resonant effects. Energy may be transferred efficiently between waves and particles. This chapter deals with the linear theory of some such waves. New effects appear among which are the phenomenon of Landau damping, various wave growth effects and a modes which do not correspond to any so far found. These can be treated by solving the Vlasov equation self-consistently with Maxwell's equations.

The introduction of such kinetic effects substantially complicates the picture of previous chapters. For this reason we only derive a general dispersion relation for some useful special cases. The first is the so-called *electrostatic approximation* in which it is assumed that the electric field is in the direction of the wave normal. The second is the case of circularly polarized waves propagated exactly along the magnetic field. These will be sufficient for application to the most important cases which occur in the magnetosphere. More general treatments are given, for example, by Stix [100] or Clemmow and Dougherty [27].

In this chapter we derive the dispersion relation in terms of a general plasma distribution function. We also describe a number of useful models of distribution functions. In chapter 7 we discuss the properties of waves in plasmas with various specified distribution functions.

6.2 Unperturbed Particle Orbits

An energetic, charged, non-relativistic particle in a uniform magnetic field $\mathbf{B}_0$ moves in a spiral about the direction of $\mathbf{B}_0$. If the component of its velocity parallel to $\mathbf{B}_0$ is $v_{\|}$, and if the component perpendicular to $\mathbf{B}_0$ has magnitude $v_\perp$, then its cyclotron frequency is $\Omega = B_0|q|/m$ and its cyclotron radius is $\rho = v_\perp/\Omega$. Note that our convention is that Ω, ρ and $v_\perp$ are positive while $v_{\|}$ may be positive or negative.

Such a motion is commonly resolved into the motion of a *guiding center*, which moves parallel to the magnetic field with speed $v_{\|}$, and *cyclotron motion*, a circular motion about the guiding center with angular velocity $\mathbf{\Omega}$. Positive ions spiral left-handedly and electrons right-handedly about the magnetic field. The *pitch angle* of the spiral is given by $\alpha = \tan^{-1} v_\perp/v_{\|}$.

We choose the z-axis to lie along the magnetic field and denote the position of the particle at time t' by x', y', z'. Then if the particle arrives at point x, y, z at time t the equation of its path is

$$x'(t') - x = \frac{v_\perp}{\Omega}\{\sin[\Omega(t'-t)+\theta] - \sin\theta\} \quad (6.1)$$

$$y'(t') - y = \pm\frac{v_\perp}{\Omega}\{\cos[\Omega(t'-t)+\theta] - \cos\theta\} \quad (6.2)$$

$$z'(t') - z = v_{\|}(t'-t) \quad (6.3)$$

The upper sign corresponds to positive ions and the lower sign to electrons. The angle θ is a phase angle. At time t particles with different phase angles arriving simultaneously at (x, y, z), have traveled along different spirals with axes which have been displaced from one another.

The velocity of such a particle is given by

$$v'_x(t') = v_\perp \cos[\Omega(t'-t)+\theta] \quad (6.4)$$

$$v'_y(t') = \mp v_\perp \sin[\Omega(t'-t)+\theta] \quad (6.5)$$

$$v'_z(t') = v_{\|} \quad (6.6)$$

6.3 Electrostatic Approximation

In the previous chapter we saw how, for a warm plasma, the equations describing waves could often be separated in such a way that one set of equations described a wave with an electric field approximately parallel to the wave normal and zero magnetic field. This was possible because the additional effects due to the energetic particles led to waves having speeds comparable to the thermal speeds of the particles. These waves were nearly longitudinal and electrostatic. They were effectively decoupled from the much faster electromagnetic waves because of their very different speeds. We anticipate that this will also be possible when we take kinetic effects into account. In this section we therefore seek electrostatic waves with $\nabla \times \mathbf{E} \equiv i\mathbf{k} \times \mathbf{E} \gg i\omega\mathbf{B} = 0$. This implies that the perturbation magnetic field is negligible. We therefore assume that the electric field in the wave can be described by a potential, so that $\mathbf{E} = -\nabla\Phi$. The only magnetic field is the uniform background field $\mathbf{B}_0$. It is assumed that the unperturbed velocity distribution function f_0 is given. At present we obtain conductivities, dielectric tensors and dispersion relations in terms of an unspecified f_0. These will then be used with specific models to discuss various phenomena. The Vlasov equation (1.23) for each species of particle may then be written

$$\frac{\partial f}{\partial t} + \mathbf{v}.\nabla f + \frac{q}{m}[\mathbf{v} \times \mathbf{B}_0 - \nabla\Phi].\nabla_v f = 0 \tag{6.7}$$

and Maxwell's equations can be replaced with Poisson's equation

$$\nabla^2\Phi = -\frac{\rho}{\epsilon_0} \tag{6.8}$$

6.3.1 Linearization of the Kinetic Equation

As usual we now assume that the wave is a small perturbation so that Φ is of first order and we can write $f = f_0 + f_1$. Substituting this in (6.7) and retaining only first order terms we can write

$$\frac{\partial f_1}{\partial t} + \mathbf{v}.\nabla f_1 + \frac{q}{m}\mathbf{v} \times \mathbf{B}_0.\nabla_v f_1 = \frac{q}{m}(\nabla\Phi).(\nabla_v f_0) \tag{6.9}$$

This is a linear relation between the first order correction to f and the perturbing field in terms of the properties of the zero order distribution function.

6.3.2 Integration Along Unperturbed Orbits

The left-hand side of (6.9) is the total derivative of f_1 following a fluid particle. Thus

$$f_1(\mathbf{r},t) = \frac{q}{m}\int_{-\infty}^{t} [\nabla'\Phi(\mathbf{r}',t')].[\nabla_{v'} f_0(\mathbf{v}')]dt' \tag{6.10}$$

We assume that the perturbation is sufficiently small so that the path along which we integrate is the unperturbed orbit of the particles. This is a suitable particular integral of the differential equation. To it can be added, however, a complementary function which is any solution to the equation

$$\frac{\partial f_1}{\partial t} + \mathbf{v}.\nabla f_1 + \frac{q}{m}\mathbf{v} \times \mathbf{B}_0.\nabla_v f_1 = 0$$

The choice of complementary function must be made in accordance with the requirements of causality. It is beyond the scope of this treatment to give a detailed discussion. The interested reader is referred to the extensive discussion by Clemmow and Dougherty [27]. Briefly, we derive an expression for the conductivity assuming that the plasma was undisturbed in the distant past and that the disturbance has grown from zero. This implies the assumption that ω has a positive imaginary part and that the complementary function can be taken as zero. The conductivity is computed for this case. The resulting expression, which is a function of ω, is then analytically continued to values of ω with negative imaginary part. Clemmow and Dougherty show that this satisfies the requirements of causality.

We proceed to find an explicit expression for f_1 by choosing coordinates so that $\mathbf{k}$ is in the x–z plane. We assume

$$\begin{aligned}
&\Phi(\mathbf{r}',t') \\
&\quad = \tilde{\varphi}\exp[i(k_\perp x' + k_\parallel z' - \omega t')] \\
&\quad = \tilde{\varphi}e^{i(k_\perp x + k_\parallel z - \omega t)}\exp i\left\{\frac{k_\perp v_\perp}{\Omega}[\sin(\Omega\tau + \theta) - \sin\theta] - (\omega - k_\parallel v_\parallel)\tau\right\}
\end{aligned}$$

where

$$\tau = t' - t$$

We can use (B.14; Appendix B) to write this in the form

$$\begin{aligned}
\Phi(\mathbf{r}',t') &= \tilde{\varphi}e^{i(k_\perp x + k_\parallel z - \omega t)} \\
&\times \sum_{l=-\infty}^{\infty}\sum_{m=-\infty}^{\infty} J_l(\zeta)J_m(\zeta)e^{-i(\omega - k_\parallel v_\parallel - l\Omega)\tau}e^{i(l-m)\theta}
\end{aligned}$$

where

$$\zeta = k_\perp v_\perp/\Omega$$

It can be seen that

$$\begin{aligned}
[\nabla'\Phi(\mathbf{r}',t')].[\nabla_{v'}f_0(\mathbf{v}')] &= i\left[k_\perp\frac{\partial f_0}{\partial v'_{x'}} + k_\parallel\frac{\partial f_0}{\partial v'_{z'}}\right]\Phi \\
&= i\left[k_\perp\frac{\partial f_0}{\partial v_\perp}\cos(\Omega\tau + \theta) + k_\parallel\frac{\partial f_0}{\partial v_\parallel}\right]\Phi
\end{aligned}$$

Then

$$\begin{aligned}
f_1(\mathbf{r},t) &= \frac{iq}{m}\tilde{\varphi}e^{i(k_\perp x + k_\parallel z - \omega t)}\sum_{l=-\infty}^{\infty}\sum_{m=-\infty}^{\infty} J_l(\zeta)J_m(\zeta)e^{i(l-m)\theta} \\
&\times \left[k_\perp\frac{\partial f_0}{\partial v_\perp}\int_{-\infty}^{0} e^{-i(\omega - l\Omega - k_\parallel v_\parallel)\tau}\cos(\Omega\tau + \theta)d\tau\right. \\
&+ \left. k_\parallel\frac{\partial f_0}{\partial v_\parallel}\int_{-\infty}^{0} e^{-i(\omega - l\Omega - k_\parallel v_\parallel)\tau}d\tau\right] \qquad (6.11)
\end{aligned}$$

Now consider the sum over l. It can be written

$$S_l = \frac{1}{2}\sum_{l=-\infty}^{\infty} J_l(\zeta)\int_{-\infty}^{0}\left\{e^{i[(l+1)-m]\theta}e^{-i[\omega-k_\parallel v_\parallel-(l+1)\Omega]\tau} + e^{i[(l-1)-m]\theta}e^{-i[\omega-k_\parallel v_\parallel-(l-1)\Omega]\tau}\right\}d\tau$$

Since it is a sum over all l we can replace $l+1$ in the first term and $l-1$ in the second term by l. Then, using (B.4), we get

$$S_l = \sum_{l=-\infty}^{\infty}\frac{l\Omega}{k_\perp v_\perp}J_l(\zeta)e^{i(l-m)\theta}\int_{-\infty}^{0} e^{-i(\omega-k_\parallel v_\parallel-l\Omega)\tau}d\tau$$

Then, if we take $f_1(\mathbf{r},t) = \tilde{f}_1\exp[i(k_\perp x + k_\parallel z - \omega t)]$,

$$\tilde{f}_1 = -\frac{q\tilde{\varphi}}{m}\sum_{l=-\infty}^{\infty}\sum_{m=-\infty}^{\infty} J_l(\zeta)J_m(\zeta)e^{i(l-m)\theta}\left\{\frac{\frac{l\Omega}{v_\perp}\frac{\partial f_0}{\partial v_\perp}+k_\parallel\frac{\partial f_0}{\partial v_\parallel}}{\omega-k_\parallel v_\parallel-l\Omega}\right\} \tag{6.12}$$

The integral has been evaluated on the assumption that ω has a positive imaginary part so that causality is preserved. When this formula is used to find the conductivity the result can be analytically continued to the case where ω has a negative imaginary part.

6.3.3 Conductivity, Susceptibility and Dielectric Constant

In the electrostatic case **J** and **E** are parallel to **k** and so, for harmonic waves, the equation of continuity gives $\rho = kJ/\omega$ for the charge density. The conductivity relationship is $J = -ik\sigma\tilde{\varphi}$ so that

$$\rho = -\frac{ik^2\sigma}{\omega}\tilde{\varphi}$$

Now the charge density perturbation is found from

$$\rho = q\int_0^{2\pi}\int_C\int_0^{\infty} f_1 v_\perp dv_\perp\, dv_\parallel\, d\theta$$

where the contour C is chosen to satisfy the requirements of causality as described below. As shown in (B.16) the integration over θ reduces the double sum to a sum over n and the expression becomes

$$\rho = -\frac{2\pi q^2\tilde{\varphi}}{m}\sum_{l=-\infty}^{\infty}\int_C\int_0^{\infty} v_\perp J_l^2(\zeta)\left\{\frac{\frac{l\Omega}{v_\perp}\frac{\partial f_0}{\partial v_\perp}+k_\parallel\frac{\partial f_0}{\partial v_\parallel}}{\omega-k_\parallel v_\parallel-l\Omega}\right\}dv_\perp\, dv_\parallel \tag{6.13}$$

The conductivity for a hot particle species is thus seen to be

$$\sigma = -\frac{2\pi i\omega q^2}{k^2 m}\sum_{l=-\infty}^{\infty}\int_C\int_0^{\infty} v_\perp J_l^2(\zeta)\left\{\frac{\frac{l\Omega}{v_\perp}\frac{\partial f_0}{\partial v_\perp}+k_{\|}\frac{\partial f_0}{\partial v_{\|}}}{\omega - k_{\|}v_{\|} - l\Omega}\right\} dv_\perp\, dv_{\|} \quad (6.14)$$

The integral over $v_{\|}$ is along the *Landau contour C* shown in Figure 6.1. To satisfy causality it was assumed that ω had a positive imaginary part

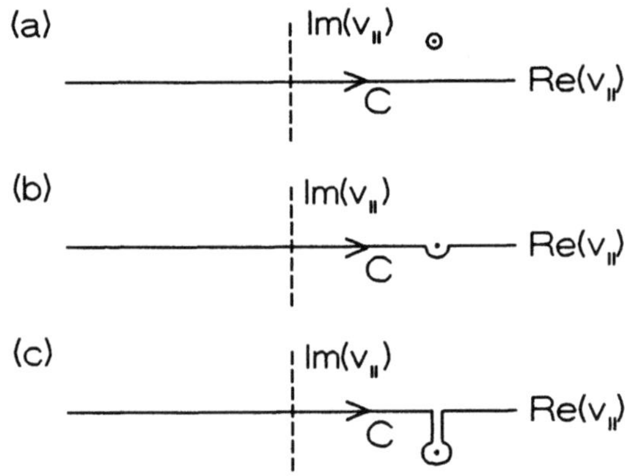

Figure 6.1: Contours for the integral over $v_{\|}$. (a) $\Im(\omega) > 0$ (b) $\Im(\omega) = 0$ (c) $\Im(\omega) < 0$. The pole occurs where $v_{\|} = (\omega \pm l\Omega)/k_{\|}$.

when evaluating the integral over the unperturbed paths of the particles. In this case the pole $v_{\|} = (\omega - l\Omega)/k$ lies above the real $v_{\|}$-axis and the integral is taken along the real axis as shown in Figure 6.1a. If the imaginary part is not positive then the charge density perturbation is found by analytic continuation. The cases for ω real and $\Im(\omega) < 0$ are shown in 6.1b and c respectively. It can be seen that when $\Im(\omega) = 0$ the integral is the principal part of the integral along the real axis minus πi times the residue at the pole. When $\Im(\omega) < 0$ it is the integral along the real axis minus $2\pi i$ times the residue at the pole.

The contribution of each particle species is summed to get the plasma conductivity.

$$\sigma = \sum_i \sigma_i$$

The expressions for a cold or a hot plasma may be used for a species as appropriate.

The scalar susceptibility for the plasma is then

$$\chi = i\sigma/\epsilon_0\omega$$

and the dielectric constant is

$$\kappa = 1 + \chi$$

6.3.4 Dispersion Relation

For electrostatic waves the dispersion relation has a simple form. Poisson's equation may be written

$$-k^2\varphi = \rho/\epsilon_0 = \frac{ik^2\sigma}{\omega}$$

Thus

$$\left(1 + \frac{i\sigma}{\omega\epsilon_0}\right)\varphi \simeq \kappa_0\varphi = 0$$

This only has a non-trivial solution if $\kappa = 0$ that is if

$$1 + \sum_{species} \frac{2\pi}{N}\frac{\omega_p^2}{k^2} \sum_{l=-\infty}^{\infty} \int_C \int_0^{\infty} v_\perp J_l^2(\zeta) \left\{ \frac{\frac{l\Omega}{v_\perp}\frac{\partial f_0}{\partial v_\perp} + k_\parallel \frac{\partial f_0}{\partial v_\parallel}}{\omega - k_\parallel v_\parallel - l\Omega} \right\} dv_\perp \, dv_\parallel = 0 \tag{6.15}$$

where ω_p is the plasma frequency for the species.

This is of the form

$$1 + \sum_{species} \chi_\parallel = 1 + \sum_{species} \frac{i\sigma_\parallel}{\omega\epsilon_0} = 0 \tag{6.16}$$

where each term in the sum is the contribution of a particular particle species with its own distribution function to the susceptibility or conductivity.

This dispersion relation relates ω and k or, equivalently, refractive index n and ω. It is not possible, as we have done in previous chapters, to survey and classify its properties in parameter space. Its properties depend on the nature of the zero order distribution function and even for a limited number of idealized models such a study would be impractically large. Fortunately such a study is not even particularly desirable. The physical difference between this and the warm plasma model is that the kinetic treatment takes into account the interaction between waves and resonant particles. This interaction allows energy exchange between waves and particles. If this interaction is small our previous treatments are quite adequate and can be used to study the propagation of waves in plasmas. It is only when we are specifically interested in wave generation or damping that a kinetic approach is desirable. In such cases one is usually concerned either with the initial growth rate of an instability or with marginal growth in a near equilibrium situation. Such a spatial perturbation can be represented by a wave with a particular wave vector $\mathbf{k}$ and the dispersion relation can be used to find the corresponding value of ω. If this has a positive imaginary part the wave is a growing wave and if it has a negative imaginary part it is a decaying wave.

6.3.5 Particle Resonance

The expression for the conductivity, and therefore the dispersion relation, involves a sum over an infinite number of terms. In each of the terms of the sum the integral over $v_{\parallel}$ has a pole. For the lth term this pole occurs where

$$\omega - k_{\parallel}v_{\parallel} - l\Omega = 0 \tag{6.17}$$

The interpretation of this *resonance condition* is that particles traveling with velocity component $v_{\parallel}$ along the field see the frequency of the wave as Doppler shifted to $\omega - k_{\parallel}v_{\parallel}$. For particles for which the resonance condition holds for $l = 0$ this Doppler shifted frequency is zero. As a result the particles remain in phase with the wave. The effect of this group of particles on the wave depends on the nature of the residue at the pole, which in turn depends on the nature of the distribution function at this value of $v_{\parallel}$. For those particles for which the resonance condition holds for $l = \pm 1$ the Doppler shifted frequency is the gyrofrequency. For higher values of l the Doppler shifted frequency is a harmonic of the gyrofrequency. In Chapter 8 we shall consider the nature of particle resonance in detail. For the present we shall note that in the following the dispersion relation is significantly affected by resonant particles and the nature of their effect is to determine the imaginary part of the frequency which describes growth or damping of the waves.

6.4 Propagation Parallel to the Magnetic Field

The problem of studying waves for which the electrostatic approximation does not apply is not conceptually harder but is more complicated algebraically. We shall consider only a simple case, that of waves propagated along the magnetic field direction with $k_{\perp} = 0$. We can follow an exactly analogous procedure to that for the electrostatic approximation. The difference in this case is that the electromagnetic force is now $\mathbf{E} + \mathbf{v} \times \mathbf{B}$ rather than $-\nabla\Phi$. The linearized kinetic equation is thus

$$\frac{\partial f_1}{\partial t} + \mathbf{v}.\nabla f_1 + \frac{q}{m}\mathbf{v} \times \mathbf{B}_0.\nabla_v f_1 = -\frac{q}{m}\left\{\mathbf{E} + \frac{\mathbf{v} \times (\mathbf{k} \times \mathbf{E})}{\omega}\right\}.\nabla_v f_0 \tag{6.18}$$

where we have used Faraday's law in the form $\mathbf{k}\times\mathbf{E}=\omega\mathbf{B}$. Then

$$f_1(\mathbf{r},t)=-\frac{q}{m}\int\left\{\mathbf{E}(\mathbf{r}',t')+\frac{\mathbf{v}'\times[\mathbf{k}\times\mathbf{E}(\mathbf{r}',t')]}{\omega}\right\}.\nabla_{v'}f_0\,dt' \tag{6.19}$$

where the integral is taken along the unperturbed orbits. Since $k_\perp=0$ we can write

$$\mathbf{E}=\tilde{\mathbf{E}}e^{ik_\parallel z}e^{-i(\omega-k_\parallel v_\parallel)\tau}$$

where $\tau=t'-t$.

6.4.1 Circularly Polarized Coordinates

Because of the cylindrical symmetry about the direction of the magnetic field it is convenient to work in the circularly polarized coordinates defined by (2.26). First we write out the integrand in Cartesian coordinates and use (6.19) to express them in circularly polarized form.

$$\begin{aligned}
&\left\{\tilde{\mathbf{E}}+\frac{\mathbf{v}'\times[\mathbf{k}\times\tilde{\mathbf{E}}]}{\omega}\right\}.\nabla_{v'}f_0\\
&= \left\{\left[\left(1-\frac{k_\parallel v_\parallel}{\omega}\right)\frac{\partial f_0}{\partial v_\perp}+\frac{k_\parallel v_\perp}{\omega}\frac{\partial f_0}{\partial v_\parallel}\right]\right.\\
&\quad\left.\times\left[\tilde{E_x}\cos(\Omega\tau+\theta)\mp\tilde{E_y}\sin(\Omega\tau+\theta)\right]+\frac{\partial f_0}{\partial v_\parallel}\tilde{E_z}\right\}e^{i(k_\parallel v_\parallel-\omega)\tau}\\
&= \left\{\frac{1}{2}\left[\left(1-\frac{k_\parallel v_\parallel}{\omega}\right)\frac{\partial f_0}{\partial v_\perp}+\frac{k_\parallel v_\perp}{\omega}\frac{\partial f_0}{\partial v_\parallel}\right]\left[\tilde{E_L}e^{\pm i\Omega\tau}e^{\pm i\theta}+\tilde{E_R}e^{\mp i\Omega\tau}e^{\mp i\theta}\right]\right.\\
&\quad\left.+\frac{1}{i(k_\parallel v_\parallel-\omega)}\frac{\partial f_0}{\partial v_\parallel}\tilde{E_\parallel}\right\}e^{i(k_\parallel v_\parallel-\omega)\tau}
\end{aligned} \tag{6.20}$$

6.4.2 Integration Along Unperturbed Orbits

The integration with respect to τ is straightforward. We get

$$\begin{aligned}
\tilde{f_1} &= -\frac{q}{m}\left\{\frac{1}{i(k_\parallel v_\parallel-\omega)}\frac{\partial f_0}{\partial v_\parallel}+\frac{1}{2}\left[\left(1-\frac{k_\parallel v_\parallel}{\omega}\right)\frac{\partial f_0}{\partial v_\perp}+\frac{k_\parallel v_\perp}{\omega}\frac{\partial f_0}{\partial v_\parallel}\right]\right.\\
&\quad\left.\times\left[\frac{E_Le^{\pm i\theta}}{i(k_\parallel v_\parallel-\omega\pm\Omega)}+\frac{E_Re^{\mp i\theta}}{i(k_\parallel v_\parallel-\omega\mp\Omega)}\right]\right\}
\end{aligned} \tag{6.21}$$

6.4.3 Conductivity and Dielectric Tensors

The conductivity relation for the species is

$$\tilde{\mathbf{J}} = q\int_{-\infty}^{\infty}\int_{0}^{\infty}\int_{0}^{2\pi} \mathbf{v}v_\perp \tilde{f}_1 d\theta\, dv_\perp\, dv_\| \tag{6.22}$$

In the circularly polarized coordinates the conductivity tensor is diagonal:

$$\begin{pmatrix} J_L \\ J_R \\ J_\| \end{pmatrix} = \begin{pmatrix} \sigma_L & 0 & 0 \\ 0 & \sigma_R & 0 \\ 0 & 0 & \sigma_\| \end{pmatrix}\begin{pmatrix} E_L \\ E_R \\ E_\| \end{pmatrix}$$

Using $v_x = v_\perp \cos\theta$, $v_y = \mp v_\perp \sin\theta$ we get

$$\begin{aligned}
\tilde{J}_L &= q\int_{-\infty}^{\infty}\int_{0}^{\infty}\int_{0}^{2\pi} v_\perp^2 \tilde{f}_1 e^{\mp i\theta} d\theta\, dv_\perp\, dv_\| \\
\tilde{J}_R &= q\int_{-\infty}^{\infty}\int_{0}^{\infty}\int_{0}^{2\pi} v_\perp^2 \tilde{f}_1 e^{\pm i\theta} d\theta\, dv_\perp\, dv_\| \\
\tilde{J}_\| &= q\int_{-\infty}^{\infty}\int_{0}^{\infty}\int_{0}^{2\pi} v_\perp v_\| \tilde{f}_1 d\theta\, dv_\perp\, dv_\|
\end{aligned} \tag{6.23}$$

and since f_1 is independent of θ we can substitute the expression for f_1 and integrate over θ to get

$$\tilde{J}_L = \frac{\pi q^2}{im}\tilde{E}_L \int_{-\infty}^{\infty}\int_{0}^{\infty} \frac{v_\perp^2}{\omega} \frac{\frac{\partial f_0}{\partial v_\perp}(\omega - k_\| v_\|) + k_\| v_\perp \frac{\partial f_0}{\partial v_\|}}{\omega - k_\| v_\| \mp \Omega} dv_\perp\, dv_\| \tag{6.24}$$

$$\tilde{J}_R = \frac{\pi q^2}{im}\tilde{E}_R \int_{-\infty}^{\infty}\int_{0}^{\infty} \frac{v_\perp^2}{\omega} \frac{\frac{\partial f_0}{\partial v_\perp}(\omega - k_\| v_\|) + k_\| v_\perp \frac{\partial f_0}{\partial v_\|}}{\omega - k_\| v_\| \pm \Omega} dv_\perp\, dv_\| \tag{6.25}$$

$$\tilde{J}_\| = \frac{2\pi q^2}{im}\tilde{E}_\| \int_{-\infty}^{\infty}\int_{0}^{\infty} \frac{v_\perp v_\| \frac{\partial f_0}{\partial v_\|}}{\omega - k_\| v_\|} dv_\perp\, dv_\| \tag{6.26}$$

The elements of the dielectric constant for the medium are then given by

$$\kappa_{L,R,\|} = 1 + \sum_{species} \frac{i\sigma_{L,R,\|}}{\epsilon_0 \omega}$$

Thus

$$\kappa_L = 1 + \sum_{species} \frac{\pi}{N}\frac{\omega_p^2}{\omega^2} \int_{-\infty}^{\infty}\int_{0}^{\infty} v_\perp^2 \frac{\frac{\partial f_0}{\partial v_\perp}(\omega - k_\| v_\|) + k_\| v_\perp \frac{\partial f_0}{\partial v_\|}}{\omega - k_\| v_\| \mp \Omega} dv_\perp\, dv_\|$$

$$\kappa_R = 1 + \sum_{species} \frac{\pi}{N}\frac{\omega_p^2}{\omega^2}\int_{-\infty}^{\infty}\int_0^{\infty} v_\perp^2 \frac{\frac{\partial f_0}{\partial v_\perp}(\omega - k_\parallel v_\parallel) + k_\parallel v_\perp \frac{\partial f_0}{\partial v_\parallel}}{\omega - k_\parallel v_\parallel \pm \Omega} dv_\perp\, dv_\parallel$$

$$\kappa_\parallel = 1 + \sum_{species} \frac{2\pi}{N}\frac{\omega_p^2}{\omega^2}\int_{-\infty}^{\infty}\int_0^{\infty} \frac{\omega v_\perp v_\parallel \frac{\partial f_0}{\partial v_\parallel}}{\omega - k_\parallel v_\parallel} dv_\perp\, dv_\parallel \tag{6.27}$$

The first two of these are the dielectric constants for the left- and right-handed circularly polarized waves propagated along the field. The third is the dielectric constant for an electrostatic wave propagated parallel to the magnetic field and is identical to the wave which is obtained by setting $k_\perp = 0$ in (6.14)[1].

It can be seen that in this case for the left and right handed waves the only resonance occurs where the resonant particles see a Doppler shifted frequency equal to the appropriate particle gyrofrequency. For the electrostatic wave they see a zero frequency. When $k_\perp$ is not zero this relation is much more complicated. In the first place it is a tensor relationship as the current density vector is not in the same direction as the applied electric field, while in the second place, as in the case of electrostatic waves, it involves an infinite sum over all the harmonics of the cyclotron frequency.

6.4.4 Dispersion Relation

The wave equation (1.35) becomes $(k^2 - \omega^2/c^2)E_{x,y} = i\omega\mu_0 J_{x,y}$, $E_z = (i/\epsilon_0\omega)J_z$. We can combine these to see that

$$\left(\frac{c^2 k_\parallel^2}{\omega^2} - \kappa_{L,R}\right) E_{L,R} = 0$$

$$\kappa_\parallel E_\parallel = 0 \tag{6.28}$$

The dispersion relations are thus

$$\frac{c^2 k_\parallel^2}{\omega^2} = \kappa_L = 1 + \frac{i\sigma_L}{\epsilon_0\omega}; \quad \frac{c^2 k_\parallel^2}{\omega^2} = \kappa_R = 1 + \frac{i\sigma_R}{\epsilon_0\omega}; \quad \kappa_\parallel = 0 \tag{6.29}$$

---

[1]To see this use $J_0(0) = 1$, $J_n(0) = 0$, $n \neq 0$ in (6.14), note that

$$\frac{\omega}{\omega - k_\parallel v_\parallel} = 1 + \frac{k_\parallel v_\parallel}{\omega} + \frac{k_\parallel^2 v_\parallel^2}{\omega - k_\parallel v_\parallel}$$

that $\int (\partial f_0/\partial v_\parallel) d^3v = 0$ and that $\frac{1}{N}\int v_\parallel (\partial f_0/\partial v_\parallel) d^3v = -1$.

6.5 Growth and Decay of Waves

6.5.1 Energy Transfer Between Particles and Waves

If a wave is propagated in a plasma characterized by conductivity σ_{jk}, or dielectric tensor $\kappa_{jk} = \delta_{jk} + i\sigma_{jk}/\epsilon_0\omega$ then the work done per unit time by the electromagnetic field on the particles of the plasma is

$$W = \mathbf{J}.\mathbf{E} = \sigma_{jk} E_j E_k = i\epsilon_0\omega(\delta_{jk} - \kappa_{jk})E_j E_k$$

The average of this over one cycle is

$$< W >= \frac{1}{2}\Re(\mathbf{J}.\mathbf{E}^*) = \frac{1}{2}\Re(\sigma_{jk} E_j E_k^*) \tag{6.30}$$

If $i\sigma_{jk}$ is Hermitian, i.e. if $\Re(\sigma_{jk}) = -\Re(\sigma_{kj})$ and thus κ_{jk} is Hermitian, then the real part of the conductivity is zero and there is no energy transfer. This has been the case for the waves considered in previous chapters. Non-Hermitian dielectric tensors imply energy transfer between wave and particles.

In this chapter we have limited ourselves to cases where the dielectric tensor is diagonal. The condition therefore implies that the real part of σ determines the energy transfer. When $\Re(\sigma) > 0$ then $\frac{1}{2}\Re(\mathbf{J}.\mathbf{E}^*)$ is positive and the wave transfers energy to the particles, decaying in the process. When $\Re(\sigma) < 0$ then $\frac{1}{2}\Re(\mathbf{J}.\mathbf{E}^*)$ is negative and the particles transfer energy to the wave, which grows in the process.

6.5.2 Nature of the Dispersion Relation for Growth and Decay of Waves

Since we have assumed that the time behavior of waves is as $\exp(-i\omega t)$ we anticipate that, if for some real value of k the dispersion relation predicts that $\Im(\omega) > 0$, the wave will grow as $\exp(|\omega_i|t$. If, on the other hand, $\Im(\omega) < 0$ it will decay as $\exp(-|\omega_i|t)$. We shall show that this is consistent with the energy considerations of the previous section. We shall limit ourselves to the simple cases considered above, although they are generally consistent.

For the left and right polarized waves the dispersion relation can be written

$$\omega - \frac{c^2k^2}{\omega} \equiv \omega - \frac{c^2k^2\omega^*}{\omega\omega^*} = -i\sigma_{L,R}/\epsilon_0$$

Thus

$$\left(1 + \frac{c^2k^2}{\omega\omega^*}\right)\omega_i = -i\sigma_r/\epsilon_0$$

for both waves so that for wave growth ω_i is positive and σ_r is negative, while for wave decay ω_i is negative and σ_r is positive as expected. For any electrostatic wave with dispersion relation $\kappa = 0$ we have

$$\omega = -i\sigma/\epsilon_0$$

and this leads to $\omega_i = -\sigma_r/\epsilon_0$. This gives the same result. We see therefore that, in these simple cases, the growth rate given by the dispersion relation is consistent with energy conservation. It is possible to show that this is true in more general cases.

6.5.3 Growth and Evanescence

In choosing to treat k as real and use the dispersion relation to find ω as a complex function of k we have assumed an initial disturbance in space as given and found how this develops in time. Clearly, if ω_i is positive it will always grow and we will have an instability. It would also be possible to treat ω as real and find complex k from the dispersion relation. This would not, however, necessarily imply instability. We have seen that complex values of k can correspond to evanescent or inhomogeneous waves with no transfer of energy between particles and wave. For example, if ω is real, $\sigma_r = 0$ and $\sigma_i > 0$ a dispersion relation of the form $c^2k^2 = \omega^2 + i\sigma\omega$ could give a negative value for k^2 and hence evanescence instead of growth. On the other hand $\sigma_r \neq 0$ always implies growth or damping and will contribute to the imaginary part of k. It is beyond the scope of this book to go into more detail of criteria for deciding between evanescence and particle-wave interaction. A full treatment is given by Clemmow and Dougherty [27, chapter 6]. To avoid ambiguity we can use either the real part of the conductivity or the imaginary part of the frequency to determine whether there is wave growth or decay.

Note that it is possible to find growth corresponding to a propagated wave which grows as it travels (convective instability) or to a non-propagated oscillation (non-convective instability) which in the limit of no growth reduces to an inhomogeneous or evanescent wave. This is shown in [27].

6.5.4 Approximations When the Growth or Decay Rate Is Small

In all the cases discussed here the general form of the dispersion relation is

$$D(\omega, k) = 0 \tag{6.31}$$

In general ω and D are both complex. If we assume that the imaginary part of ω is small, $\omega_i \ll \omega_r$, and that D is analytic in the region of interest, then we may write the dispersion relation in the form

$$D_r(\omega_r, k) + i\omega_i \left(\frac{\partial D_r}{\partial \omega}\right)_{\omega=\omega_r} + iD_i(\omega, k) + O(\omega_i^2) = 0 \qquad (6.32)$$

Since both real and imaginary parts of D must be zero in order to satisfy the dispersion relation it is clear that D_i must have been of the order of ω_i in order for our assumption of small ω_i to be justified. To this order of approximation $D_i(\omega, k) = D_i(\omega_r, k)$. The first approximation to the dispersion relation is then

$$D_r(\omega_r, k) = 0 \qquad (6.33)$$

which allows us to find ω_r as a function of k, giving an undamped, unamplified wave. The next approximation gives an explicit expression for the growth rate or decay rate:

$$\omega_i = -\frac{D_i(\omega_r, k)}{(\partial D_r/\partial \omega)_{\omega=\omega_r}} \qquad (6.34)$$

This gives a positive or negative value for ω_i corresponding to growth or decay. The cases of growth and decay which we treat in Chapter 7 are all of this type.

6.6 The Equilibrium Distribution Function – The Maxwellian

Whatever the initial state of a plasma, after a time that is long compared with the collision time it will come to an equilibrium for which the entropy is maximum. It is shown for example in [23] that the equilibrium state for a particle species with temperature T is the Maxwellian

$$f_0(\mathbf{r}, \mathbf{v}, t) = N(\mathbf{r}, t)\left(\frac{m}{2\pi KT}\right)^{3/2} e^{-mv^2/2KT} \qquad (6.35)$$

This has been normalized so that the integral of f over velocity space gives the number density N. We shall assume that the medium is uniform so that there is no dependence on $\mathbf{r}$. It is useful to express f_0 as a function of $v_\perp$ and $v_\parallel$. Since $v^2 = v_\perp^2 + v_\parallel^2$ we can write

$$f_0(v_\perp, v_\parallel) = \frac{1}{2\pi} f_\perp(v_\perp) f_\parallel(v_\parallel) \qquad (6.36)$$

where

$$f_\perp(v_\perp) = \frac{m}{KT} e^{-mv_\perp^2/2KT}; \quad f_\parallel(v_\parallel) = N\left(\frac{m}{2\pi KT}\right)^{1/2} e^{-mv_\parallel^2/2KT} \qquad (6.37)$$

The normalization has been chosen so that

$$\int_{-\infty}^{\infty} f_{\|}\, dv_{\|} = N; \quad \int_{0}^{\infty} v_{\perp} f_{\perp}\, dv_{\perp} = 1 \tag{6.38}$$

and the factor $1/2\pi$ arises to account for the integration over θ.

6.7 Non-Equilibrium Distribution Functions

In the plasmas of the magnetosphere the collision time may be much longer than the time scales of interest in the problem. Wave-particle interactions are often more important in bringing the plasma to equilibrium. Nevertheless, on time scales that are long compared with the characteristic times for the wave-particle interactions, the final equilibrium state will still be expected to be such a Maxwellian. These time scales for the wave-particle interactions also may be much longer than the time scales of interest in the problem. The hot plasma populations of the magnetosphere are often far from equilibrium and this is an abundant source of free energy for the generation of waves.

6.7.1 Quasi-Equilibrium States

On the road to equilibrium there are often quasi-equilibrium states for which we can find constants of the motion which satisfy the Vlasov equation. For example, as we have seen, in a uniform magnetic field $v_{\|}$ and $v_{\perp}$ are such constants. It follows that any f which is a function only of such constants of the motion is a solution to the Vlasov equation. Since f does not then evolve further in time it represents a quasi-equilibrium state. Whether it is stable to small perturbations is, of course, a matter for further investigation. We present a number of useful models of such distribution functions. It is emphasized that these are mathematical models with features chosen for analytical convenience. They do not model real distributions in detail. Nevertheless they are chosen to have those features of real distributions which are important for understanding the generation of waves. In a uniform plasma permeated by a uniform magnetic field $\mathbf{B}_0$ the distribution function does not vary in space and time and the unperturbed Vlasov equation (1.23) may be written

$$(\mathbf{v} \times \mathbf{B}_0).\nabla_v f_0 = 0 \tag{6.39}$$

Then $v_{\perp}$ and $v_{\|}$ are constants of the motion. Because $Dv_{\perp}/Dt = 0$ and $Dv_{\|}/Dt = 0$ *any* function of these constants of the motion, $f_0(v_{\perp}, v_{\|})$ is a suitable zero order distribution function. This is only true if the collision term on the right hand side of the Boltzmann equation can be neglected because the collision times are long compared with time scales of interest in the problem. In more general problems other constants of the motion can be found and the distribution function can be expressed as a function of these.

6.7.2 Separable Distribution Functions

It is often assumed, for mathematical convenience but without much physical justification, that f_0 is separable so that

$$f_0(v_\perp, v_\parallel) = \frac{1}{2\pi} f_\perp(v_\perp) f_\parallel(v_\parallel)$$

The factor $1/2\pi$ is introduced because f_0 is normalized such that

$$N = \int_{-\infty}^{\infty} \int_0^{\infty} \int_0^{2\pi} f_0(v_\perp, v_\parallel) v_\perp \, d\theta \, dv_\perp \, dv_\parallel$$

Thus we have normalized $f_\perp$ and $f_\parallel$ such that, on integration over θ,

$$N = \int_{-\infty}^{\infty} f_\parallel(v_\parallel) \, dv_\parallel \int_0^{\infty} f_\perp(v_\perp) \, v_\perp \, dv_\perp$$

There are many models of distribution functions which are of this type and which can be made to show properties close to those of observed distribution functions. Some such models are given below.

6.7.3 Bi-Maxwellian Distribution

The bi-Maxwellian distribution is a simple analytical model for an anisotropic distribution. It is assumed that the parallel and perpendicular components of the distribution are independently in thermal equilibrium with different temperatures $T_\parallel$ and $T_\perp$ respectively.

$$f(v_\perp, v_\parallel) = N \left(\frac{m}{2\pi k}\right)^{3/2} T_\perp^{-1} T_\parallel^{-1/2} e^{-mv_\perp^2/2kT_\perp} e^{-mv_\parallel^2/2kT_\parallel} \tag{6.40}$$

This distribution is separable and, as in the case of the isotropic distribution, we define

$$f_0(v_\perp, v_\parallel) = \frac{1}{2\pi} f_\perp(v_\perp) f_\parallel(v_\parallel) \tag{6.41}$$

where

$$f_\perp(v_\perp) = \frac{m}{KT_\perp} e^{-mv_\perp^2/2KT_\perp}; \quad f_\parallel(v_\parallel) = N \left(\frac{m}{2\pi KT_\parallel}\right)^{1/2} e^{-mv_\parallel^2/2KT_\parallel} \tag{6.42}$$

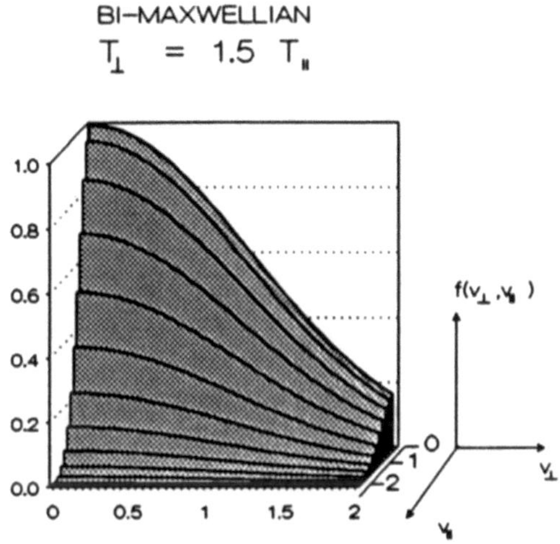

Figure 6.2: Bi-Maxwellian distribution with $T_\perp > T_\parallel$. The distribution function is plotted in the $v_\parallel$–$v_\perp$ plane.

The normalization has again been chosen so that

$$\int_{-\infty}^{\infty} f_\parallel \, dv_\parallel = N; \quad \int_0^\infty v_\perp f_\perp \, dv_\perp = 1 \tag{6.43}$$

In Figure 6.2 we show a three-dimensional plot of the bi-Maxwellian distribution function in the $v_\parallel$–$v_\perp$ plane.

6.7.4 Loss-Cone Distributions

In the earth's magnetic field energetic particles, gyrating around the magnetic field, bounce back and forth between opposite hemispheres. A detailed discussion of their behavior is given, for example, by Roederer [90]. Briefly their magnetic moment is the current $q\Omega$ multiplied by the area of a cyclotron orbit and is conserved. The cyclotron radius is proportional to $v_\perp/B_0$ and Ω is proportional to B_0. Thus $v_\perp/B_0$ is conserved as the guiding center of the particle moves along a field line. The kinetic energy of the particle, $\frac{1}{2}m(v_\perp^2 + v_\parallel^2)$, is also conserved. As the particle moves from the equator along the field line the magnetic field, and hence $v_\perp$, increases and as a result $v_\parallel$ decreases until at some point it becomes zero and the particle is reflected. The smaller the pitch angle, $\arctan v_\perp/v_\parallel$, the further down the field line is the reflection. For pitch angles near zero the particle penetrates deep enough into the atmosphere to collide with neutral molecules and be lost. There is thus a range of equatorial pitch angles round zero for which there are no particles in the distribution. This range of angles with small $v_\perp$ defines the *loss-cone.* Pitch angle diffusion ensures that, as the particles progreess along the field the loss-cone becomes partly filled. There are two useful ways of simulating such a distribution.

Subtracted Maxwellian Distribution

Here a bi-Maxwellian distribution is used as the basis. The perpendicu-

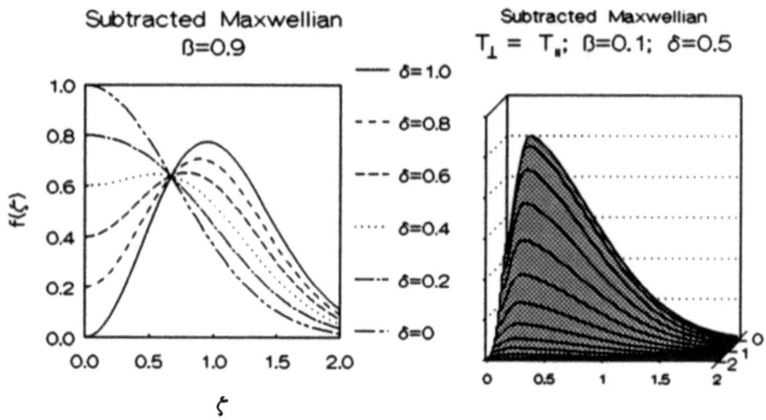

Figure 6.3: Subtracted Maxwellian distribution

lar distribution function is then modified by having another Maxwellian subtracted from it.

$$f_\perp(v_\perp) = \frac{m}{KT_\perp}\left[\left(1+\frac{\beta\delta}{1-\beta}\right)e^{-mv_\perp^2/2KT_\perp} - \frac{\delta}{1-\beta}e^{-mv_\perp^2/2\beta KT_\perp}\right] \tag{6.44}$$

Here $\delta/(1-\beta)$ represents the amplitude of the subtracted particles. When $\delta = 1$ the distribution has no particles with $v_\perp = 0$. When $\delta = 0$ the distribution has no loss-cone; it reduces to a bi-Maxwellian. The parameter β is a measure of the width of the loss-cone relative to the width of the distribution. The reason for choosing $\delta/(1-\beta)$ rather than δ as the amplitude is so that the distribution remains positive for all positive values of β and δ. An example of such a distribution is shown in Figure 6.3.

Modified Maxwellian Distribution

We can also simulate a loss-cone distribution by multiplying $f_\perp$ in the bi-Maxwellian distribution by an even power of $v_\perp$, getting

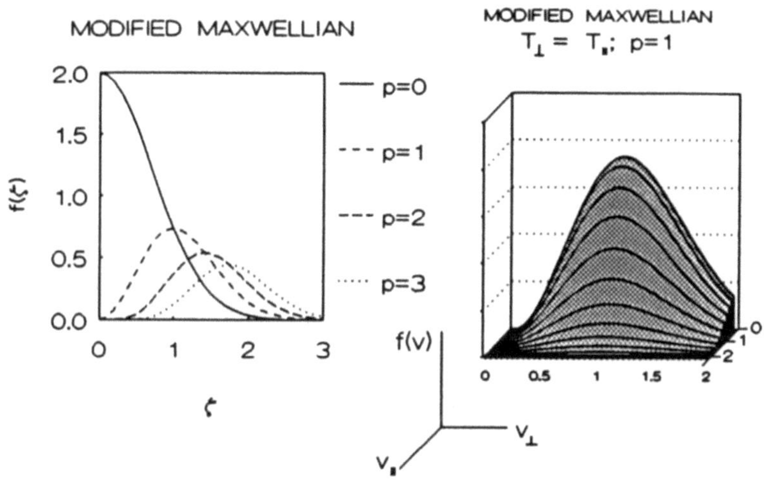

Figure 6.4: Normalized loss-cone distribution. The left-hand panel shows a three-dimensional plot of $f_{\|} f_{\perp}$ for $p = 1$. The right-hand panel shows $f_{\perp} = 2\zeta^{2p}\exp[-\zeta^2]/p!$ for various values of p. The curve for $p = 0$ is a Maxwellian distribution.

$$f_{\perp}(v_{\perp}) = \frac{2}{p!}\left(\frac{m}{2KT}\right)^{p+1} v_{\perp}^{2p} e^{-mv_{\perp}^2/2KT} \tag{6.45}$$

Again the normalization is chosen so that $\int_0^{\infty} v_{\perp} f_{\perp}\, dv_{\perp} = 1$. The distribution is shown in Figure 6.4.

6.7.5 Particle Beams

It is possible that a particular species of particle may have an average velocity relative to other species which define the observational frame of reference. For example a bunch of particles bouncing between hemispheres would move through the background along the magnetic field. Such a species constitutes a particle beam. We shall only consider the case of a beam of particles moving along the magnetic field with speed V_B. If we assume that they have a Maxwellian spread of the component

$v_{\parallel}$ about this mean then we can model $f_{\parallel}$ as

$$f_{\parallel} = N \left(\frac{m}{2\pi K T_{\parallel}} \right)^{1/2} e^{-m(v_{\parallel} - V_B)^2 / 2KT_{\parallel}} \tag{6.46}$$

6.7.6 Composite Distributions

The dispersion relation for any plasma involves a sum over all the particle species. In practical cases we model the plasma as a combination of

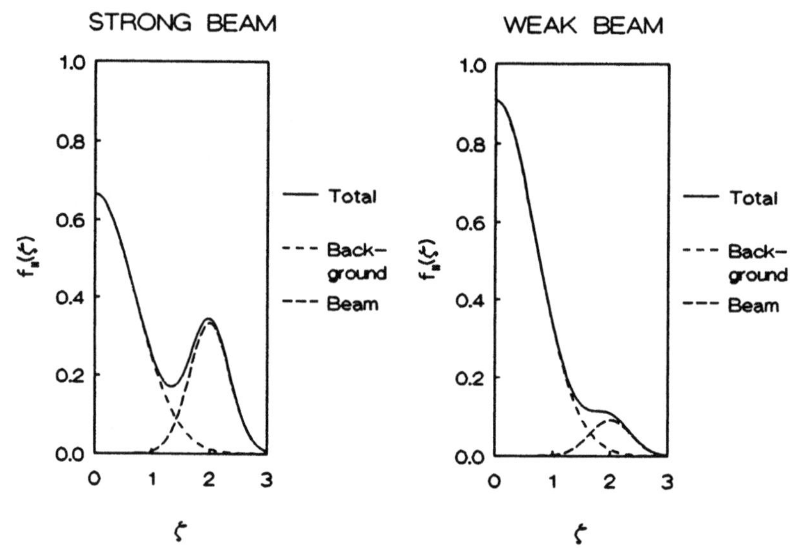

Figure 6.5: Superposition of a Maxwellian and a shifted Maxwellian to represent the parallel distribution function for a beam passing through a background population of particles.

several such distributions. For example we might have a plasma consisting of anisotropic ions, a population of electrons with different temperatures, and a beam of electrons with yet another temperature. In Figure 6.5 we show two examples of $f_{\parallel}$ as the superposition of a simple Maxwellian and a particle beam. Depending on the values of the parameters this can result in a sharply defined two-humped distribution or one with a slight inflection. The distribution illustrated could, for example, be the electron distribution and there could be a backgound of positive ions. In this case there would be a current flowing in the plasma. If there was to be no current we could impose a streaming velocity on the background electrons in such a way as to make the momentum zero. This would require that $N_{\text{background}} v_{\text{background}} = N_{\text{beam}} v_{\text{beam}}$. The current density associated with the electrons would then be zero.

When treating practical cases we shall attempt to model the distribution by such suitable superpositions of particle species.

6.8 Summary

- If the thermal speed of a significant number of particles is comparable with the phase speed of a wave then kinetic effects become important. The constitutive relation for the medium may then be found by using the Vlasov equation for the particle distribution to average the effects of thermal particles on the wave.

- The unperturbed path of a particle in a uniform medium with constant magnetic field is a helix about the field direction. A linear theory can be developed by assuming the existence of a harmonic wave and evaluating its effect on the particle by integrating the force exerted by the wave fields along the unperturbed path of the particle from $t = -\infty$ to t.

- In the electrostatic approximation the kinetic equation can be linearized assuming a small electric field which can be described by a scalar potential. This leads to an expression for the first order perturbation of the velocity distribution.

- The charge density perturbation can be found by integrating this over velocity space. This leads to a formal expression for the conductivity. This expression is an infinite sum. Each term in the sum involves an integral over derivatives of the zero order distribution function. Actual evaluation of the expression requires a knowledge of the distribution function. The integral has a pole corresponding to resonance. The contour of integration near the pole is determined in a manner that satisfies the requirements of causality. The resonance occurs for particles whose parallel velocity is the same as the speed of phase advance along the magnetic field or if the Doppler shifted frequency seen by a particle is equal to a harmonic of the gyrofrequency.

- When the electrostatic approximation is not valid the conductivity is a tensor each of whose elements is a sum of a similar type to the scalar conductivity in the electrostatic approximation. Its computation [49] is beyond the scope of this book. Here it is computed only in the special case of propagation along the field. In rotating coordinates there is a conductivity expression for the left-

and right-handed circularly polarized components of the fields and for the electrostatic component parallel to the magnetic field. Each has a pole corresponding to resonance. The left- and right-handed waves only have a single resonance where the particles see a Doppler shifted frequency equal to their gyrofrequency. The electrostatic wave has a resonance when it sees a zero Doppler shifted frequency.

- Dispersion relations can be computed for all these cases.

- Energy considerations show that wave growth or decay is related to the sign of the real part of the conductivity. If this is positive there is decay, if negative there is growth. The real part of the conductivity arises from the poles in the integrals in the conductivity relation. Growth and decay are thus related to the resonant particles. It can be shown that positive real conductivity corresponds to a negative imaginary part of ω and *vice versa.*

- When the growth rate is small it can be ignored in a zero order approximation and the real part of ω found from the dispersion relation. This can then be used to find a first order approximation for the imaginary part of ω.

- In order to study the propagation of waves in a hot plasma it is necessary to have a suitable model for the zero order distribution function. In plasmas which have reached thermal equilibrium this is a Maxwellian distribution.

- In the magnetosphere plasmas are often not at thermal equilibrium although they may be near other quasi-equilibria. Such distributions are often modeled in terms of separable distributions in which the distribution function is a product of two factors, one of which depends on $v_\perp$ only and the other on $v_\parallel$ only. Examples of such distributions are anisotropic Maxwellians in which the perpendicular and parallel components have different temperatures, loss-cone distributions which may be modeled by modified Maxwellians or by superposed Maxwellians and particle beams which may be modeled by a component with a bulk velocity along the field superimposed on a background Maxwellian.

Chapter 7

Waves in a Hot Plasma – 2. Equilibrium and Non-Equilibrium Distributions

7.1 Introduction

The dispersion relations found in Chapter 6 are expressed in terms of the zero order distribution function. While some general deductions can be made from the dispersion relations described in Chapter 6 the detailed nature of the waves can only be discussed when the actual distribution function is known. We therefore, in this chapter, discuss examples of waves in a plasma with a specified distribution function.

We first choose a plasma which is in thermal equilibrium so that the zero order distribution function is Maxwellian. The specific examples studied are (i) Bernstein modes, which are new modes which only appear when kinetic effects are taken into account, (ii) the Landau damping of electrostatic plasma waves which is also a kinetic effect and (iii) cyclotron damping of whistlers propagated along the direction of the magnetic field.

Because of the long collision time the plasmas in the magnetosphere are often not in thermal equilibrium. They can, however, exist in quasi-equilibrium states with non-Maxwellian distribution functions. There is then available free energy which may be dissipated through various instabilities which lead to wave growth.

If a plasma is far from thermal equilibrium rapid growth of waves to large amplitudes takes place with consequent non-linear effects occurring. The plasma may become turbulent. The waves react on the particles, and may change the distribution function so that it approaches an equilibrium or quasi-equilibrium state. Fully non-linear processes such as turbulence are of great intrinsic interest; they are, however, beyond the scope of this book and we shall not discuss them in any detail. The effects of waves on particles are discussed in a quasi-linear way in the next chapter. In this chapter we shall discuss examples of the linear theory of how waves are generated or amplified in a plasma which is not very far from equilibrium.

We shall not attempt full generality. Instead we shall take some specific examples of non-equilibrium zero order distribution functions and investigate the instabilities associated with them. When we require specific models we shall make use of the distribution functions which we have presented in Section 6.7. These are all close relatives of the Maxwellian distribution. We can thus use a restricted set of mathematical techniques to illustrate the physical principles. For a treatment of a much wider set of phenomena the reader should consult the book by Hasegawa [49].

7.2 Waves in Plasmas in Thermal Equilibrium

7.2.1 Bernstein Modes

An example of the rich variety of additional modes that appear when kinetic effects are included is the class of Bernstein [11] modes. These are electrostatic modes. We shall consider the case where the wave normal is exactly perpendicular to the magnetic field so that $k_{\|} = 0$. We neglect the effects of ions and use the dispersion relation (6.15). The Maxwellian distribution is used in the form (6.37) and the integration over $v_{\|}$ carried out to get

$$1 - \frac{\omega_p^2}{k_\perp^2} \sum_{l=-\infty}^{\infty} \frac{l\Omega}{\omega - l\Omega} \frac{m^2}{K^2T^2} \int_0^\infty v_\perp J_l\left(\frac{k_\perp v_\perp}{\Omega}\right) e^{-mv_\perp^2/2KT} dv_\perp = 0 \quad (7.1)$$

The integral is Weber's second exponential integral (B.20) with $a = b = k_\perp/\Omega$, $p^2 = m/2KT$. This gives

$$\frac{\omega_p^2}{\Omega} \sum_{l=-\infty}^{\infty} \frac{l}{\omega - l\Omega} \frac{e^{-\xi}}{\xi} I_l(\xi) = 1 \quad (7.2)$$

where $\xi = k_\perp^2 KT/m\Omega^2$. The $l = 0$ term is zero and, since $I_{-l} = I_l$, we can write

$$\omega_p^2 \sum_{l=1}^{\infty} \frac{2l^2}{\omega^2 - l^2\Omega^2} \frac{e^{-\xi}}{\xi} I_l(\xi) = 1 \quad (7.3)$$

which is the desired dispersion relation connecting ω and $k_\perp$. It can only be evaluated numerically. Before presenting numerical results we consider the limiting cases $\xi \to 0$ and $\xi \to \infty$.

The case $\xi = 0$ corresponds either to a cold plasma ($T = 0$) or infinite wavelength ($k_\perp = 0$). We note that $I_l(\xi) \to \xi^l/2l!$ in this case. Then

$$\omega_p^2 \sum_{l=1}^{\infty} \frac{l\xi^{l-1}}{(l-1)!(\omega^2 - l^2\Omega^2)} = 0 \quad (7.4)$$

If $\omega \neq l\Omega$ all the terms in the sum are zero except the first and we are left with

$$\omega^2 = \omega_p^2 + \Omega^2$$

which is an oscillation at the upper hybrid resonance. In addition, if $l > 1$, we can satisfy the dispersion relation by

$$\omega = l\Omega$$

There are therefore oscillations at all the harmonics of the gyrofrequency except the first as well as the upper hybrid resonance.

The Larmor radius, ρ_T, of an electron with the r.m.s. thermal speed is given by $\rho_T^2 = 2KT/m\Omega$. Thus $\xi = k_\perp^2 \rho_T^2 = 2\pi^2\rho_T^2/\lambda^2$. The case $\xi \to \infty$ corresponds to wavelengths much shorter than the average radius of gyration of the particles. The first term of the asymptotic expansion of I_l for large ξ is $e^\xi/\sqrt{2\pi\xi}$; each term of the sum is of order $\xi^{-1/2}$. Thus, as $\xi \to \infty$, the dispersion relation can be satisfied by $\omega = l\Omega$ at all the harmonics of the gyrofrequency, including the first. The question arises of how these modes connect for intermediate values of $k_\perp$. This requires numerical calculation such as that carried out by Crawford [29] and illustrated in Figure 7.1. It can be seen from this diagram that there are

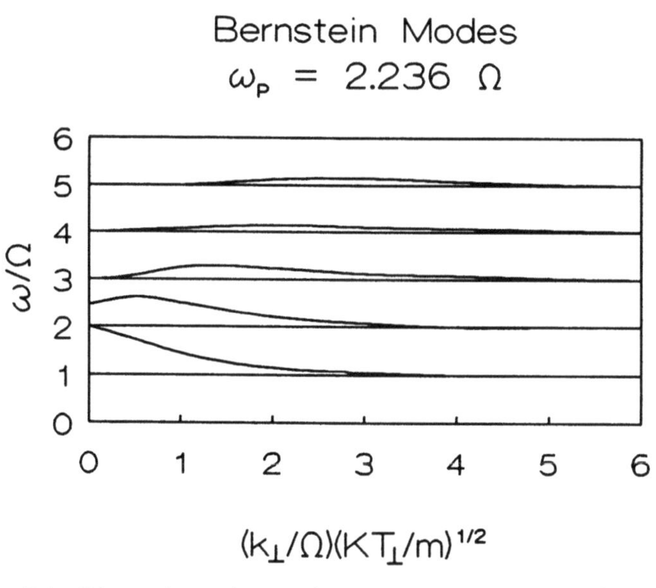

Figure 7.1: Dispersion relation for the Bernstein modes (after Crawford [29])

forbidden frequency bands separating the modes. If $l\Omega > (\omega_p^2 + \Omega^2)^{1/2}$ the curve starting at $\omega = l\Omega$ for small $k_\perp$ becomes $\omega = l\Omega$ for large k, with a maximum at intermediate values of $k_\perp$. The curve starting at the upper hybrid frequency when $k_\perp = 0$ rises to a maximum and then approaches the value of the harmonic immediately below the hybrid frequency. Curves starting at harmonic $l\Omega$ below the gyrofrequency decrease and tend to the value $(l-1)\Omega$ at large $k_\perp$.

We shall not consider cases where the wave is not exactly perpendicular to the magnetic field. We expect modes analogous to the Bernstein modes to exist but, because $k_\parallel \neq 0$ we anticipate that resonant particles may give rise to damping.

7.2.2 Landau Damping of Electrostatic Waves

For waves propagating exactly along the field $k_\perp$ is zero. Let us consider frequencies much greater than the ion plasma frequency so that ions are unaffected by the wave and only the electrons contribute to the dispersion relation. The dielectric constant is then

$$\kappa = 1 + \frac{i\sigma}{\epsilon_0 \omega}$$

where σ follows from (6.14) for electrons and f_0 is the Maxwellian distribution. Then

$$\kappa = 1 + \frac{\omega_p^2}{N k_\parallel^2} \int_0^\infty v_\perp f_\perp \, dv_\perp \int_C \frac{k_\parallel \frac{\partial f_\parallel}{\partial v_\parallel}}{\omega - k_\parallel v_\parallel} dv_\parallel$$

The integral with respect to $v_\perp$ is unity and

$$\frac{\partial f_\parallel}{\partial v_\parallel} = -\frac{2N}{\sqrt{\pi}} \left(\frac{m}{2KT}\right)^{\frac{3}{2}} v_\parallel e^{-mv_\parallel^2/2KT}$$

Thus

$$\kappa = 1 + \frac{\omega_p^2 m}{k_\parallel^2 KT} \left\{ 1 - \left(\frac{m}{2\pi KT}\right)^{\frac{1}{2}} \int_C \left(\frac{\omega}{\omega - k_\parallel v_\parallel}\right) e^{-mv_\parallel^2/2KT} dv_\parallel \right\}$$

This can be expressed in terms of the plasma dispersion function whose properties are outlined in Appendix B.2. The dispersion relation for electrostatic waves has the form $\kappa = 0$ so for this case it is

$$1 + \frac{\omega_p^2 m}{k_\parallel^2 KT} \left[1 + \frac{\omega}{\sqrt{2} k_\parallel} \sqrt{\frac{m}{KT}} Z\left(\frac{\omega}{\sqrt{2} k_\parallel} \sqrt{\frac{m}{KT}}\right) \right] = 0 \qquad (7.5)$$

The plasma dispersion function is complex. If $k_\parallel$ is specified the equation will therefore give a complex value of ω. On physical grounds, since the plasma is in thermal equilibrium, we can anticipate that this will be negative so that the wave is damped. Generally it is necessary to perform a numerical computation. The cases of physical interest however are those for which the damping is small. We argue, initially on physical grounds, that this will occur when the thermal speed is small compared to the wave speed, i.e. $KT/m \ll \omega/k_{\|}$. This means that there can be little transfer of energy to the particles through resonance. In this case the argument of the Z-function is large and we can use the asymptotic approximation (B.28). We assume on physical grounds that omega has a negative imaginary part. This will prove to be the case when we obtain the final formula. The result is

$$1-\frac{\omega_p^2}{\omega^2}\left\{1+\frac{3k_{\|}^2 KT}{m\omega^2}-i\sqrt{2\pi}\frac{\omega^3}{k_{\|}^3}\left(\frac{m}{KT}\right)^{\frac{3}{2}}\exp\left(-\frac{\omega^2 m}{2k_{\|}^2 KT}\right)+\ldots\right\}=0 \tag{7.6}$$

The first approximation to this is $\omega_p^2 = \omega^2$. We can substitute this in the small second term and if we neglect the last term which arises from the particle resonance this can be written in the form

$$n^2 = \frac{1-X}{U^2} \tag{7.7}$$

where $n = ck_{\|}/\omega$ is the refractive index, $X = \omega_p^2/\omega^2$ as usual, and $U^2 = 3KT/mc^2$. This is exactly the same formula as we found for the case $\theta = 0$ using warm plasma theory, U being the normalized sound speed with $\gamma = 3$. The effect of the particle resonance is to introduce a damping term. To find this we set $\omega = \omega_r + i\omega_i$. We use the approximation $\omega = \omega_p$ in the small imaginary part to get

$$\omega_i = -\sqrt{\frac{\pi}{2}}\frac{\omega_p}{(k_{\|}\lambda_D)^3}e^{-1/2k_{\|}^2\lambda_D^2} \tag{7.8}$$

where $\lambda_D = \sqrt{\epsilon_0 KT/Nq^2}$ is the Debye length. As assumed ω_i is negative and so the wave is damped. This is an example of linear *Landau damping*. It arises because of interaction between the wave and resonant plasma particles which travel at the wave speed. For these waves the damping is small provided the wave has a frequency close to the plasma frequency. The first rigorous treatment of the initial value problem was carried out by Landau [73]. A physical picture of how the damping occurs in the linear case is given by Chen [25, §§7.5, 7.6].

When the wave is exactly along the magnetic field ($k_\perp = 0$) there are no resonances at the harmonics of the cyclotron frequency. This is because the wave electric field is exactly parallel to $\mathbf{B}_0$ and therefore cannot affect the perpendicular component of the particle velocity. When the wave is not exactly along the magnetic field ($k_\perp \neq 0$) the dispersion relation is much more complicated, including an infinite sum of resonances at the harmonics of the Doppler shifted cyclotron frequency. There is now a component of electric field perpendicular to $\mathbf{B}_0$ and thus the perpendicular component of the velocity can be affected by the wave. We shall not work out this case in detail, but shall note that in the case of waves in the electrostatic approximation the strongest resonance is the Landau resonance which dominates the damping. This is because for such waves the electric field is approximately parallel to the wave normal. Particles traveling in phase with the wave experience a steady electric field parallel to their motion. This does work on them and changes their energy and therefore the energy of the wave. Cyclotron resonances are important for transverse waves as discussed in the next section.

7.2.3 Cyclotron Damping of Whistlers

Purely transverse waves do not suffer Landau damping because the longitudinal motion of the particles is at right angles to the direction of the electric field. In this case work can only be done on the particles if the transverse component of their velocity remains in phase with the electric field. Because the particles are gyrating this means that they can have the same angular frequency as the Doppler shifted frequency of a cicularly polarized wave.

We have seen (6.27) that waves propagated exactly along the magnetic field separate into two oppositely circularly polarized waves and the electrostatic plasma wave that we have already considered. In Chapter 3 it has been shown that, in cold plasmas, whistler waves propagating along the field are of this type. They are right hand polarized waves when propagated in the positive $\mathbf{B}_0$ direction and left hand polarized waves when propagated in the opposite direction. They occur for conditions corresponding to regions 7 and 8 of the CMA diagram. They are unaffected by ions. We consider the modifications produced to the cold plasma results by kinetic theory.

Consider a plasma in thermal equilibrium in which a right hand polarized wave is propagated along the magnetic field. Assume that the frequency is much greater than the ion gyrofrequency so that the motion of positive ions can be neglected. The distribution function is Maxwellian and represented by (6.37). In the expression for κ_R given by (6.27) we choose the lower sign, corresponding to electrons, set $k_\perp = 0$, perform the trivial integrations over θ and $v_\perp$ and express the integral over $v_\parallel$ in terms of the plasma dispersion function (Appendix B.2) to get

$$\kappa_R = 1 + \frac{\omega_p^2}{\omega k_\parallel}\sqrt{\frac{m}{2KT}}\,Z\left(\frac{\omega - \Omega}{k_\parallel}\sqrt{\frac{m}{2KT}}\right) \tag{7.9}$$

As in the case of electrostatic waves a full investigation of this expression requires a numerical study. We can, however, make approximations which are appropriate in almost the whole region of practical interest.

The argument of the Z-function can be written in the form

$$\frac{V_P}{v_T}\frac{\omega - \Omega}{\omega}$$

where $V_P = \omega/k_\parallel$ is the phase speed and $v_T = \sqrt{2KT/m}$ is the thermal speed of the particles. If the thermal speed of the particles is large we expect heavy damping and so we consider the case $v_T \ll V_P$. Whistlers are propagated at frequencies comparable with Ω. There are thus two extreme cases, one where the frequency is not particularly close to the cyclotron frequency so that $|\omega - \Omega| \gg k_\parallel v_T$ and the other close to the cyclotron frequency so that $|\omega - \Omega| \ll k_\parallel v_T$. In the first case we apply the asymptotic expansion of Z given by (B.28) and in the second the series expansion given by (B.27).

If we take the asymptotic form of the Z-function the dispersion relation $c^2k_\parallel^2/\omega^2 = \kappa_R$ may be written

$$\frac{c^2k_\parallel^2}{\omega^2} = 1 - \frac{\omega_p^2}{(\omega - \Omega)\omega} + 2i\sqrt{\pi}\frac{\omega_p^2}{k_\parallel v_T\omega}e^{-\frac{(\omega-\Omega)^2}{k_\parallel^2 v_T^2}} + \dots \qquad (7.10)$$

If we ignore the small imaginary term this can be written in the notation of Chapter 3 as

$$n^2 = 1 - \frac{X}{1 - Y}$$

which is the formula for the refractive index of a whistler propagated along the magnetic field in a cold plasma. To this order of approximation the plasma is cold. The imaginary term arises from the pole at $\omega - k_\parallel v_\parallel = \Omega$. It therefore corresponds to Doppler shifted cylotron resonance with particles of the appropriate energy. We can proceed in the same way as for electrostatic waves to find an expression for ω_i. It again gives a negative value for ω_i, showing that our implicit assumption that the wave is damped is correct. The expression for ω is complicated and not worth quoting. It turns out to be very small in all practical cases.

The other extreme case occurs in the immediate neighbourhood of the cyclotron frequency. The series expansion (B.27) is valid in the range $\Omega - k_\parallel v_T < \omega < \Omega + k_\parallel v_T$. The dispersion relation is then

$$\frac{c^2k_\parallel^2}{\omega^2} = 1 + \frac{\omega_p^2}{\omega k_\parallel v_T}\left[i\sqrt{\pi} - \frac{2V_P}{v_T}\left(\frac{\omega - \Omega}{\omega}\right) + \dots\right] \qquad (7.11)$$

The imaginary term is thus large. There is strong damping near the cyclotron frequency. We thus see that, in general, for whistlers propagated along the magnetic field the cold plasma theory is generally valid. In a high temperature plasma kinetic effects produce small Doppler shifted cyclotron frequency damping. Near the cyclotron frequency the cold plasma theory breaks down and there is strong absorption of the wave.

7.3 Longitudinal Waves Excited by a Particle Beam

Consider a plasma which consists of a cold ion background and a hot electron population of the type shown in Figure 6.5 in which a beam of electrons moving along the magnetic field is superposed on a background distribution. If current arising from the drift velocity of particles in the beam is not desired a small drift velocity can be imposed on the background electrons or on the cold ions in such a direction and of such a magnitude as to make the net current zero. There is a wide variety of instabilities related to such a situation. For example various cases include ion beams or current driven instabilities involving the relative drift of electrons and ions. We shall illustrate this type of instability by considering a weak electron beam parallel to the magnetic field.

Such a beam will interact with electrostatic waves propagated in any direction. The strongest interaction will, however, be for those waves propagated in the magnetic field direction parallel to the beam. In the electrostatic approximation, since $k_\perp = 0$ and $k_\| = k$, the dispersion relation is substantially simplified. The arguments of the Bessel functions are all zero and thus $J_0 = 1$, and $J_l = 0$ when $l \neq 0$. The dispersion relation (6.15) for this case will thus have the form

$$1 + \chi_\| \equiv 1 + \frac{i\sigma_\|}{\epsilon_0 \omega} = 0$$

where

$$\sigma_\| = -\frac{2\pi i \epsilon_0}{N} \frac{\omega_p^2 \omega}{k} \int_C \int_0^\infty \frac{v_\perp \frac{\partial f_0}{\partial v_\|}}{\omega - k_\| v_\|} \, dv_\perp \, dv_\| \tag{7.12}$$

and C is the Landau contour. The distribution function f_0 is assumed to be separable and we can thus immediately integrate over $v_\perp$ to get

$$\sigma = -i\frac{\epsilon_0}{N} \frac{\omega_p^2 \omega}{k^2} \int_C \frac{f'(v)}{\omega/k - v} \, dv \tag{7.13}$$

where $f'(v) = \partial f_\| / \partial v_\|$ and we have dropped the subscript $\|$ from k, f and v.

7.3.1 Condition for Growth

We shall illustrate simple linear growth for this case only. We shall assume that the growth is small so that ω has a small positive imaginary part. In this case the pole lies a little above the real v-axis and the Plemelj condition (B.29) allows us to replace the integral with one in which there is a pole on the real axis. Thus

$$\sigma = i\frac{\epsilon_0}{N}\frac{\omega_p^2\omega}{k^2}\left\{P\int_{-\infty}^{\infty}\frac{f'(v)}{v-\omega_r/k}\,dv + i\pi f_0'(\frac{\omega_r}{k})\right\} \tag{7.14}$$

where k is positive. When the real part of σ is negative the conditions for growth are satisfied because the particles do work on the wave. In this case

$$\Re(\sigma) = -\frac{\epsilon_0\pi}{N}\frac{\omega_p^2\omega}{k^2}f_0'(\frac{\omega_r}{k})$$

It can immediately be seen that a necessary condition for instability is that $f'(\omega_r/k) > 0$ when the phase velocity of the wave is positive. The slope of the distribution function which we have chosen is positive over part of the range and only waves with phase velocities in this range can be unstable. This is a necessary, but not sufficient, criterion.

7.3.2 Phase and Group Speeds

It will be useful to find a general relationship between the phase and group speeds. The dispersion relation has the form

$$\kappa_{||}(\omega,k) = 0$$

where

$$\kappa_{||}(\omega,k) = 1 + \frac{\omega_p^2}{Nk^2}\int_C\frac{f'(v)}{V_P - v}\,dv, \tag{7.15}$$

where $V_P = \omega/k$ and is the (complex) phase speed.

If we regard this relationship as expressing ω as a function of k we can take the total derivative with respect to k getting

$$\frac{\partial\kappa_{||}}{\partial\omega}\frac{d\omega}{dk} + \frac{\partial\kappa_{||}}{\partial k} = 0$$

Obviously

$$\frac{\partial\kappa_{||}}{\partial\omega} = \frac{1}{k}\frac{\partial\kappa_{||}}{\partial V_P}$$

and

$$\begin{aligned}\frac{\partial\kappa_{||}}{\partial k} &= -\frac{2}{k}(\kappa_{||}-1) - \frac{\omega}{k^2}\frac{\partial\kappa_{||}}{\partial V_P}\\ &= -\frac{2}{k}(\kappa_{||}-1) - \frac{\omega}{k}\frac{\partial\kappa_{||}}{\partial\omega}\end{aligned} \tag{7.16}$$

But $\kappa_{||} = 0$. Thus

$$\frac{d\omega}{dk} - \frac{\omega}{k} = \frac{2}{k\,\partial\kappa_{||}/\partial\omega} \tag{7.17}$$

If losses or growth are very small so that ω is real then this can be written

$$V_G - V_P = \frac{2}{k\,\partial\kappa_{||}/\partial\omega_r} \tag{7.18}$$

7.3.3 Approximation to the Dispersion Relation

We have assumed that the growth rate is small. We shall now make some further realistic approximations which allow us to remove some mathematical complexities. We shall consider the case of a weak beam in an electron distribution with r. m. s. speed much smaller than the phase speed of the wave. The beam is thus a small perturbation on the tail of a Maxwellian distribution. The dispersion relation can be written in the approximate form

$$\kappa_{||}(\omega, k) = P[\kappa_{||}(\omega, k)] - \frac{i\pi\omega_p^2}{Nk^2} f'(\omega/k) \tag{7.19}$$

where

$$P[\kappa_{||}(\omega, k)] = 1 + \frac{\omega_p^2}{N\omega k} P \int_{-\infty}^{\infty} \frac{f'(v)}{1 - kv/\omega}\, dv \tag{7.20}$$

Assume the following ordering:

$$|\omega_r/k| \gg |v| \gg |\omega_i/k| \tag{7.21}$$

We start by assuming that growth or loss can be completely neglected. Then ω is real and the principal part of the dispersion relation is also real. We can take it that the only significant contribution to the principal value of the integral is for values of v much less than ω_r/k. Thus we can expand the integral in powers of kv/ω_r and then integrate by parts getting

$$\begin{aligned} P \int_{-\infty}^{\infty} \frac{f'(v)}{1 - kv/\omega_r}\, dv &= \int_{-\infty}^{\infty} f'(v) \left[1 + \frac{kv}{\omega_r} + \left(\frac{kv}{\omega_r}\right)^2 + \ldots\right] dv \\ &= \frac{k}{\omega_r} \int_{-\infty}^{\infty} f(v) \left[1 + 2\frac{kv}{\omega_r} + 3\left(\frac{kv}{\omega_r}\right)^2 + \ldots\right] dv \\ &= \frac{k}{\omega_r} \left\{ N + \int_{-\infty}^{\infty} f(v) \left[2\frac{kv}{\omega_r} + 3\left(\frac{kv}{\omega_r}\right)^2 + \ldots\right] dv \right\} \end{aligned}$$

The dispersion relation is thus

$$1 - \frac{\omega_p^2}{\omega_r^2} - \frac{\omega_p^2}{N\omega_r^2} \int_{-\infty}^{\infty} f(v) \left[2\frac{kv}{\omega_r} + 3\left(\frac{kv}{\omega_r}\right)^2 + \ldots\right] dv = 0 \tag{7.22}$$

This can be solved by successive approximations. The first approximation is just

$$\omega = \omega_p$$

The second approximation is then obtained by replacing ω by ω_p in the remaining terms:

$$\omega_r^2 - \omega_p^2 - \frac{\omega_p^2}{N}\int_{-\infty}^{\infty} f(v)\left[2\frac{kv}{\omega_p} + 3\left(\frac{kv}{\omega_p}\right)^2 + \ldots\right] dv = 0 \tag{7.23}$$

The first integral is proportional to the current density in the plasma. In many cases it is zero if there is no net current. The second integral gives $3k^2KT_{\text{eff}}/m$. To this order the dispersion relation is essentially the same as that for plasma waves in a warm plasma. Indeed, if the beam is regarded as a perturbation on a Maxwellian background, and its contribution is entirely neglected to this order, the analysis is exactly the same as we carried out for the case of Landau damping. The difference will arise in the next order of approximation when we allow ω to have a small imaginary part.

7.3.4 Growth Rate

The introduction of the effect of the beam means that ω is no longer real. If it has a small imaginary part we can write

$$\kappa_{\|}(\omega, k) = \kappa_{\|}(\omega_r, k) + i\omega_i\left(\frac{\partial\kappa_{\|}}{\partial\omega}\right)_{\omega=\omega_r} - \frac{i\pi\omega_p^2}{Nk^2}f'(\omega_r/k) = 0 \tag{7.24}$$

We have already ensured that the real part is zero. We can use (7.18) to find the derivative of $\kappa_{\|}$. The imaginary part must also be zero which gives

$$\omega_i = \frac{\pi}{2}\frac{\omega_p^2}{Nk}f'(\omega_r/k)(V_P - V_G) \tag{7.25}$$

Note that the analysis has been carried out for waves propagated in the positive z-direction. The opposite sign will be obtained if V_P is negative. Another point to be noted is that this shows that a positive slope for f', while a necessary condition for instability, is not sufficient. We also require that $V_P > V_G$. In Figure 7.2 we show a computation of the dispersion relation and growth rate in a simple case. We have taken the model distribution such as that shown in Figure 6.5 with a weak beam, of normalized amplitude 0.05, centered on $v = 3\sqrt{2KT_{\text{background}}/m}$. The left hand panel shows normalized frequency and growth rate as a function of normalized wave number and growth rate. The straight line is drawn

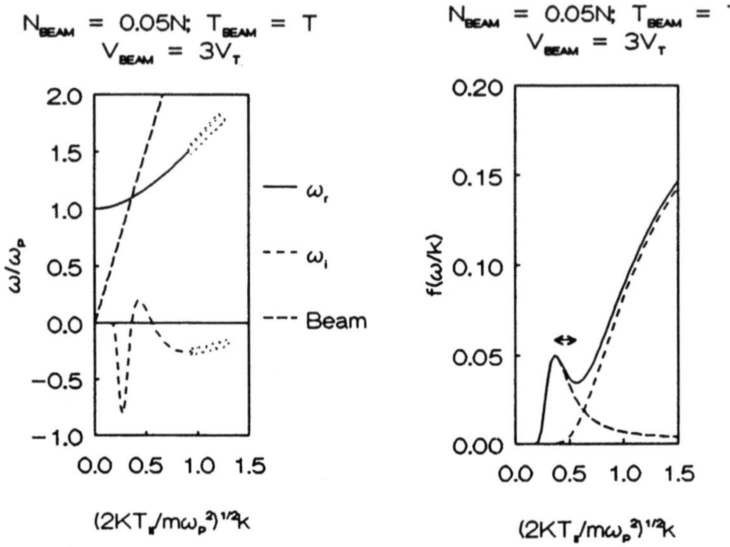

Figure 7.2: Dispersion relation for a plasma in which a weak electron beam is superimposed on the background. For explanation see text.

with a slope corresponding to a speed equal to the mean speed of the particles in the beam. The shaded region shows where the approximations, which we have used, break down. When $k = 0$ the wave frequency is equal to the plasma frequency and the damping is negligible. This is just the plasma oscillation of cold plasma theory at $X = 1$. As k increases ω also increases and it is clear from the slope of the tangent to the curve that the group speed is always much less than the phase speed.

The right-hand panel shows the distribution function plotted, not as a function of v but as a function of k. The slope of this curve is

$$\frac{df}{dk} = f'(\omega_r/k)\frac{d}{dk}\left(\frac{\omega(k)}{k}\right) = -\frac{1}{k}(V_P - V_G)f'(\omega_r/k)$$

and is thus negative when there is growth and positive when there is damping.

In can be seen in the left-hand panel that as k increases from zero damping is at first negligible for phase speeds very much greater than the drift speed of the beam. As k increases Landau damping increases as kinetic effects become appreciable. As k further increases there is a range of wave numbers for which growth occurs. This range corresponds to the range marked by the double arrow in the right-hand panel where the slope is negative (positive $f'[v]$). The maximum growth occurs where the slope of the distribution function is greatest. As k increases Landau damping takes over and will become larger as the phase speed approaches the r. m. s. speed of the background plasma. There is no point in attempting a more accurate calculation in the shaded portion of the graph because Landau damping will ensure that the wave decays rapidly to zero.

It must be emphasized that our detailed calculations apply to a special case. Many other cases can be envisaged. Nevertheless it epitomizes many cases which can be treated by linear theory. There is a background plasma distribution which can support waves, in this case the electron plasma waves of warm plasma theory. Superimposed on this distribution is a small perturbation which is a source of free energy, in this case the streaming plasma. The particles associated with the perturbation interact with the waves to cause growth. The physical picture is of a set of resonant particles interacting with a wave in such a way as to transfer energy to it. More complicated cases can be envisaged where this simple picture is not true. In such cases the interaction may be with the whole distribution rather than with a subset of resonant particles.

7.4 Electrostatic Waves Associated with Anisotropic Distributions

When the wave normal is not parallel to the magnetic field and the distribution function is anisotropic, either because of unequal $T_{\|}$ and $T_{\perp}$ or because of a loss-cone, new types of instability appear [67]. We shall study these using the subtracted Maxwellian distribution (6.44). Special cases of interest can then be studied by setting β and δ appropriately. Some of the more interesting cases occur when there is a cold background plasma and this will be included when necessary.

7.4.1 General Expression for the Conductivity

When $k_{\perp} \neq 0$ the conductivity takes the form of an infinite sum (6.14). Assume that the distribution function is separable. The lth term of this sum is then

$$\sigma_l = \frac{i\epsilon_0 \omega \omega_p^2}{N k^2} \int_C \int_0^\infty v_\perp J_l^2\left(\frac{k_\perp v_\perp}{\Omega}\right) \times \frac{\frac{l\Omega}{k_\| v_\perp} f_\|(v_\|) f'_\perp(v_\perp) + f'_\|(v_\|) f_\perp(v_\perp)}{v_\| - \frac{\omega - l\Omega}{k_\|}} \, dv_\perp \, dv_\| \tag{7.26}$$

We now substitute the subtracted loss-cone distribution (6.44) for $f_\perp$ and a Maxwellian for $f_\parallel$. These can be explicitly differentiated to find $f'_\parallel$ and $f'_\perp$. In particular we note that

$$\frac{f'_\parallel}{v_\parallel - \frac{\omega - l\Omega}{k_\parallel}} = \frac{m v_\parallel}{KT_\parallel} \frac{f_\parallel}{v_\parallel - \frac{\omega - l\Omega}{k_\parallel}} = \frac{m}{KT_\parallel}\left[1 + \frac{\frac{\omega - l\Omega}{k_\parallel}}{v_\parallel - \frac{\omega - l\Omega}{k_\parallel}}\right] f_\parallel$$

The double integral then can be expressed as the sum of several products of single integrals. The integrals over $v_\parallel$ can be expressed in terms of the plasma dispersion function, Z, (B.24) and Weber's second exponential integral (B.20) can be used to express the integrals over $v_\perp$ in terms of modified Bessel functions of the first kind. The result is

$$\begin{aligned}
\sigma_l &= -\frac{2i\epsilon_0\omega\omega_p^2}{k^2 V_\parallel^2} \\
&\times \left\{\left[1 + \frac{\beta\delta}{1-\beta}\right] G_l\left(\frac{k_\perp^2 V_\perp^2}{2\Omega^2}\right)\left[1 + \frac{\omega - l\Omega\left(1 - \frac{T_\parallel}{T_\perp}\right)}{k_\parallel V_\parallel} Z(\zeta_l)\right]\right. \\
&- \left.\frac{\beta\delta}{1-\beta} G_l\left(\frac{\beta k_\perp^2 V_\perp^2}{2\Omega^2}\right)\left[1 + \frac{\omega - l\Omega\left(1 - \frac{T_\parallel}{\beta T_\perp}\right)}{k_\parallel V_\parallel} Z(\zeta_l)\right]\right\}
\end{aligned} \tag{7.27}$$

where

$$V_{\parallel,\perp} = \sqrt{2KT_{\parallel,\perp}/m} \tag{7.28}$$

$$G_l(\xi^2) = \exp(-\xi^2) I_l(\xi^2) \tag{7.29}$$

and

$$\zeta_l = (\omega - l\Omega)/k_\parallel V_\parallel \tag{7.30}$$

This expression for the conductivity can be used to model the behavior of different particle species in a wide range of plasmas. For example, for electrons with a cold ion background, if $T_\parallel = T_\perp$, $\delta = 0$ and $k_\parallel = 0$ we obtain the conductivity for the Bernstein modes which we have already studied. If $k_\parallel \neq 0$ it corresponds to Bernstein modes which are not at right angles to the magnetic field. In its most general form we can use it to model plasmas with temperature anisotropy or with a loss-cone distribution. We shall discover that such distributions can lead to instability.

It should not be forgotten that the expression represents only one term in a sum. When studying growth we must study the competing effects of growth and damping in all the terms.

7.4.2 General Conditions for Growth

In this section we study some requirements for growth by considering the sign of the real part of the conductivity. A wide and confusing variety of special cases can be investigated but we shall first discuss some general considerations before going into detail. We restrict ourselves to the case for which growth and decay are small. In these circumstances ω_i is small and we can write

$$\sigma_l(\omega, k_\parallel, k_\perp) \simeq \Re\sigma_l(\omega_r, k_\parallel, k_\perp) + \Im\sigma_l(\omega_r, k_\parallel, k_\perp)$$

For real ω only the plasma dispersion function contributes to $\Re\sigma_l$ through its imaginary part, which from (B.31) is

$$Z_i\left(\frac{\omega - l\Omega}{k_\parallel V_\parallel}\right) = i\sqrt{\pi}\exp\left[-\left(\frac{\omega - l\Omega}{k_\parallel V_\parallel}\right)^2\right]$$

If this is used in the expression for σ_l the expression for the real part is

$$\begin{aligned}(\sigma_l)_r &= \frac{2\sqrt{\pi}\epsilon_0\omega\omega_p^2}{k^2 V_\parallel^2}\exp\left[-\left(\frac{\omega - l\Omega}{k_\parallel V_\parallel}\right)^2\right] \\ &\times \left\{\left[1 + \frac{\beta\delta}{1-\beta}\right] G_l\left(\frac{k_\perp^2 V_\perp^2}{2\Omega^2}\right)\left[\frac{\omega - l\Omega\left(1 - \frac{T_\parallel}{T_\perp}\right)}{k_\parallel V_\parallel}\right]\right. \\ &- \left.\frac{\beta\delta}{1-\beta} G_l\left(\frac{\beta k_\perp^2 V_\perp^2}{2\Omega^2}\right)\left[\frac{\omega - l\Omega\left(1 - \frac{T_\parallel}{\beta T_\perp}\right)}{k_\parallel V_\parallel}\right]\right\}\end{aligned} \tag{7.31}$$

In general it is possible for this expression to be negative which is a necessary condition for growth. We consider some special cases.

Isotropic Distribution

If $T_\parallel = T_\perp$ and $\delta = 0$ then the expression becomes

$$(\sigma_l)_r = \frac{2\sqrt{\pi}\epsilon_0\omega\omega_p^2}{k^2 V_\parallel^2}\exp\left[-\left(\frac{\omega - l\Omega}{k_\parallel V_\parallel}\right)^2\right] G_l\left(\frac{k_\perp^2 V_\perp^2}{2\Omega^2}\right)\frac{\omega}{k_\parallel V_\parallel} \tag{7.32}$$

Clearly this is positive unless $k_\parallel$ is zero. Thus any wave arising in such a plasma will be damped unless its wave normal is at right angles to the magnetic field. As $k_\parallel \to 0$ the real part of the conductivity tends to zero because $\lim_{x\to\infty} xe^{-x} = 0$. This applies to the Bernstein waves discussed above and justifies our statement that they are damped unless the wave normal is at right angles to the magnetic field.

Temperature Anisotropy

If $T_{\|} \neq T_{\perp}$ and $\delta = 0$ then there is no loss-cone but the distribution is anisotropic. The real part of the conductivity is then

$$(\sigma_l)_r = \frac{2\sqrt{\pi}\epsilon_0\omega\omega_p^2}{k^2V_{\|}^2} \exp\left[-\left(\frac{\omega - l\Omega}{k_{\|}V_{\|}}\right)^2\right] G_l\left(\frac{k_{\perp}^2V_{\perp}^2}{2\Omega^2}\right) \frac{\omega - l\Omega\left(1 - \frac{T_{\|}}{T_{\perp}}\right)}{k_{\|}V_{\|}} \tag{7.33}$$

It is clear that, provided $T_{\|} < T_{\perp}$ it is conceivable that this is negative. It must, however, be borne in mind that this expression is only a single term in an infinite sum. Whether or not there is growth or damping at a particular frequency depends on the competing effects of all the terms in the sum. It must also be borne in mind that the values of ω and $\mathbf{k}$ used in this expression must satisfy the real part of the dispersion relation; negative real conductivity is necessary but not sufficient. It is useful to illustrate the behavior of the real part of the conductivity with a specific numerical example.

In Figure 7.3 the result of such a computation shown. For the parameters shown the argument of the function G is large and it can be evaluated using an asymptotic approximation which shows that $G_l(x) \simeq 1/\sqrt{2\pi x}$ [1, eq. 9.7.1]. It thus has the same value for each term in the sum. The contribution of each term thus becomes proportional to

$$\omega \exp\left[-\left(\frac{\omega - l\Omega}{k_{\|}V_{\|}}\right)^2\right] \frac{\omega - l\Omega\left(1 - \frac{T_{\|}}{T_{\perp}}\right)}{k_{\|}V_{\|}}$$

It is clear that, for $l = 0$, the real part of the conductivity is positive and this term cannot contribute to growth. If we focus on a single component for which $l \neq 0$, say σ_2 in Figure 7.3, we see that the exponential factor ensures that is negligible except near $\omega = 2\Omega$. The width over which it is appreciable depends on the value of $\Omega/k_{\|}V_{\|}$. Here we have chosen values so that this is quite large. This is possible because, although the thermal speed may be quite large, the angle θ is close to 90^0 and $V_{\|}$ is small. For all values of the frequency satisfying

$$\omega < l\Omega\left(1 - \frac{T_{\|}}{T_{\perp}}\right)$$

this term is negative so that it contributes to growth. In the frequency band $\Omega < \omega < 2\Omega$ the term σ_1 is appreciable and positive. The contribution of the two terms is comparable in the middle of the band near $\omega = \frac{3}{2}\Omega$. The total conductivity thus has a negative real part in the band $\frac{3}{2}\Omega < \omega < 2\Omega\left(1 - \frac{T_{\|}}{T_{\perp}}\right)$. Similar arguments apply for the other

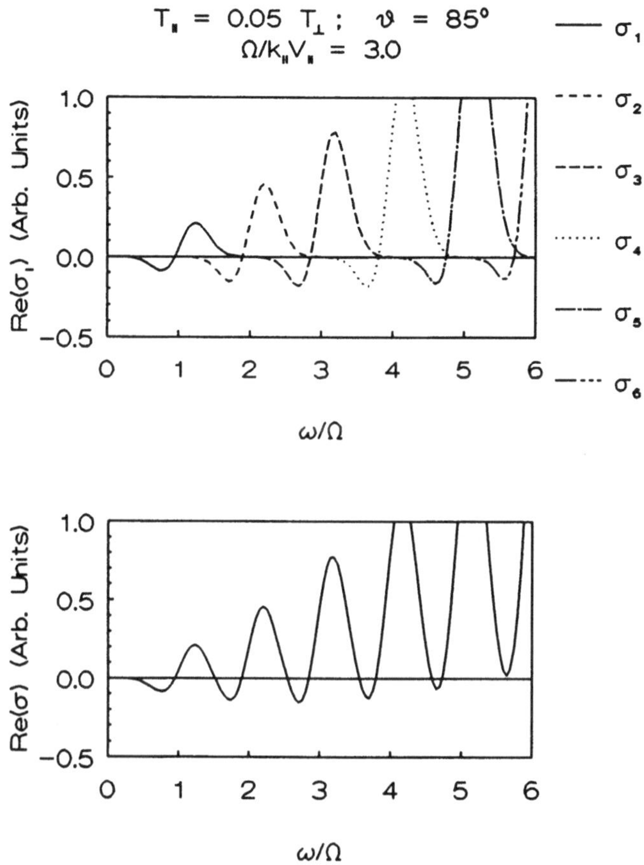

Figure 7.3: Real part of conductivity for distribution with strong temperature anisotropy. $T_{||}/T_{\perp} = V_{||}/V_{\perp} = 0.05$; $\theta = \arctan k_{\perp}/k_{||} = 85^0$; $\Omega/k_{||}V_{||} = 3.0$. The upper panel shows the contribution of each term in the sum. The lower panel shows the real part of the resultant conductivity.

components so there is a set of such bands given by

$$(l - \frac{1}{2})\Omega < \omega < l\Omega\left(1 - \frac{T_{\|}}{T_{\perp}}\right) \tag{7.34}$$

Because of increasing l each successive graph for σ_l is displaced upwards, and because of increasing ω the amplitude is increasing. The net effect is that the bands of negative conductivity only appear below a particular harmonic of the gyrofrequency. An upper limit for this harmonic follows directly from (7.34). The condition is

$$\frac{T_{\|}}{T_{\perp}} < \frac{1}{2l} \tag{7.35}$$

It can be seen that this implies a high degree of anisotropy for instability to be possible.

Loss-Cone Distributions

In the case of a beam the free energy required for the instability arose because of the positive slope of the parallel distribution function; because of this slope there was a maximum in the distribution function with a peak located where $v_{\|} = \sqrt{2KT_{\|}/m}$. In the case of anisotropic temperature the source of free energy was the differing perpendicular and parallel thermal energies. The distribution function was monotonically decreasing from the origin in velocity space and there was no local peak in the distribution function. In the case of a loss-cone distribution there is a local peak in the perpendicular distribution function because of the loss of particles in the region of velocity space near $v_{\perp} = 0$. The most general case has $T_{\perp} \neq T_{\|}$. There are no additional features introduced by considering this case so we shall illustrate the features introduced by the existence of a loss-cone by setting $T_{\perp} = T_{\|}$ and thus $V_{\perp} = V_{\|} = V$ in (7.31).

$$\begin{aligned}(\sigma_l)_r \quad = \quad & \frac{2\sqrt{\pi}\epsilon_0\omega\omega_p^2}{k^2V^2} e^{-[(\omega - l\Omega)/k_{\|}V)^2]} \left\{ \frac{\omega}{k_{\|}V} G_l\left(\frac{k_{\perp}^2V^2}{2\Omega^2}\right) \right. \\ + \quad & \frac{\beta\delta}{1-\beta}\frac{\omega}{k_{\|}V}\left[G_l\left(\frac{k_{\perp}^2V^2}{2\Omega^2}\right) - G_l\left(\frac{\beta k_{\perp}^2V^2}{2\Omega^2}\right)\right] \\ - \quad & \left. \delta\frac{l\Omega}{k_{\|}V}G_l\left(\frac{\beta k_{\perp}^2V^2}{2\Omega^2}\right)\right\}\end{aligned} \tag{7.36}$$

Of the three major terms within the outermost braces the first is clearly positive. The second is also positive if $\beta < 1$ because $G_l(\xi)$ is

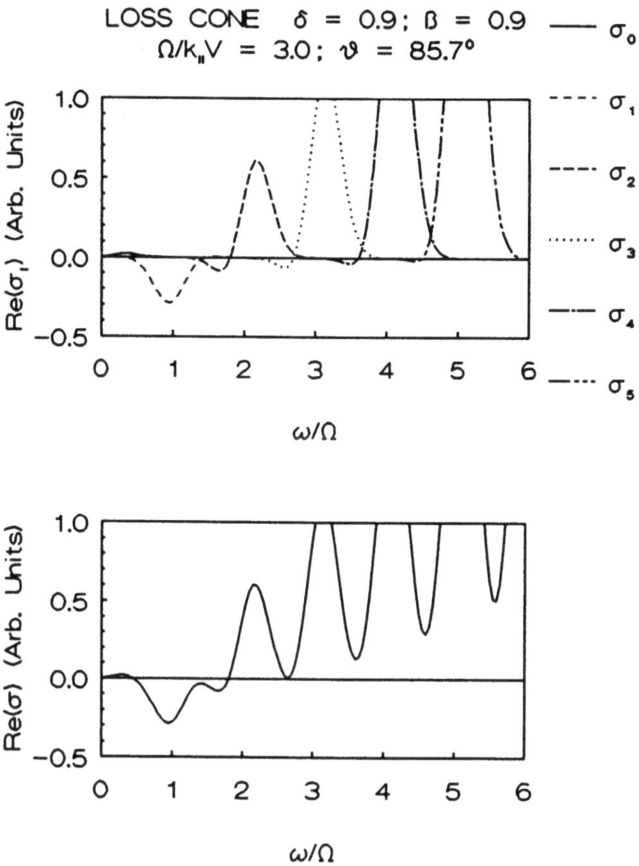

Figure 7.4: Real part of conductivity for model loss-cone distribution. $T_{\|}/T_{\perp} = V_{\|}/V_{\perp} = 1$; $\theta = \arctan k_{\perp}/k_{\|} = 85.7^0$; $\delta = 0.9$; $\beta = 0.9$; $\Omega/k_{\|}V_{\|} = 3.0$. The upper panel shows the contribution of each term in the sum. The lower panel shows the real part of the resultant conductivity.

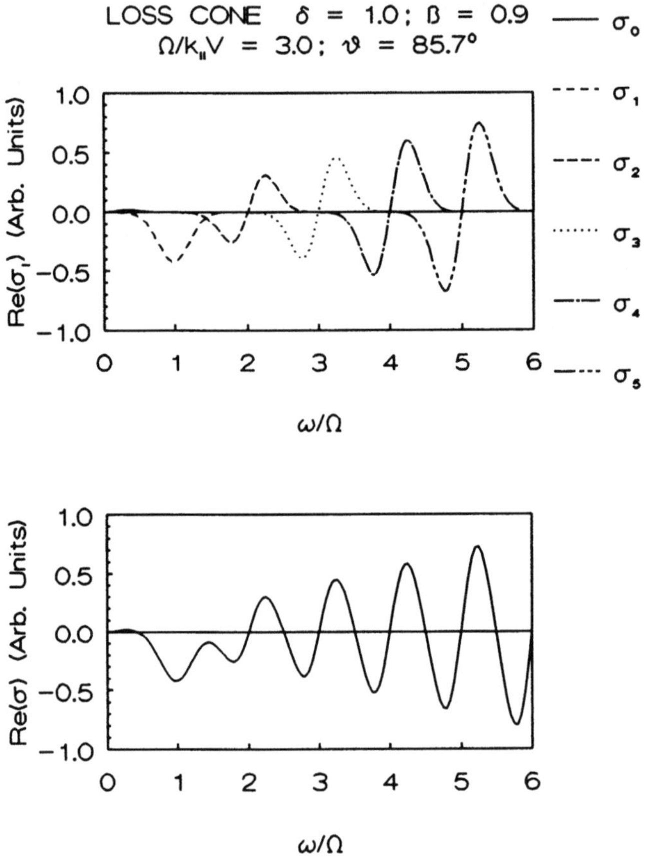

Figure 7.5: Real part of conductivity for model loss-cone distribution. $T_{\|}/T_{\perp} = V_{\|}/V_{\perp} = 1$; $\theta = \arctan k_{\perp}/k_{\|} = 85.7^0$; $\delta = 1.0$; $\beta = 0.9$; $\Omega/k_{\|}V_{\|} = 3.0$. The upper panel shows the contribution of each term in the sum. The lower panel shows the real part of the resultant conductivity.

a monotonically decreasing positive function of ξ. The third term is, however, negative. Depending on the relative sizes of the terms regions may be found where the real part of σ is negative and growth is possible. Of course it is still necessary to solve the real part of the dispersion relation to establish whether waves can exist with the appropriate values of ω.

Just as in the case of temperature anisotropy we show in Figures 7.4 and 7.5 the real part of the conductivity for a particular model of a loss-cone distribution. The parameters chosen in the first case correspond to a partly filled and fairly gentle loss-cone and in the second case to an empty but gentle loss-cone. While we do not claim that these parameters cover all cases the curves do show a feature which is not exhibited by distributions with temperature anisotropy. Instability is in principle possible at the gyrofrequency ($l = 1$) and also in the region $\omega < 1.5\Omega$. In the case of temperature anisotropy we have seen that no instability can occur in this region. For the partly filled loss-cone there is no instability at higher frequencies; for the empty loss-cone bands of possible instability appear between $(l - \frac{1}{2})\Omega$ and Ω.

To make further progress it is necessary to consider the nature of the distribution function.

7.4.3 Growth Rate for Various Models of the Particle Distribution

It is not possible to provide a comprehensive set of calculations of growth rates in a representative set of models. We can only hope to provide a few typical examples. One feature of the magnetosphere of importance is that the plasma frequently consists of a mixture of hot and cold plasmas. For this reason we first consider the contribution to the dispersion relation of a cold plasma component.

Electrostatic Oscillations in a Cold Plasma

If the plasma is cold we can find the conductivity for the case $\delta = 0$, $\beta = 0$, $T_{\|} = T_{\perp} \to 0$. The arguments of the G-functions tend to zero, $\lim_{\xi\to 0} G_0(\xi) \to 1$, $\lim_{\xi\to 0} G_{\pm 1}(\xi) \to \xi/2$, and the other terms in the sum are zero. The argument of the plasma dispersion function tends to infinity and we can use its asymptotic approximation (B.28). The resulting value of σ is used in the dispersion relation $\kappa_{\|} \equiv 1 + \sum_s i\sigma_s/\epsilon_0\omega = 0$ where s is the label for each particle species. The result is

$$1 + \sum_s \left\{ \frac{\omega_{p,s}^2}{\omega^2} \cos^2\theta + \frac{\omega_{p,s}^2}{\omega^2 - \Omega_s^2} \sin^2\theta \right\} = 0 \qquad (7.37)$$

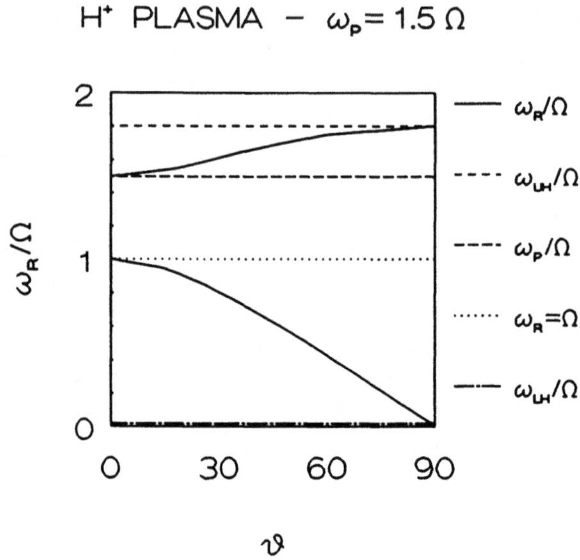

Figure 7.6: Cold plasma resonance frequency as a function of the angle θ between the wave normal and magnetic field.

How is this related to the theory of cold plasmas discussed earlier in this book? It is immediately obvious that this relation can be written in the form

$$A = \kappa_1 \sin^2\theta + \kappa_3 \cos^2\theta = 0$$

where A is given by (2.59). The dispersion relation which we have derived corresponds to the condition that the quadratic for the refractive index should have an infinite root. It is the condition for resonance in subsection 2.7.5.

The dispersion relation is independent of the magnitude of the wave number k. The resonant frequency is a function only of the plasma parameters and the direction of the wave normal. It is illustrated in Figure 7.6 for a plasma frequency greater than the gyrofrequency. It can be seen that, as the wave normal direction changes, there are two bands of possible resonance frequencies. One is between the upper hybrid resonance frequency (2.105) and the electron plasma frequency. The other is between the lower hybrid resonance frequency (2.106) and the electron gyrofrequency. There is a third band below the ion gyrofrequency which is not shown; it is not of interest here.

It must be remembered that these resonances are not isolated resonances. The upper resonance band represents the behavior of the Z-mode at the resonance cone and the lower represents the behavior of the whistler mode on its resonance cone. If there is any non-uniformity in the plasma an oscillation generated on the resonance cone can be refracted to a region where it becomes a progressive transverse wave. In considering waves in a warm plasma we have seen how a transverse electromagnetic wave can couple to an electrostatic plasma wave at a resonance cone. In the warm plasma case there are two separate modes which, near the resonance cone, have identical speeds and polarizations and therefore are coupled. In the case of a mixture of a cold and hot plasma considered here there is a similar effect. The real parts of the cold and hot contributions to the plasma dispersion relation combine to give a propagated wave having the resonance frequency. The imaginary part of the hot plasma contribution determines growth or decay.

Upper Hybrid Waves in a Plasma with a Loss-Cone Distribution

We consider one case of electrostatic waves generated by interaction by the free energy of a loss-cone distribution. A systematic study would require a prohibitive amount of space. One of the most comprehensive studies has been carried out by Kennel and Ashour-Abdalla and is reviewed in [67]. Our example is one of the cases which they consider [7]. A similar problem was first studied by Young *et al.* [120].

Consider a plasma which consists of a hot electron component, a cold electron component and a cold ion background. The distribution function for the hot electrons is a subtracted Maxwellian with an empty loss-cone ($\delta = 1$) and with $T_\perp = T_\parallel$. We consider waves with frequencies near the upper hybrid frequency so that the cold ions do not contribute to the dispersion relation. The dispersion relation is then of the form

$$1 + \chi_c + \chi_h = 0 \tag{7.38}$$

where the cold ion susceptibility is found from (7.37)

$$\chi_c = \frac{\omega_p^2}{\omega^2}\frac{(\omega^2 - \Omega^2\cos^2\theta)}{(\omega^2 - \Omega^2)} \tag{7.39}$$

and the hot ion susceptibility is

$$\chi_h = \sum_{l=-\infty}^{\infty} \chi_l = \frac{i}{\epsilon_0\omega}\sum_{l=-\infty}^{\infty} \sigma_l \tag{7.40}$$

Thus, from (7.27)

$$\chi_l \;=\; \frac{2\omega_p^2}{k^2V_\parallel^2}$$

$$\times \left\{ \frac{1}{1-\beta} G_l \left(\frac{k_\perp^2 V_\perp^2}{2\Omega^2} \right) \left[1 + \frac{\omega}{k_\| V_\|} Z(\zeta_l) \right] \right.$$
$$\left. - \frac{\beta}{1-\beta} G_l \left(\frac{\beta k_\perp^2 V_\perp^2}{2\Omega^2} \right) \left[1 + \frac{\omega - l\Omega \left(1 - \frac{1}{\beta}\right)}{k_\| V_\|} Z(\zeta_l) \right] \right\} \tag{7.41}$$

When growth or decay rates are small we can use the approximate form of the Z-function (B.31). The only complex quantity in the expression for χ arises from Z. The real part of Z can be found from tables or computed using either the series expansion or asymptotic expansion, depending on the value of ζ[1]. We have already found an expression for the real part of σ and hence for the imaginary part of χ. The growth rate is given by (6.34). It is

$$\omega_i = \frac{\chi_{hi}}{/} \frac{\partial \chi_c}{\partial \omega_r} + \frac{\partial \chi_{hr}}{\partial \omega_r} \tag{7.42}$$

The dispersion relation can be differentiated with respect to ω by using (B.26).

In Figure 7.7 we show some examples of results extracted from the calculations presented by Ashour-Abdalla and Kennel [7] who have performed a very extensive survey of parameter space in such a model. The model is that of a hot electron distribution with a gentle loss-cone ($\beta =. 0.9$) which is empty at $v_\perp = 0$ ($\delta = 1$), and with $T_\perp = T_\| = T$, $V^2 = 2KT/m$. The hot and cold electron densities can be varied and are specified by the plasma frequencies ω_{pc} and ω_{ph}. The angle of the wave normal is defined by the ratio of the parallel and perpendicular components of the wave vector, $k_\|$ and $k_\perp$, normalized in terms of the r. m. s. gyroradius $\rho = V/\Omega$. It is assumed that θ, the angle between wave normal and magnetic field is near 90^0 so that $k_\perp \gg k_\|$ and $k^2 \simeq k_\perp^2$. In any set of computations $k_\| V/\Omega$ is held fixed, the real part of ω (subscript r) as a function of $k_\perp V/\Omega$, and hence of the wave normal angle, is computed from the real part of the dispersion relation. The imaginary part of ω (subscript i) is then found from these.

The left hand panel of Figure 7.7 shows plots of $k_\perp V/\Omega$ and ω_i/Ω against ω_r/Ω. The full line represents that portion of the curve $k_\perp V/\Omega$ for which the growth rate is negative. The dashed line shows that portion for which growth is positive and the dotted line shows the growth rate when growth is possible. It can be seen that, for a given frequency there are in general two possible directions of the wave normal. One of these directions makes a comparatively small angle with the magnetic

[1]Satisfactory accuracy can be obtained on the real axis by using the series expansion for $|\zeta| \leq 4$ and using the asymptotic expansion for $\zeta > 4$.

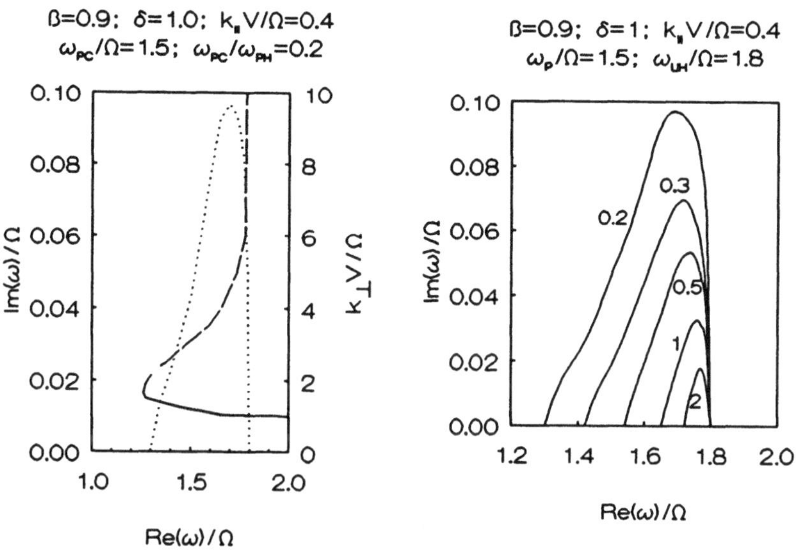

Figure 7.7: Resonant frequencies and growth rates. (a) The curve plotted against the left hand axis shows the growth rate as a function of frequency for fixed $k_{||}$. The curve plotted against the right hand axis shows the corresponding value of $k_\perp$. (b) Growth rate as a function of frequency for differing proportions of hot and cold plasma.

field ($k_\perp$ relatively small) and is always stable with a negative growth rate. The other direction is more nearly at right angles to the field ($\tan\theta \equiv k_\perp/k_{||} > 4$) and has a band of frequencies which is unstable. The upper cut-off frequency of the unstable band is the upper hybrid frequency and maximum growth occurs for a frequency slightly below this. Similar behavior is shown for a wide range of conditions.

The actual growth rate depends on the density of the hot electrons and also on the relative proportions of hot and cold electrons. The cold electrons are necessary, defining the hybrid oscillation which can be driven by the unstable hot electron distribution. The right hand panel of Figure 7.7 shows the growth rate for varying proportions of cold and hot plasma with the cold plasma density fixed. As is to be expected the growth rate falls and the width of the unstable band decreases as the hot electron density decreases.

The number of possible situations is enormous and a systematic description is not possible here. One factor which has not been taken into account and which may be important is allowing for a finite temperature of the cold plasma. This leads to the possibility of a non-convective instability with a large spatial growth rate at points where the dispersion relation gives zero group velocity [7].

7.4.4 Whistler Growth

Finally we consider a type of instability which is fundamental to the understanding of VLF wave generation. It is a resonant particle instability in which the particles resonate with a transverse, circularly polarized wave at the Doppler-shifted cyclotron resonance frequency. We consider a uniform plasma in which there is both a cold and a hot electron component. The hot electron distribution function is represented by a subtracted Maxwellian. For frequencies well above the lower hybrid frequency the ions do not contribute to the dispersion relation. We assume that the density of cold electrons is much larger than that of hot electrons so that the real part of the dispersion relation is dominated by the cold electrons. The hot electron distribution determines the imaginary part of the dispersion relation.

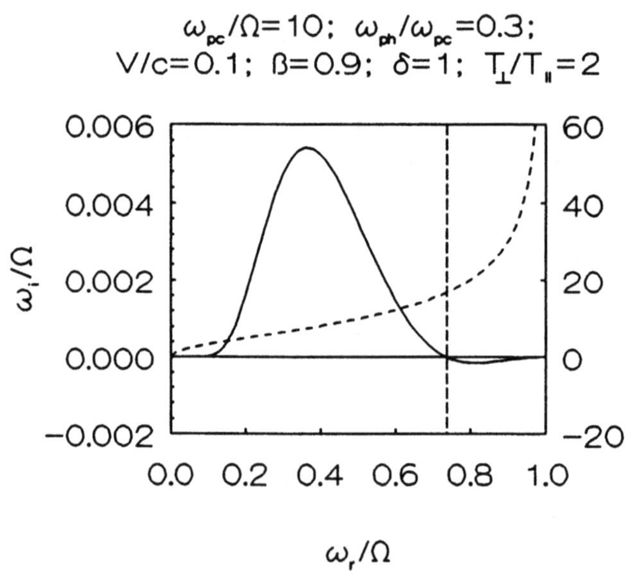

Figure 7.8: Whistler dispersion and growth rate. The dashed curve (right axis) is the whistler dispersion relation. The solid curve is the growth rate. The parameters are chosen to be typical of the region of the magnetosphere in which whistler propagation is most common.

We confine ourselves to waves propagated parallel to the magnetic field with $k_{\|} = k$, $k_{\perp} = 0$. The dispersion relation for the right hand polarized wave can then be written

$$\frac{c^2k^2}{\omega^2} = 1 - \frac{\omega_{pc}^2}{\omega(\omega - \Omega)} + \frac{i}{\epsilon_0 \omega}\Re(\sigma_h) \tag{7.43}$$

where the contribution to the conductivity σ of the hot electrons is found from the expression relating J_R and $\tilde{E}_R$ in (6.25) by using a Maxwellian distribution for $f_{\|}$ and a subtracted Maxwellian for $f_{\perp}$. If this is done the integration over $v_{\perp}$ is easy. The integral over $v_{\perp}$ can then be expressed in terms of the plasma dispersion function Z. The result is

$$\begin{aligned} \frac{i}{\epsilon_0 \omega}\Re(\sigma_h) &= \frac{\omega_{ph}^2}{\omega^2}\left\{\frac{\omega}{kV_{\|}} Z\left(\frac{\omega - \Omega}{kV_{\|}}\right)\right. \\ &- \left.\left[1 - \frac{T_{\perp}}{T_{\|}}(1 + \beta\delta)\right]\left[1 + \frac{\omega - \Omega}{kV_{\|}} Z\left(\frac{\omega - \Omega}{kV_{\|}}\right)\right]\right\} \end{aligned} \tag{7.44}$$

For small growth this quantity is small and ω may be replaced by its real part ω_r. Only the imaginary part of the expression (the real part of σ) contributes to the dispersion relation. All the quantities are real except Z. If the imaginary part of Z is found from (B.31) we can write for the imaginary part of the dispersion relation

$$D_i = \sqrt{\pi}\frac{\omega_{ph}^2}{\omega^2}\left\{\frac{\omega}{kV_{\|}} - \left[1 - \frac{T_{\perp}}{T_{\|}}(1 + \beta\delta)\right]\left[\frac{\omega - \Omega}{kV_{\|}}\right]\right\} e^{-(\omega - \Omega)^2/k^2V_{\|}^2} \tag{7.45}$$

From (6.34) the growth rate, ω_i is

$$-\frac{D_i}{\partial D_r / \partial \omega_r}$$

Since

$$D_r \equiv 1 - \frac{c^2k^2}{\omega^2} - \frac{\omega_{pc}^2}{\omega(\omega - \Omega)}$$

we obtain

$$\frac{\partial D_r}{\partial \omega_r} = \frac{2}{\omega}\left[1 + \frac{2\omega_{ph}^2}{(\omega - \Omega)^2}\right]$$

The result of some computations for parameters typical of whistler propagation in the magnetosphere is shown in Figure 7.8. It can be seen that maximum growth occurs for rather low frequencies, well below the ion gyrofrequency. In the region immediately below the gyrofrequency the wave is damped. Examination of the formula for the growth rate shows that the boundary of the region where growth takes place is determined by the point at which the coefficient of the exponential changes sign. There is thus growth for frequencies for which

$$\omega_r \leq \frac{1 + \beta\delta - T_{\|}/T_{\perp}}{1 + \beta\delta} \tag{7.46}$$

The form of the dependence of the magnitude of the growth on frequency is strongly influenced by the exponential factor. This is of Gaussian form, centered on the gyrofrequency, of width $kV_{\|}$, which is determined by the energy of the hot particles. The energy must be high enough so that there can be significant growth at frequencies well below the gyrofrequency as shown. The magnitude of the growth rate is directly proportional to the ratio of the hot to cold electron densities $\omega_{ph}^2/\omega_{pc}^2 \equiv N_h/N_c$. It is important to note that, in these circumstances, a wide band of frequencies can be unstable.

7.5 Summary

- For plasmas in thermal equilibrium kinetic effects are of two kinds. (i) For wave normals making large angles with the magnetic field the gyroradius may be comparable with the wavelength. This introduces a whole menagerie of waves which are not predicted by cold or warm plasma theory. They are typified by the Bernstein modes. (ii) Resonance between particles and waves can lead to damping. We consider two cases, Landau damping of longitudinal electrostatic waves and cyclotron damping of whistlers.

- If the plasma is not in thermal equilibrium many instabilities leading to wave growth become possible. The following are the cases which are presented as examples.

- Longitudinal electrostatic waves can be excited by a plasma in which an energetic beam of particles exists.

- Electrostatic waves at arbitrary angles to the field can be excited by anisotropic distributions. Cases of importance are temperature anisotropy and loss cone distributions. These can lead to bands of noise which are organized in terms of the gyro-harmonics as well as near the upper hybrid frequency.

- Loss cone distributions are of great importance for whistler growth.

Chapter 8

The Effect of Wave Fields on Energetic Particles

8.1 Introduction

In Chapter 6 we have seen how wave growth and decay is dominated by the behavior of the particle distribution function near the velocity, $v_{\parallel} = (\omega - l\Omega)/k_{\parallel}$. Integrals over velocity space occurring in the expressions for the conductivity or the dispersion relation have poles at this velocity; the residue at this pole determines the imaginary part of the frequency or growth rate. The physics of this situation is that the particles near this velocity remain approximately in phase with the wave as it is propagated. They can thus exchange energy efficiently with the wave. This can lead to growth or decay depending on the relative proportions of the particles which transfer energy to the wave and which gain energy from the wave.

In this chapter we study the effect of this resonant energy exchange on the particle orbits in velocity space. Resonant particle effects are an efficient means of scattering particles in the velocity space ($v_{\perp}$–$v_{\parallel}$). As a result quasi-equilibrium distributions may be changed. Wave particle interactions are important mechanisms by which such plasmas ultimately relax towards equilibrium. They are also a key means by which particles which would otherwise be stably trapped in the magnetosphere are lost to the atmosphere.

In Chapters 6 and 7 we considered the linear theory of growth and decay of waves. In this theory the background distribution function f_0 was regarded as unchanging and the wave was a small perturbation on it. This is a valid approach for a small decaying wave. In the case of wave growth it is clear that it can only describe the initial stages of an instability. As the wave grows the perturbation in the distribution function becomes larger and ultimately the disturbance can no longer be regarded as a small perturbation. For a rapidly growing wave there is no escape from a fully non-linear treatment of the problem. Fortunately, in magnetospheric plasma physics, this is usually unnecessary. Magnetospheric plasmas are often not far from a quasi-equilibrium state for which the growth rate is small. If this is so a quasi-linear approach may be appropriate. For example the plasma could consist of a cold background with a low density anisotropic hot plasma providing the source of energy for an instability, or a Maxwellian hot distribution might have a weak beam superimposed as the energy source.

The basis of a complete quasi-linear theory is that, as a result of the action of the wave, the zero order distribution function changes with time. The time scale on which this happens is long compared with the period of the wave. The wave field and the perturbation of the distribution function are computed linearly on the assumption that the change in the zero order distribution function does not change significantly over a wave period. These perturbations are then inserted into the non-linear terms of the equation for the time evolution of the distribution function. The resulting equation has the form of a diffusion equation in velocity space.

The approach we adopt is first to consider the effect of a pre-existing wave field on a test particle with a typical orbit in configuration space [16, 35]. Under the influence of the wave the velocity of a resonating particle is changed. The path in velocity space along which it moves is called the single wave characteristic. We deduce its shape. A qualitative picture of diffusion along this path is easy to obtain. A more quantitative picture is difficult and we present a limited treatment.

8.2 Particle Resonance

8.2.1 Reference Frames

We usually consider waves in the *plasma frame* , a reference frame which is at rest with respect to the background plasma. It is often useful to use other frames of reference. One of these is the *guiding center frame* in which the guiding center of a test particle is at rest. Another is the *wave frame* in which the wave is at rest. In this section we shall find it illuminating to change from frame to frame in order to consider different situations.

8.2.2 Resonance Seen from the Guiding Center Frame

Consider a plane wave and choose axes so that the wave vector lies in the x–z plane of the plasma frame. Field components vary in space and time as $\exp[i(k_\perp x + k_\parallel z - \omega t)]$. If we observe this wave from the guiding-center frame of a test particle with parallel velocity component $v_\parallel$ its frequency will be Doppler shifted to a frequency $\omega - k_\parallel v_\parallel$. The test particle has zero $v_\parallel$ in the guiding center frame. It will gyrate about the magnetic field with the gyrofrequency Ω. In general the effect of the wave on the test particle in the plasma will be different in each successive cycle and will average to zero over many gyroperiods. If, however, the frequency of rotation of the particle or one of its harmonics matches the Doppler shifted frequency of the wave then the particle may see a field with constant phase and we can expect the same amount of energy to be transferred between wave and particle on each cycle. This results in strong resonance effects. Such resonance occurs for a test particle with parallel velocity obeying the resonance condition:

$$\omega - k_{\parallel} v_{\parallel} - l\Omega = 0 \tag{8.1}$$

where l is an integer, $-\infty < l < \infty$.

It is sometimes important to include relativistic effects, even for relatively low particle energies. In such cases the resonance condition is modified by replacing the particle rest mass m by its relativistic mass

$$\frac{m}{\sqrt{1 - v^2/c^2}}$$

where v is the particle speed. The cyclotron frequency is then replaced by $\Omega(1 - v^2/c^2)^{1/2}$ and the resonance condition becomes

$$\omega - k_{\parallel} v_{\parallel} - l\Omega(1 - v^2/c^2)^{1/2} = 0 \tag{8.2}$$

8.2.3 Resonance seen from the Rest Frame of the Wave

It is always possible to transform to the guiding center frame of an energetic particle. There are, however, advantages to working in the *wave frame* in which the wave is at rest and the wave fields will be independent of time. This is, of course, only possible if the phase speed of the wave is less than the speed of light as is often the case with plasma waves. A little thought shows that there is no unambiguous wave rest frame. In an isotropic medium an obvious frame to use is one which moves relative to the plasma frame with the phase velocity. However any other frame moving with a velocity relative to this frame which is parallel to the wave fronts is also a rest frame. In an anisotropic medium, because of the background magnetic field, a transformation into a frame in which the plasma is moving with velocity $\mathbf{V}$ will introduce an electrostatic field $\mathbf{V} \times \mathbf{B_0}$. Only if $\mathbf{V}$ is parallel to $\mathbf{B_0}$ is this field zero. For this reason we restrict our definition of the wave rest frame to a frame moving parallel to the magnetic field with phase speed $\omega/k_{\parallel}$ which is assumed to be much less than the speed of light in this treatment.

In this frame the component of the particle velocity parallel to the magnetic field is $v'_{\parallel} = v_{\parallel} - \omega/k_{\parallel}$. In the relativistic case it is $v'_{\parallel} = (v_{\parallel} - \omega/k_{\parallel})/(1 - \omega v_{\parallel}/k_{\parallel}c^2)$. The distance it moves during one gyroperiod is $2\pi v'_{\parallel}/\Omega$. If this distance is an integral number of wavelengths in either the positive or negative direction then resonance occurs. The condition for this is the same as before: $\omega - k_{\parallel} v_{\parallel} - l\Omega = 0$ where l is an integer, $-\infty < l < \infty$ or, in the relativistic case, $\omega - k_{\parallel} v_{\parallel} - l\Omega(1 - v^2/c^2)^{1/2} = 0$.

8.2.4 Location of Resonant Particles in Velocity Space

Consider the location of resonant particles in the velocity space ($v_\perp$–$v_\parallel$) as seen in the plasma frame. For a slow wave with frequency ω and with the component of the wave vector in the magnetic field direction equal to $k_\parallel$ the condition (8.1) defines a set of straight lines parallel to the $v_\perp$ axis. In each case the resonance is independent of $v_\perp$. This is because, for a non-relativistic particle, the gyrofrequency is independent of the gyroradius. The $l = 0$ resonance is located at $v_\parallel = \omega/k_\parallel = V_P$ and is called the *Čerenkov resonance* . There are gyroresonances at $v_\parallel = (\omega \pm \Omega)/k_\parallel$ and resonances at harmonics of the gyrofrequency, $v_\parallel = (\omega \pm l\Omega)/k_\parallel$, $l = 2, 3, \ldots$. For these resonances it is possible for the particle to be traveling along the magnetic field in either the same direction or in the opposite direction to the phase velocity of the wave. Gyroresonance is also possible if the particle is traveling faster than the wave.

In the relativistic case matters are more complicated. Equation 8.2 can be rearranged in the form

$$A(u_\parallel - B)^2 + l^2Y^2u_\perp^2 = C \tag{8.3}$$

where

$$A = n_\parallel^2 + l^2Y^2; \quad B = n_\parallel/A: \quad C = AB^2 + l^2Y^2 - 1 \tag{8.4}$$

In these expressions $u_{\parallel,\perp} = v_{\parallel,\perp}/c$, are the velocity components normalized to the speed of light, $n_\parallel = ck_\parallel/\omega$ is the parallel component of the refractive index vector and $Y = \Omega/\omega$ is the usual Appleton parameter.

The relativistic resonance condition thus describes an ellipse in velocity space with its center located at $u_\parallel = B$; $u_\perp = 0$, and with semi-axes of length $\sqrt{C}/lY$ in the $u_\perp$ direction and $\sqrt{C/A}$ in the $u_\parallel$ direction. We see that particles resonant with a particular wave no longer all have the same value of $v_\parallel$.

Figure 8.1 shows a number of examples of resonance ellipses. There are various ways of parameterizing such diagrams. We assume that, in each diagram, Y is fixed. This implies that for a given magnetic field strength the frequency is fixed. Each contour is then chosen to correspond to a particular value of $n_\parallel$, that is to a particular phase velocity in the magnetic field direction. In a uniform medium the refractive index is a function of θ, the angle between wave vector and magnetic field. Fixing $n_\parallel$ does not ensure that a propagated wave exists. These contours therefore simply show where there is the possibility of resonance with a wave of given frequency and wave normal direction. In any actual problem it is necessary to know the real part of the dispersion relation of plasma to see if waves with the assumed $k_\parallel$ exist.

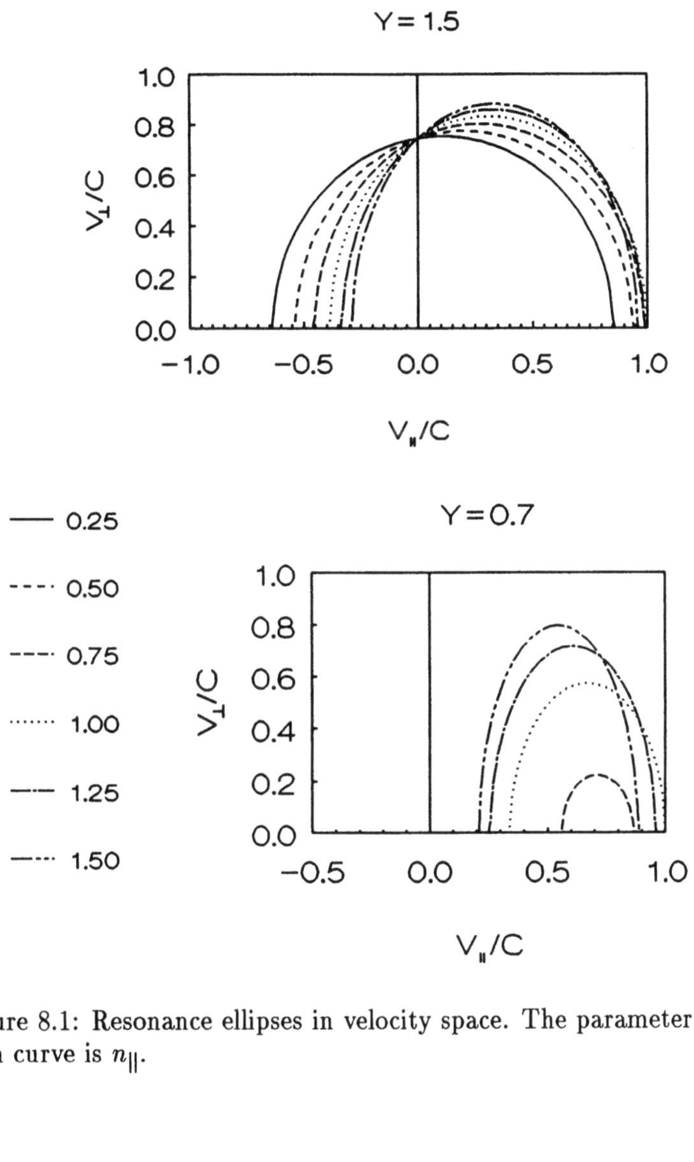

Figure 8.1: Resonance ellipses in velocity space. The parameter defining each curve is $n_{\|}$.

8.3 Trajectories of Resonant Particles in Velocity Space

In a uniform plasma with a uniform magnetic field a charged particle travels in a spiral path with constant $v_\perp$ and $v_\parallel$. It therefore remains at the same point in velocity space. In general, the effect of any wave on the particle averages to zero over one cycle of the wave. When, however, the particle is in resonance with the wave it remains in phase with the wave field and both $v_\parallel$ and $v_\perp$ may be changed. In this case the particle moves along a path in velocity space which is called the *single wave characteristic*. In this section we find the characteristics in various situations.

For waves with phase speeds less than the speed of light the single wave characteristic is most easily found by considering the situation in the rest frame of the wave and then transforming to the plasma frame. We do this both for non-relativistic and relativistic particles.

If the phase speed of the wave is greater than the speed of light this approach is no longer possible. In this case a quantum mechanical argument, first introduced by Kennel and Engelmann [68], for a non-relativistic case can be used. This approach has the advantage of both brevity and elegance but, in what is essentially a classical process, is not particularly illuminating. For this reason we use a longer approach which applies generally to all cases and which gives insight into the physical process of resonance.

Because of the gyration of the particle about the background magnetic field the average momentum of the particle is in the direction of the magnetic field. If the wave changes the average momentum and energy of a resonant particle the average momentum change is also in the magnetic field direction. Before and after the wave-particle interaction the particle is gyrating about the magnetic field so the perpendicular momentum component always averages to zero over one gyroperiod. The change in average energy due to the wave field can be written

$$\left\langle \frac{dW}{dt} \right\rangle = q\langle \mathbf{v}.\mathbf{E} \rangle$$

and the change in average parallel component of momentum is

$$\left\langle \frac{dp_\parallel}{dt} \right\rangle = \hat{\mathbf{e}}.q\langle [\mathbf{E} + \mathbf{v} \times \mathbf{B}] \rangle$$

where $\hat{\mathbf{e}}$ is a unit vector in the direction of the background magnetic field, the averages are taken over many gyroperiods and the differentiation with respect to time is taken along the particle path. This immediately allows the determination of $\langle dW/dp_\parallel \rangle$ and since W and $p_\parallel$ are functions of $v_\perp$

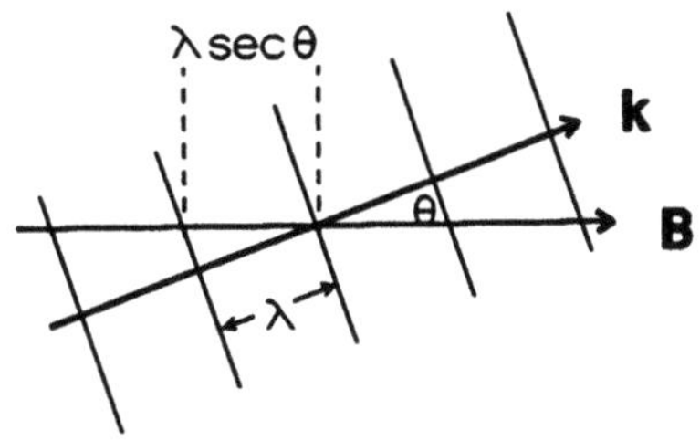

Figure 8.2: Orientation of wave.

and $v_{\|}$ the single wave characteristic can be found by integrating this equation.

8.3.1 Single Wave Characteristics for Waves with $\omega/k_{\|} \ll c$ – Non-Relativistic Particles

Consider a slow harmonic wave in the plasma frame as shown in Figure 8.2. By "slow" we mean that its phase speed is much less than the speed of light. Examples of such waves are whistlers with $\omega_p^2 \gg \omega\Omega$ or electron plasma waves in the electrostatic approximation. Transform to the wave frame. The electric field in the wave frame is electrostatic and can be expressed in terms of a potential $\mathbf{E} = -\nabla\phi$. In the special case of a transverse wave such as a whistler or ion cyclotron wave the electric field in the wave frame is zero.

Now consider a resonant particle moving in this field. Its unperturbed path is a helix along which the particle remains in phase with the wave fields. After each complete cycle of the helix it returns to a point with the same potential as at the beginning of the cycle. In this frame of reference there can be no net transfer of energy to the particle. Thus in the rest frame, if its velocity components are $v'_\perp$ and $v'_{\|}$, conservation of kinetic energy requires

$$v'^2_{\|} + v'^2_{\perp} = V_0^2$$

where V_0 is a constant. The effect of the wave on the particle is to change the relative sizes of $v'_\perp$ and $v'_{\|}$, and thus the pitch angle, without changing the kinetic energy in this frame.

In the plasma frame this condition becomes

$$\left(v_{\parallel} - \frac{\omega}{k_{\parallel}}\right)^2 + v_{\perp}^2 = V_0^2 \tag{8.5}$$

which is the equation for the single wave characteristic. For the case of Čerenkov resonance where $v_{\parallel} = \omega/k_{\parallel}$ the single wave characteristics are straight lines parallel to the $v_{\parallel}$-axis. For resonance at the gyrofrequency or any of its harmonics they are circles centered on $v_{\parallel} = \omega/k_{\parallel}$.

Single wave characteristics for both types of resonance are shown in Figure 8.3 It can be seen that in the plasma frame a resonant particle moving along a single wave characteristic gains or loses energy depending on the direction of motion. This direction in turn depends on the phase of the particle rotation relative to the wave field.

8.3.2 Single Wave Characteristics for Waves with $\omega/k_{\parallel} < c$ – Relativistic Particles

If a wave has a phase speed close to, but not greater than the speed of light it is still possible to consider resonance in the wave rest frame. In such circumstances we can carry out a relativistic transformation. It is not necessary to describe the fields in detail. In the wave frame the field is still electrostatic and thus there can be no energy transfer to the wave. We must allow for relativistic energy effects. The relativistic energy mc^2, where m is the relativistic mass, must be constant. This still requires

$$v'^2_{\perp} + v'^2_{\parallel} = V_0^2 \tag{8.6}$$

where V_0 is a constant.

For frames moving with a relative velocity V_P the relativistic transformation of the particle velocity is

$$\begin{aligned} v'_{\parallel} &= \frac{v_{\parallel} - V_P}{1 - v_{\parallel}V_P/c^2} \\ v'_{\perp} &= \frac{\sqrt{1-(v_{\parallel}^2 + v_{\perp}^2)/c^2}}{1 - v_{\parallel}V_P/c^2} v_{\perp} \end{aligned} \tag{8.7}$$

When these expressions are substituted in (8.6) we obtain an expression for the single wave characteristic along which relativistic particles are constrained to move. After some manipulation we get the result

$$v_{\parallel}^2\left(1 - \frac{V_0^2V_P^2}{c^4}\right) - 2V_{\parallel}V_P\left(1 - \frac{V_0^2}{c^2}\right) + v_{\perp}^2\left(1 - \frac{V_P^2}{c^2}\right) = V_0^2 - V_P^2 \tag{8.8}$$

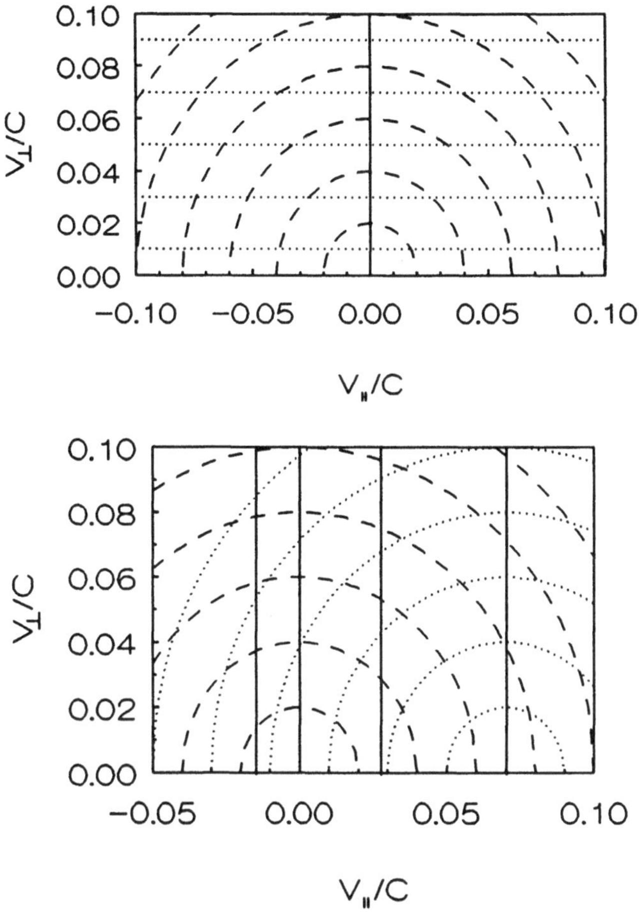

Figure 8.3: Single wave characteristics in velocity space for resonant particles. The circles centered on the origin are surfaces of constant energy. Upper panel: Čerenkov resonance. Lower panel: Resonance at the gyrofrequency or its harmonics.

Now write $V_P = c/n_{\|}$, $u_0 = V_0/c$, $u_{\perp,\|} = v_{\perp,\|}/c$. This relation can then be written in the form

$$F(u_{\|} - G)^2 + (n_{\|}^2 - 1)u_{\perp}^2 = H \tag{8.9}$$

where $F = n_{\|}^2 - u_0^2$, $G = n_{\|}(1 - u_0^2)/F$ and $H = u_0^2(n_{\|}^2 - 1)^2/F$. The single wave characteristic of a relativistic particle resonating with the wave is therefore, in general, an ellipse rather than a circle in velocity space. The special case of Čerenkov resonance has the particle moving with the phase speed so that $v_{\|}' = 0$. In this case the single wave characteristics are straight lines $v_{\perp} = V_0$ parallel to the $v_{\|}$-axis as in the non-relativistic case.

An example of gyroresonance of a relativistic particle with a wave having $\omega/k_{\|} < c$ is shown in the upper panel of Figure 8.4.

8.3.3 General Picture of Resonant Wave-Particle Interaction

In the most general case, for which the particles may be relativistic and the phase change perpendicular to the field may be significant over distances of the order of a gyroradius, the picture of the previous subsection is still useful. We can still resolve the wave field into longitudinal and two oppositely rotating circularly polarized components. The particle still resonates with one of the three components and the effect of the others averages to zero. In this case we must, however, allow for the fact that the particle may see a phase variation because of its motion perpendicular to the field as well as due to its motion along the field. We study the wave-particle interaction in the plasma frame.

Components of the Wave Fields

Consider the nature of the polarization of a wave which is propagated at some angle to the field. There is no loss of generality in choosing coordinates with magnetic field in the z-direction and wave normal in the x-z plane. We use the real wave components rather than exponential notation. The most general polarization can be represented as the sum of a right-handed circularly polarized wave with field amplitudes denoted by subscript R, a left-handed circularly polarized wave with field amplitudes denoted by subscript L, and a wave linearly polarized parallel to the magnetic field with field components denoted by subscript $\|$. In general each of these has a different amplitude and phase. The situation can be somewhat simplified if we assume that, to zero order, there is no dissipation. The symmetry of the situation then demands that the projection of the fields on the x-y plane is an ellipse with its axes aligned with the x- and

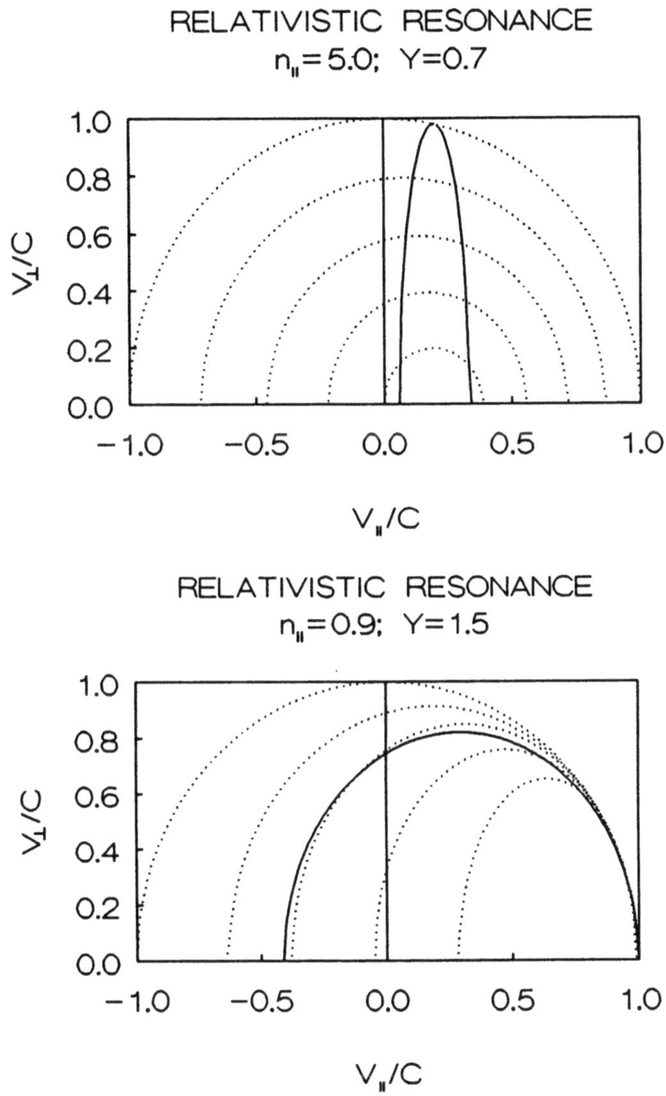

Figure 8.4: Single wave characteristics in velocity space for relativistic resonant particles. The dotted ellipses are the single wave characteristics. The solid ellipse is the locus of resonant particles for the parameters shown. Upper panel: $\omega/k_{\|} < c$. Lower panel: $\omega/k_{\|} > c$.

y-axes. Thus the x- and y-components of the wave differ in phase by $\pi/2$. The condition $\nabla.\mathbf{B} \equiv k_\perp B_x + k_\| B_z = 0$ shows that the x- and z-components of $\mathbf{B}$ are in phase or in antiphase. The combination of these requirements with Faraday's law requires that x- and z-components of $\mathbf{E}$ are also in phase or in antiphase. The x- and y-components may then be written

$$\begin{aligned}
B_x &= (B_R + B_L)\sin(\omega t - k_\| z - k_\perp x) \\
B_y &= (B_R - B_L)\cos(\omega t - k_\| z - k_\perp x) \\
B_z &= -(k_\perp / k_\|) B_x \\
E_y &= -\frac{\omega}{k_\|}(B_R + B_L)\sin(\omega t - k_\| z - k_\perp x) \\
E_x &= \left[\frac{\omega}{k_\|}(B_R - B_L) + \frac{k_\perp}{k_\|} E_\|\right]\cos(\omega t - k_\| z - k_\perp x) \\
E_z &= E_\| \cos(\omega t - k_\| z - k_\perp x)
\end{aligned} \tag{8.10}$$

These expressions are consistent with the symmetry and with Faraday's law. We choose B_R positive. The other amplitudes may be positive or negative.

Fields Experienced by a Test Particle Moving along its Unperturbed Path

Consider a positive test particle with gyroradius $\rho_\perp = v_\perp/\Omega$. At time $t = 0$ its guiding center is at $z = z_0$ and its gyroradius makes an angle ψ, measured in the positive sense, with the x-axis. It rotates in the left-handed sense about the z-axis. The equations for its path and velocity are therefore

$$\begin{aligned}
x &= \frac{v_\perp}{\Omega}\sin(\Omega t + \psi) \\
y &= \frac{v_\perp}{\Omega}\cos(\Omega t + \psi) \\
z &= v_\| t + z_0 \\
v_x &= v_\perp \cos(\Omega t + \psi) \\
v_y &= -v_\perp \sin(\Omega t + \psi) \\
v_z &= v_\|
\end{aligned} \tag{8.11}$$

It is easy to see that, in the following, we can deal with a negative test particle simply by changing the sign of Ω wherever it appears.

The rate at which the wave does work on the particle is

$$\frac{dW}{dt} = q\mathbf{v}.\mathbf{E}$$

and the rate at which the parallel momentum changes is

$$\frac{dp_{\|}}{dt} = q(v_x B_y - v_y B_x + E_z)$$

Substitute the expressions for the fields and the position and velocity of the particle in these and group the terms to get

$$\begin{aligned}
\frac{1}{q}\frac{dW}{dt} = & \\
& \frac{v_\perp \omega}{k_\|} B_R \cos\left\{(\omega - k_\| v_\| + \Omega)t - \frac{k_\perp v_\perp}{\Omega}\sin(\Omega t + \psi) + \psi - k_\| z_0\right\} \\
- & \frac{v_\perp \omega}{k_\|} B_L \cos\left\{(\omega - k_\| v_\| - \Omega)t - \frac{k_\perp v_\perp}{\Omega}\sin(\Omega t + \psi) - \psi - k_\| z_0\right\} \\
+ & E_\| \left\{\frac{k_\perp}{k_\|} v_\perp \cos(\Omega t + \psi) + v_\|\right\} \times \\
& \cos\left\{(\omega - k_\| v_\|)t - \frac{k_\perp v_\perp}{\Omega}\sin(\Omega t + \psi) - k_\| z_0\right\}
\end{aligned} \tag{8.12}$$

and

$$\begin{aligned}
\frac{1}{q}\frac{dp_\|}{dt} = & \\
& v_\perp B_R \cos\left\{(\omega - k_\| v_\| + \Omega)t - \frac{k_\perp v_\perp}{\Omega}\sin(\Omega t + \psi) + \psi - k_\| z_0\right\} \\
- & v_\perp B_L \cos\left\{(\omega - k_\| v_\| - \Omega)t - \frac{k_\perp v_\perp}{\Omega}\sin(\Omega t + \psi) - \psi - k_\| z_0\right\} \\
+ & E_\| \cos\left\{(\omega - k_\| v_\|)t - \frac{k_\perp v_\perp}{\Omega}\sin(\Omega t + \psi) - k_\| z_0\right\}
\end{aligned} \tag{8.13}$$

Each of these expressions has three major terms, corresponding to interaction between the particle and the left-handed circularly polarized component, right-handed circularly polarized component and the parallel component.

Time of Interaction of a Wave with a Test Particle

The wave-particle interaction does not last for ever. We have assumed a harmonic wave interacting with a particle moving on its unperturbed path in a uniform medium. The time τ is a measure of the period for which this assumption holds. After this time the resonance condition will fail. In general τ is long compared with a gyroperiod and our assumption is that the growth rate is small so that τ is short compared with the time scale on which growth or decay of the wave or modification of the distribution function occurs. The factors determining τ depend upon the nature of the problem. Possible cases are:

- A wave packet which originates in a remote part of the plasma passes through the region where wave-particle interaction takes place. Let the component of its group velocity along the magnetic field be V_G and its length along the magnetic field be $N\lambda_{\|}$. Then

$$\tau = \frac{N\lambda_{\|}}{|V_G - v_{\|}|} \tag{8.14}$$

- A noise-like wave form has a statistical character. In this case τ is a suitably defined coherence time, related to the autocorrelation width of the spectrum. For a gyroresonance a similar formula to (8.14) can be found by the following argument. Suppose the wave spectrum is centered on $k_{\|}$ and the width of its spectrum is $\Delta k_{\|}$. This will be related to the the autocorrelation width of a noise-like spectrum but the same argument applies to the previous case. Consider a resonant particle with $v_{\|} = (\omega - \Omega)/k_{\|}$. A wave with wave number $k \pm \frac{1}{2}\delta k_{\|}$ has frequency $\omega \pm \frac{1}{2}\Delta k_{\|}\, \partial\omega/\partial k_{\|} = \omega \pm \frac{1}{2}\Delta k_{\|}\, V_G$. In the particle rest frame the particle gyrates with angular frequency Ω and it sees the frequency of the wave Doppler shifted by an amount $-(k_{\|} \pm \frac{1}{2}\Delta k_{\|})v_{\|} = -\omega + \Omega \mp \Delta k_{\|} v_{\|}$. After time t the phase difference between wave and particle has changed by $[\omega \pm \frac{1}{2}\Delta k_{\|}\, V_G - (k_{\|} \pm \frac{1}{2}\Delta k_{\|})v_{\|} - \Omega]t = \mp\frac{1}{2}\Delta k_{\|}(V_G - v_{\|})$. After a time τ this phase difference will have changed by ± 1 radian and the wave-particle coherence will have been lost. This occurs for

$$\tau = \frac{2}{\Delta k_{\|}|V_G - v_{\|}|} \tag{8.15}$$

  This is exactly the same as (8.14) with $N\lambda_{\|} = 2/\Delta k_{\|}$.

- If the magnetic field is not uniform, phase coherence between wave and particle will eventually be lost because of the change in the gyrofrequency of the particle and because of the associated change in the wave phase speed. These changes give a value for τ of the order

$$\tau \simeq \left(\frac{\pi}{v_{\|}[(d\Omega/dz) + \omega(dn_{\|}/dz)]} \right)^{\frac{1}{2}}$$

Average Effect of the Wave Fields on Resonant Particles

We wish to average the wave field effects (8.12) and (8.13) over the time τ. Let

$$\begin{aligned}
(k_\perp v_\perp/\Omega) &= \zeta \\
\Omega t + \psi &= \theta \\
(\omega - k_\| v_\| + \eta_{R,L,\|}\Omega)t &= \nu_{R,L,\|}\Omega t \\
&= \nu_{R,L,\|}(\theta - \psi) \\
\chi_{R,L,\|} &= (\nu_{R,L,\|} - \eta_{R,L,\|})\psi + k_\| z_0
\end{aligned}$$

where $\eta_R = +1$, $\eta_L = -1$ and $\eta_\| = 0$. Suppose that, over the wave-particle interaction time τ the particle completes N gyrations. The average values of dW/dt and $dp_\|/dt$ over the wave-particle interaction time are then $1/2\pi N$ times the integral with respect to θ between the limits 0 and 2π. Thus

$$\begin{aligned}
\frac{1}{q}\left\langle \frac{dW}{dt} \right\rangle &= \frac{1}{2\pi N}\int_0^{2\pi N} \left\{ \frac{v_\perp \omega}{k_\|} B_R \cos[\nu_R\theta - \zeta\sin\theta - \chi_R] \right. \\
&- \frac{v_\perp \omega}{k_\|} B_L \cos[\nu_L\theta - \zeta\sin\theta - \chi_L] \\
&+ \left. E_\| \left[\frac{k_\perp}{k_\|} v_\perp \cos\theta + v_\| \right] \cos[\nu_\|\theta - \zeta\sin\theta - \chi_\|] \right\} d\theta
\end{aligned} \tag{8.16}$$

and

$$\begin{aligned}
\frac{1}{q}\left\langle \frac{dp_\|}{dt} \right\rangle &= \frac{1}{2\pi N}\int_0^{2\pi N} \{ v_\perp B_R \cos[\nu_R\theta - \zeta\sin\theta - \chi_R] \\
&- v_\perp B_L \cos[\nu_L\theta - \zeta\sin\theta - \chi_L] + E_\| \} \, d\theta
\end{aligned} \tag{8.17}$$

There are just two forms of integral to evaluate. They are

$$\begin{aligned}
I_1 &= \frac{1}{2\pi N}\int_0^{2\pi N} \cos[\nu\theta - \zeta\sin\theta - \chi]\, d\theta \\
I_2 &= \frac{1}{2\pi N}\int_0^{2\pi N} \cos\theta \cos[\nu\theta - \zeta\sin\theta - \chi]\, d\theta
\end{aligned}$$

They can be transformed into standard integrals which can be evaluated in terms of Bessel functions. We may write

$$I_1 = \frac{1}{2\pi N}\sum_{p=0}^{N-1}\int_0^{2\pi} \cos[\nu\theta - \zeta\sin\theta - \chi + 2\pi p\nu]\, d\theta$$

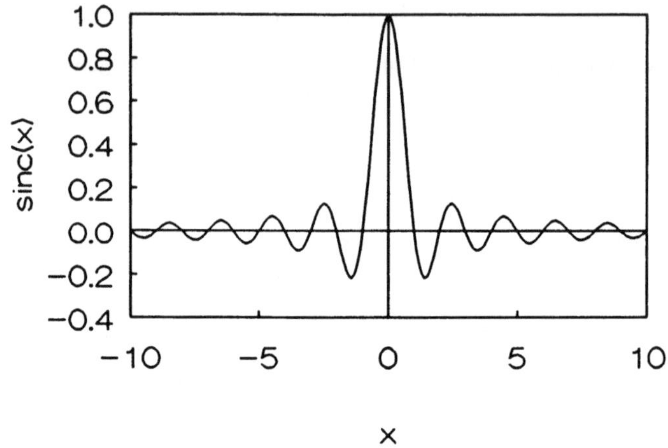

Figure 8.5: The function $\operatorname{sinc} x \equiv \frac{\sin \pi x}{\pi x}$.

$$= \frac{1}{2\pi N} \sum_{p=0}^{N-1} \int_0^{2\pi} \{\cos[\nu\theta - \zeta \sin\theta - \chi] \cos[2\pi p \nu]$$
$$- \sin[\nu\theta - \zeta \sin\theta - \chi] \sin[2\pi p \nu]\} \; d\theta$$
$$= \frac{1}{N} \left\{ \sum_{p=0}^{N-1} \cos(2\pi p \nu) \right\} \left\{ \int_0^{2\pi} \cos[\nu\theta - \zeta \sin\theta - \chi] \, d\theta \right\}$$
$$- \frac{1}{N} \left\{ \sum_{p=0}^{N-1} \sin(2\pi p \nu) \right\} \left\{ \int_0^{2\pi} \sin[\nu\theta - \zeta \sin\theta - \chi] \, d\theta \right\}$$

Now consider the first sum. If ν is equal to an integer then its value is unity. Let $\nu = l + \eta + \delta\nu$, where l is an integer chosen such that $-\frac{1}{2} < \delta\nu < \frac{1}{2}$. Then the sum may be written

$$S_1 = \frac{1}{N} \sum_{p=0}^{N-1} \cos(2\pi \, \delta\nu \, p)$$

If N is large the sum may be approximated by an integral. If we write $\alpha = p/N$ and $\delta\xi = \delta p/N = 1/N$, then

$$S_1 \simeq \frac{1}{2\pi} \int_0^1 \cos(2\pi \, N \, \delta\nu \, \xi) \, d\xi = \frac{\sin(2\pi \, N \, \delta\nu)}{2\pi \, N \, \delta\nu}$$

This is the well known function $\operatorname{sinc} x$ shown in Figure 8.5. It is strongly peaked at the origin. The successive maxima and minima have magnitude

$1/(l+\frac{1}{2})\pi$. For large $N\,\delta\nu$ it dies away very rapidly. There is little contribution except from the central peak. The width to the first zero is given by $\delta\nu = 1/2N$. Thus, if N is large, there is a very narrow band of frequencies about the integral values of ν which makes any significant contribution to resonance. In the same way the second sum may be approximated by an integral and becomes

$$S_2 = \frac{1-\cos(2\pi N\,\delta\nu)}{2\pi N\,\delta\nu}$$

Near $\delta\nu = 0$ it is of order $\delta\nu$ and tends to zero as $\delta\nu \to 0$. Elsewhere it is of order $(\pi N\,\delta\nu)^{-1}$ and is negligible when $\delta\nu > N^{-1}$. In addition, from (B.23), the integral multiplying this sum is zero when $\delta\nu = 0$ and of order $\delta\nu$ in this neighbourhood. Thus when N is large we may use this result and (B.22) to write

$$\begin{aligned} I_1 &= \frac{1}{2\pi}\int_0^{2\pi} \cos[(l+\eta)\theta - \zeta\sin\theta - \chi]\,d\theta \\ &= \frac{1}{2\pi}\int_0^{2\pi} \cos[(l+\eta)\theta - \zeta\sin\theta]\cos\chi\,d\theta \\ &= J_{l+\eta}(\zeta)\cos\chi \end{aligned} \tag{8.18}$$

We may treat I_2 in exactly the same way. We find that

$$\begin{aligned} I_2 &= \frac{1}{2}\left\{\frac{1}{2\pi}\int_0^{2\pi} \cos[(l+1)\theta - \zeta\sin\theta - \chi]\,d\theta \right. \\ &\quad \left. + \frac{1}{2\pi}\int_0^{2\pi}\frac{1}{2\pi}\int_0^{2\pi} \cos[(l-1)\theta - \zeta\sin\theta - \chi]\,d\theta\right\} \\ &= \frac{1}{2}[J_{l+1}(\zeta) + J_{l-1}(\zeta)]\cos\chi \\ &= \frac{n}{\zeta}J_l(\zeta)\cos\chi \end{aligned} \tag{8.19}$$

The last result follows from (B.4).

When these results are substituted in (8.16) and (8.17) we find that

$$\begin{aligned} \left\langle\frac{dp_\parallel}{dt}\right\rangle &= q\left\{v_\perp B_R J_{l+1}\left(\frac{k_\perp v_\perp}{\Omega}\right) - v_\perp B_L J_{l-1}\left(\frac{k_\perp v_\perp}{\Omega}\right)\right. \\ &\quad \left. + E_\parallel J_l\left(\frac{k_\perp v_\perp}{\Omega}\right)\right\}\cos(l\psi + k_\parallel z_0) \end{aligned} \tag{8.20}$$

and

$$\left\langle\frac{dW}{dt}\right\rangle = \frac{\omega}{k_\parallel}\left\langle\frac{dp_\parallel}{dt}\right\rangle \tag{8.21}$$

Thus

$$\left\langle \frac{dW}{dp_{\|}} \right\rangle = \frac{\omega}{k_{\|}} \tag{8.22}$$

is a general result applying to all types of interaction between waves and particles. From it can be found the single wave characteristics for any case.

8.3.4 Single Wave Characteristics in the General Case

Equation 8.22 may be integrated to obtain

$$W - W_0 = \omega p_{\|}/k_{\|} \tag{8.23}$$

where W_0 is a constant. This is a general expression for the single wave characteristic. The actual characteristics can be found by expressing W and $p_{\|}$ in terms of $v_{\|}$ and $v_{\perp}$.

In the non-relativistic case it is convenient to define a different constant V_0 by $V_0^2 = (\omega/k_{\|})^2 + W_0/m$. In this case $W = \frac{1}{2}m(v_{\|}^2 + v_{\perp}^2)$ and $p_{\|} = mv_{\|}$. The single wave characteristic becomes

$$\left(v_{\|} - \frac{\omega}{k_{\|}} \right)^2 + v_{\perp}^2 = V_0^2 \tag{8.24}$$

which is the same expression as previously obtained.

In the relativistic case we define

$$\frac{V_0^2}{c^2} = 1 - \frac{m^2c^4}{W_0^2}\left(1 - \frac{\omega^2}{c^2k_{\|}^2}\right)$$

The expressions for energy and parallel momentum are

$$W = \frac{mc^2}{\sqrt{1 - (v_{\|}^2 + v_{\perp}^2)/c^2}}; \; p_{\|} = \frac{mv_{\|}}{\sqrt{1 - (v_{\|}^2 + v_{\perp}^2)/c^2}}$$

If these are substituted in the general expression for the single wave characteristic we get

$$v_{\|}^2\left(1 - \frac{V_0^2V_P^2}{c^4}\right) - 2V_{\|}V_P\left(1 - \frac{V_0^2}{c^2}\right) + v_{\perp}^2\left(1 - \frac{V_P^2}{c^2}\right) = V_0^2 - V_P^2 \tag{8.25}$$

which again is the same as that obtained previously. In this case there is, however, an important difference. The previous derivation required that we work from the wave frame; this implied that the expression was valid only for waves with $\omega/k_{\|} < c$. No such restriction has been made in this derivation. The expression for the single wave characteristic is valid no matter what the phase velocity along the magnetic field. An example of the single wave characteristics for $\omega/k_{\|} > c$ is shown in the lower panel of Figure 8.4.

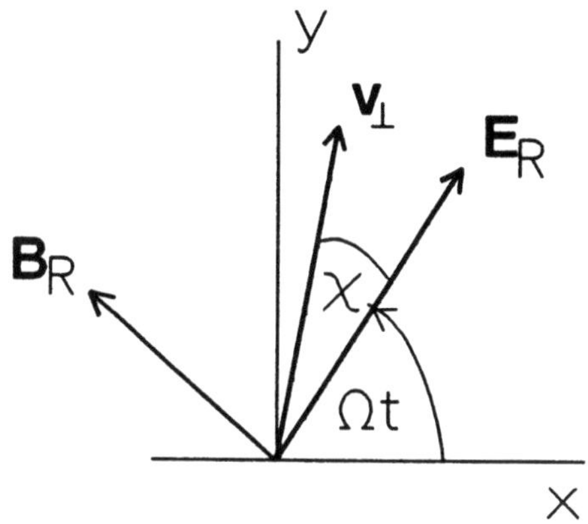

Figure 8.6: Phase difference between particle and wave as seen in the guiding center frame.

8.4 Diffusion in Velocity Space

Under the action of a wave a resonant particle moves in velocity space along the single wave characteristic. The rate and sense of the motion depend on the phase difference between particle and wave; all the expressions in the previous section have a factor $\cos\chi$. The situation is shown in Figure 8.6. This shows the situation for a negatively charged particle which rotates about the field in a right-handed sense and resonates with the right-hand circularly polarized wave. In the guiding center frame the wave has a Doppler shifted frequency equal to the gyrofrequency. The wave field vectors $\mathbf{E}_R$ and $\mathbf{B}_R$ rotate as shown with angular velocity Ω. The angle χ between $\mathbf{E}_R$ and $\mathbf{v}_\perp$ is constant. The force exerted by the electric field of the wave on the particle is clearly $qE_R\cos\chi$ in the direction of $\mathbf{E}_R$. The force exerted by the magnetic field of the wave is $qv_\perp B_R\cos\chi$ along the background magnetic field direction. Both these forces are proportional to $\cos\chi$; they are zero when $\chi = \pm\pi/2$ and maximum when $\chi = 0$ or π.

The zero order distribution function $f_0(v_\perp, v_\parallel)$ is independent of the particle phase. Quantities such as $\langle dW/dt\rangle$ and $\langle dp_\parallel/dt\rangle$ which must be averaged over velocity space do depend on the phase. Thus we must not use $2\pi f_0$, the phase integrated version of the distribution function, but must integrate over all three dimensions of velocity space with volume element $v_\perp\, d\chi\, dv_\perp\, dv_\parallel$.

8.4.1 Net Flux of Resonant Particles in Velocity Space

Consider a point O located at $(v_\perp, v_\|)$ in velocity space for which the particles are in resonance with a wave of angular frequency ω. Let s be the distance (of course having the dimensions of velocity) measured along the single wave characteristic from O. Let r be the distance along the resonance curve from O. An example is shown in Figure 8.7. This figure corresponds to the case of a gyroresonance in the non-relativistic case. The top panel shows the location of O in velocity space. The resonance is defined by a straight line parallel to the $v_\perp$-axis, located at $v_\| = V_R \equiv (\omega - \Omega)/k_\|$. The distance along this line, measured from O, is r. The single wave characteristic through O for a wave with phase speed $V_P = \omega/k_\|$ is a circle centered on V_P. The phase coordinate of velocity space is perpendicular to the plane of the paper. The second panel shows detail of the region near O. The arrow s is the direction of the single wave characteristic, making angle β with the direction of the $v_\|$-axis. A "tube" of velocity space is shown. In the direction perpendicular to the diagram it has width $d\chi$. In the plane of the diagram its width is $PQ = \delta v_\perp \cos\beta$ where $\delta v_\perp = OQ$. The bottom panel shows a related space where s, the distance measured along the single wave characteristic is plotted against particle phase χ. In other cases the details would be different. For Čerenkov resonance the single wave characteristics would be parallel to the $v_\|$-axis and $s = v_\| - V_R$. For the relativistic case the resonance condition would define an ellipse.

Near the resonance, in the region defined by $|v_\| - V_R| < \tau |m^{-1}| dp_\|/dt|$, the particle moves along the single wave characteristic at a rate

$$\frac{ds}{dt} = \frac{1}{m\cos\beta}\left\langle\frac{dp_\|}{dt}\right\rangle = \Theta_l \cos\chi$$

where Θ_l depends on the nature of the resonance and has the dimensions of acceleration. The angle $\chi = l\psi + k_\| z_0$ is the phase difference between particle and wave. For the case shown in Figure 8.7 from the geometry it can be seen from (8.20) that

$$\begin{aligned}\Theta_l &= \frac{q}{m\cos\beta}\left\{v_\perp B_R J_{l+1}\left(\frac{k_\perp v_\perp}{\Omega}\right) - v_\perp B_L J_{l-1}\left(\frac{k_\perp v_\perp}{\Omega}\right)\right. \\ &\quad \left. + E_\| J_l\left(\frac{k_\perp v_\perp}{\Omega}\right)\right\} \qquad (8.26)\end{aligned}$$

This also applies to the case of Čerenkov resonance if we set $\beta = 0$. The relativistic case requires some generalization of this formula because the curve on which resonance occurs is not parallel to the $v_\perp$-axis.

The rate ds/dt is positive in the range $-\pi/2 < \chi < \pi/2$ and negative in the range $\pi/2 < \chi < 3\pi/2$. During the wave-particle interaction time

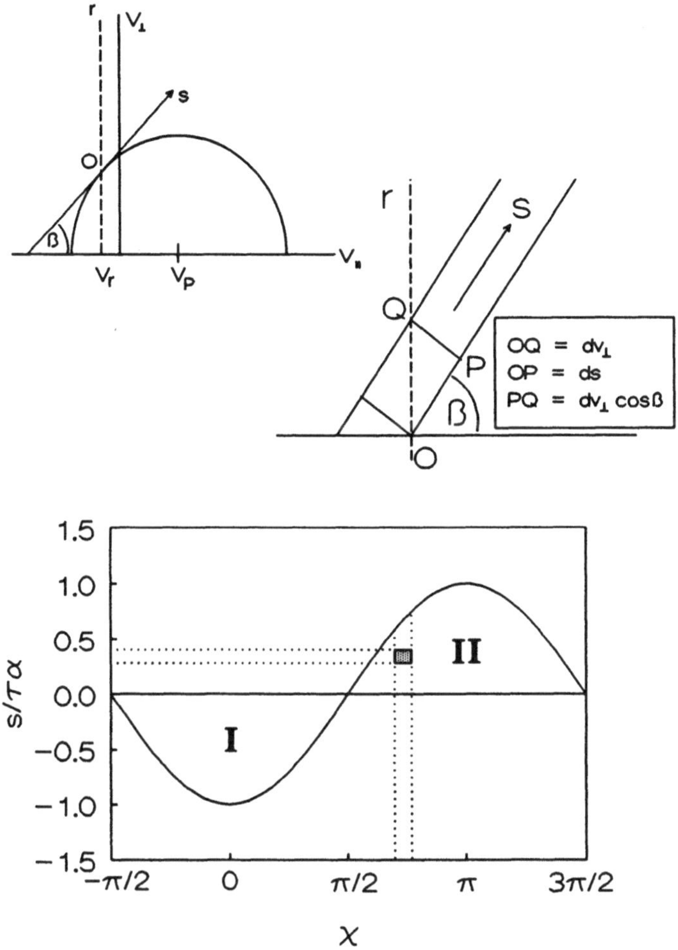

Figure 8.7: Diffusion in velocity space. For explanation see text.

τ the particle moves a distance $\Theta_l \tau \cos \chi$ along the single wave characteristic. In the bottom panel of Figure 8.7 we plot a region of the velocity space in the neighbourhood of the resonance. In this case we choose to use coordinates s and the wave-particle phase difference χ. The $v_\perp$-axis is perpendicular to this diagram. The horizontal line defines the center of the resonance located at O. Then two regions of velocity space are defined by the curve $s = \Theta_l \tau \cos \chi$. All the particles in region I are moving in the $+s$-direction and all those in region II in the $-s$-direction. Thus all the particles in regions I and II, and only these particles, cross the point O in time τ. The difference between the number of particles in region I and the number in region II is proportional to the net flux of particles in velocity space across the resonance due to the action of the wave.

The zero-order distribution function is a function of s and $v_\perp$. In the neighbourhood of the resonance it is approximately given by

$$f_0(s, v_\perp) \simeq f_0(0, v_\perp) + s \left.\frac{\partial f_0}{\partial s}\right|_{s=0}$$

From the diagram it can be seen that $dv_\parallel = ds \cos \beta$. The element of volume in the space $(v_\perp, s, \chi)$ is thus $v_\perp \cos \beta \, d\chi \, ds \, dv_\perp$. The number of particles in the element of volume shown shaded in Figure 8.7 and having width $\delta v_\perp$ in the $v_\perp$-direction is thus

$$v_\perp \cos \beta \left\{ f_0(0, v_\perp) + s \left.\frac{\partial f_0}{\partial s}\right|_{s=0} \right\} d\chi \, ds \, \delta v_\perp$$

We wish to find flux at O in velocity space, that is to say the number of particles per second per unit "area" of velocity space, crossing the resonance curve at O. From Figure 8.7 the area is $\delta v_\perp \int_0^\pi v_\perp \, d\chi \, dr = \pi v_\perp \, \delta v_\perp$. The positive flux is thus

$$\begin{aligned} &\frac{v_\perp \cos \beta \, \delta v_\perp}{\pi \tau v_\perp \, \delta v_\perp} \int_{-\pi/2}^{\pi/2} \int_{-\tau \Theta_l \cos \chi}^{0} \left\{ f_0 + s \frac{\partial f_0}{\partial s} \right\} d\chi \, ds \\ &= \frac{\cos \beta}{\pi \tau} \int_{-\pi/2}^{\pi/2} \tau \Theta_l \cos \chi \left[f_0 - \frac{1}{2} \tau \Theta_l \cos \chi \frac{\partial f_0}{\partial s} \right] d\chi \end{aligned}$$

In exactly the same way the negative flux is

$$\begin{aligned} &\frac{v_\perp \cos \beta \delta v_\perp}{\pi \tau v_\perp \, \delta v_\perp} \int_{\pi/2}^{3\pi/2} \int_{0}^{-\tau \Theta_l \cos \chi} \left\{ f_0 + s \frac{\partial f_0}{\partial s} \right\} d\chi \, ds \\ &= \frac{\cos \beta}{\pi \tau} \int_{\pi/2}^{3\pi/2} \tau \Theta_l \cos \chi \left[-f_0 + \frac{1}{2} \tau \Theta_l \cos \chi \frac{\partial f_0}{\partial s} \right] d\chi \end{aligned}$$

The net flux is the difference between these

$$\Pi = \frac{\Theta_l \cos \beta}{\pi} \int_{-\pi/2}^{3\pi/2} \left[f_0 \cos \chi - \frac{1}{2} \tau \Theta_l \cos^2 \chi \frac{\partial f_0}{\partial s} \right] d\chi = -\frac{1}{2} \tau \Theta_l^2 \cos \beta \frac{\partial f_0}{\partial s} \tag{8.27}$$

This shows that the flux of particles in velocity space is locally along the single wave characteristics in the direction of decreasing particle density in velocity space. Depending on the nature of the distribution function this may be in the direction of increasing or decreasing particle energy. If the net flux is in the direction of increasing particle energy the particles gain energy from the wave which is therefore damped. If it is in the direction of decreasing particle energy the waves gain energy and grow.

It can immediately be seen that, since Θ_l is linearly related to the field amplitudes through (8.26) and (8.20), this flux is a non-linear function of the field amplitudes. This is the basis of a quasi-linear theory. The wave amplitudes are calculated linearly on the assumption that the distribution function does not change significantly over the wave-particle interaction time. This can be done by whatever method is appropriate to solve the problem; in different circumstances cold plasma theory, warm plasma theory or linear kinetic theory may be appropriate. It is assumed that the rate of change in the distribution function is of second order and can be non-linearly calculated from the first order solutions. Non-linear effects on the waves are, however, neglected.

8.4.2 Quasi-Linear Diffusion

It remains only to find the rate at which the zero-order distribution function changes with time. Conservation of particles requires that f_0 obeys a conservation equation in velocity space of the form

$$\frac{\partial f_0}{\partial t} + \boldsymbol{\nabla}_v \,.\, \boldsymbol{\Pi} = 0$$

The flux is along the single wave characteristics and thus the divergence operator can be written

$$\boldsymbol{\nabla}_v = \hat{\mathbf{s}} \frac{1}{\cos\beta} \frac{\partial}{\partial s}$$

This leads to

$$\frac{\partial f_0}{\partial t} = \frac{1}{\cos\beta} \frac{\partial}{\partial s} \left(\frac{1}{2} \tau \Theta_l^2 \cos\beta \frac{\partial f_0}{\partial s} \right) \tag{8.28}$$

This can immediately be recognized as a diffusion equation. It is a simple form of the *quasi-linear diffusion equation.* The diffusion coefficient is $\frac{1}{2}\tau\Theta_l^2$. The factors $\cos\beta$ arise from the fact that the s- and $v_\perp$-axes are not orthogonal.

The approach here has very significant limitations. It is developed by considering the case of a single almost monochromatic wave. A particle located at a particular point in velocity space may be resonant with more than one wave mode or with a number of gyroharmonics of a single mode. In general a broad wave spectrum may exist and it may be necessary

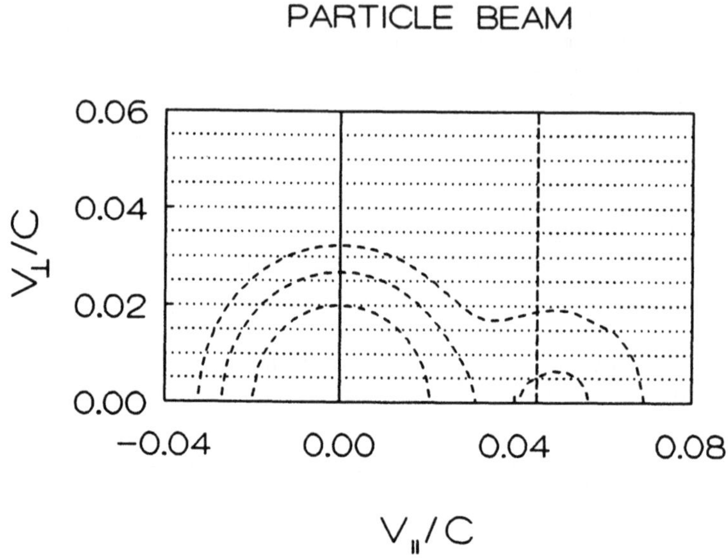

Figure 8.8: Diffusion in velocity space for a Maxwellian distribution with a superimposed beam. The dashed curves are contours of f_0. The horizontal dotted lines are single wave characteristics. The vertical dashed line is the Čerenkov resonance.

to consider its statistics. A great deal of physics may be involved in establishing the value of the wave-particle interaction time τ. The reader requiring a more rigorous approach should study the specific cases treated by Swanson [106].

8.4.3 Examples of Diffusion in Velocity Space for Specific Distributions

Čerenkov Resonance for Plasma Beam Distribution

Consider a plasma with a low temperature Maxwellian distribution on which a particle beam such as that shown in Figure 6.5 is superimposed. Some contours of the distribution function are shown in Figure 8.8. The distribution function has a maximum at the origin and a subsidiary maximum due to the beam can be seen at $v_{\|} = 0.045c$. We study the Čerenkov resonance of a wave with parallel phase velocity $\omega/k_{\|} = V_R$ where V_R is defined by the resonance line. The vertical dashed line defines those particles which are resonant with the wave. The single wave characteristics are the horizontal dotted lines.

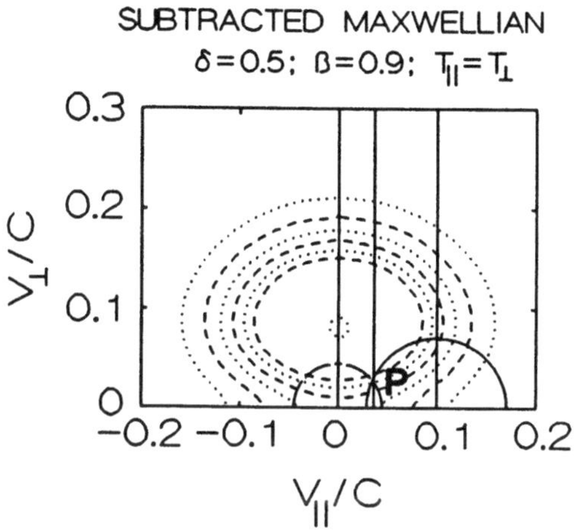

Figure 8.9: Diffusion in velocity space for a loss cone distribution. The dashed curves are contours of f_0. The parallel phase speed of the wave is $\omega/k_{\parallel} = 0.1$. The vertical line through P represents resonance. The solid circle is a single wave characteristic and the dashed circle is a contour of constant particle kinetic energy.

The resonant particles diffuse along the single wave characteristics. The direction of the diffusion is from high particle density to low particle density. Thus resonant particles at the point defined by the intersection of the inner contour and the resonance line diffuse to the left, the direction in which f_0 decreases. In the process $v_{\parallel}$ would decrease and so would the mean particle energy. Particles with large values of $v_{\perp}$ would diffuse to the right, increasing their mean energy. The nature of the distribution function is that the density is largest along that part of the resonance line on which particles lose energy. Thus, if the energy transferred to resonant particle is summed over all the resonant particles, there is a net loss of particle energy. This particle energy is transferred to the wave with the consequence that it grows. This wave growth arises because of the positive value of $\partial f_0/\partial v_{\parallel}$ for most of the resonant particles. This aids our physical intuition of the linear growth processes discussed in Chapters 6 and 7.

8.4.4 Gyroresonance for a Loss Cone Distribution

Suppose now that we have a plasma with a loss cone distribution represented by the subtracted Maxwellian such as that shown in Figure 6.3. Examples of contours for such a distribution are shown in Figure 8.9. The point P lies on the resonance line. A single wave characteristic for gyroresonance has been drawn through P. It is a circle centered on $v_\parallel = \omega/k_\parallel$. The distribution is a half filled subtracted Maxwellian with a maximum at the point represented by a small circle. The net diffusion of particles at P will be in the direction of decreasing f_0. For the case shown they diffuse in an anticlockwise direction along the single wave characteristic. The dashed circle centered on the origin is a contour of constant particle kinetic energy. The diffusion means that particles cross this circle from outside to inside so that their energy is reduced. The wave energy will be correspondingly increased. The most important feature of the diffusion is, however, that the pitch angle of the particles is changed; while $v_\perp$ decreases $v_\parallel$ increases. This is an example of *pitch angle diffusion* under the action of the wave. The direction of this diffusion is such that the particles tend to enter the loss cone thus filling it and reducing the pitch angle anisotropy which is the source of the instability. We have considered a single wave which acts only on a limited number of resonant particles. A broad band spectrum of waves acts on many particles resonating with different frequencies and is an important mechanism in driving the plasma towards equilibrium.

8.5 Some Non-Linear Effects

8.5.1 Non-Linear Čerenkov Resonance

The simplest intuitive treatment of Landau damping envisages a particle which is trapped in the potential well formed by the trough of an electrostatic wave. Trapped particles traveling more slowly than the wave are accelerated by it and extract energy from it; those traveling more quickly lose energy to it. If there are more slower particles than faster ones as is the case for a Maxwellian distribution there is net damping of the wave. The process is often compared to a surfer being accelerated by a wave. This is an inherently non-linear process. The paths of the trapped particles are crucial to this explanation. They in no way approximate the unperturbed paths which the particles are assumed to follow in the linear treatment of Chapter 7.

In the quasi-linear treatment of instability in the previous section it was assumed that the wave could be treated by linear theory. The particle was assumed to travel along its unperturbed path and the effect of the wave on the particle evaluated to find the change in parallel momentum and energy. This depended on second order terms in the fields and was thus non-linear. It still did not depend on particle trapping.

In this section we study qualitatively some non-linear effects which do depend on trapped particles. As an example we consider the non-linear development of the instability excited by a particle beam [34].

Suppose we have a sinusoidal electron plasma wave propagated in the z-direction. Assume that the stability is marginal so that the wave is of constant amplitude. The acceleration of the particles is proportional to the electric field

$$\frac{dv_{\|}}{dt} \propto \sin kz \tag{8.29}$$

In the rest frame of the wave the field can be described in terms of a sinusoidal electric potential. Figure 8.10a shows the potential energy

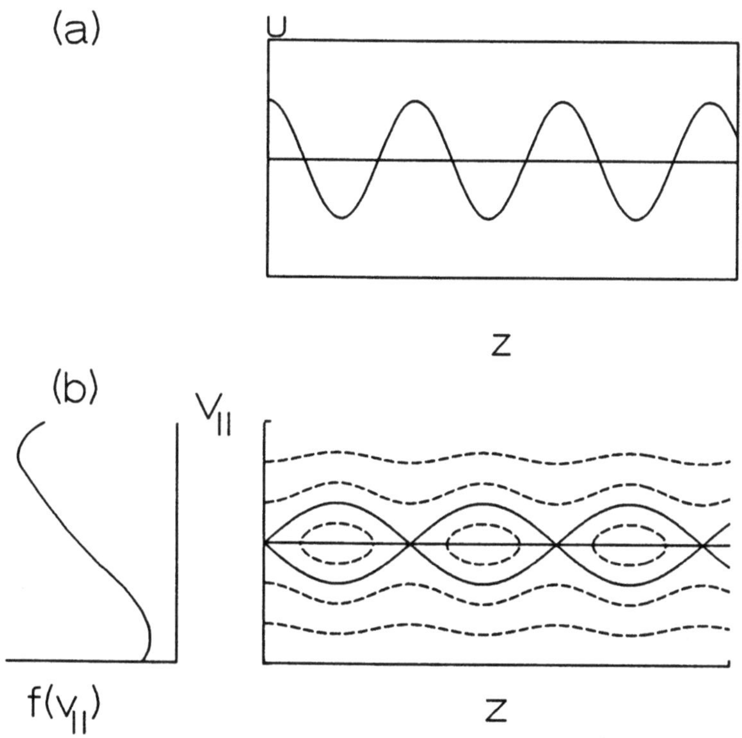

Figure 8.10: Particle trapping. (a) Potential energy in the rest frame of an electrostatic wave. (b) Trajectories of near-resonant particles in phase space. The left hand portion of the diagram shows a portion of the distribution function which would lead to growth.

$U = q\Phi == e\Phi$ of an electron in the field of the wave. There is a series of potential energy wells. Conservation of the electron energy requires that they move in such a way that $U(z) + \frac{1}{2}mv_{\|}^2 = \mathcal{E} - \frac{1}{2}mv_{\perp}^2 =$ constant where $\mathcal{E}$ is the total energy. Electrons near resonance have small $v_{\|}$ in the rest frame of the wave. Those with sufficiently small kinetic energy will be trapped in the potential wells. Those with larger kinetic energies have a fluctuating $v_{\|}$ as they move in the z-direction. Their paths in phase space are illustrated on the right hand side of Figure 8.10b. The untrapped particles with $v_{\|} > 0$ move to the right. Those with $v_{\|} < 0$ move to the left. The trapped particles move clockwise in closed loops. The width of the region in phase space in which trapping takes place depends on the amplitude of the wave. The greater the amplitude the greater is $\Delta v_{\|}$, the region over which trapping takes place. Even when the wave is not constant in amplitude, if the growth rate is small the amplitude does not vary very much over a wave cycle and this picture is a useful one.

In the case of linear Landau damping or growth the wave amplitude is assumed to be very small. The particles which are trapped form a negligible fraction of the particles which resonate with the wave. The unperturbed paths of all the other particles are horizontal straight lines in phase space. The effect of the wave is to start to move the particles away from this path and the wave grows or decays depending on whether more particles lose or gain energy in this process. A detailed discussion is given by Chen [25, §§7.5, 7.6].

In the quasi-linear theory the wave is still given by linear theory. As it grows it moves the particles away from their unperturbed paths. The interaction time, however, is assumed to be relatively short. The bandwidth of the wave $\Delta k_{\|}$ is thus sufficiently broad such that particles resonating with waves in the spectrum at $k_{\|} = k_0 \pm \frac{1}{2}\Delta k_{\|}$ do not get out of phase during the wave interaction time. They thus move only a small distance along their trajectories in Figure 8.10b.

Non-linear effects become important when a significant number of particles become trapped by the wave. Consider a situation in which the distribution function has a weak beam on the tail of a Maxwellian distribution. The distribution function in the neighborhood of the resonance is shown on the left in Figure 8.10b. This will lead to wave growth. We assume that the growth is slow and that the bandwidth of the unstable wave is small so that the wave-particle interaction time is very long. Consider Figure 8.11. In the top panel the shaded region represents a band of particles which previously were untrapped. These particles formed part of the beam and their density is greater than the background. The amplitude of the wave has grown to trap them and they now follow new trajectories represented by the dashed lines. If the potential well were parabolic all the trapped particles would have the same angular velocity in phase space. Because the well is not exactly parabolic the angular velocity depends on the contour on which the particle is located. After one half period the trapped particles will have rotated to the position shown in the middle panel and those which were not trapped will be car-

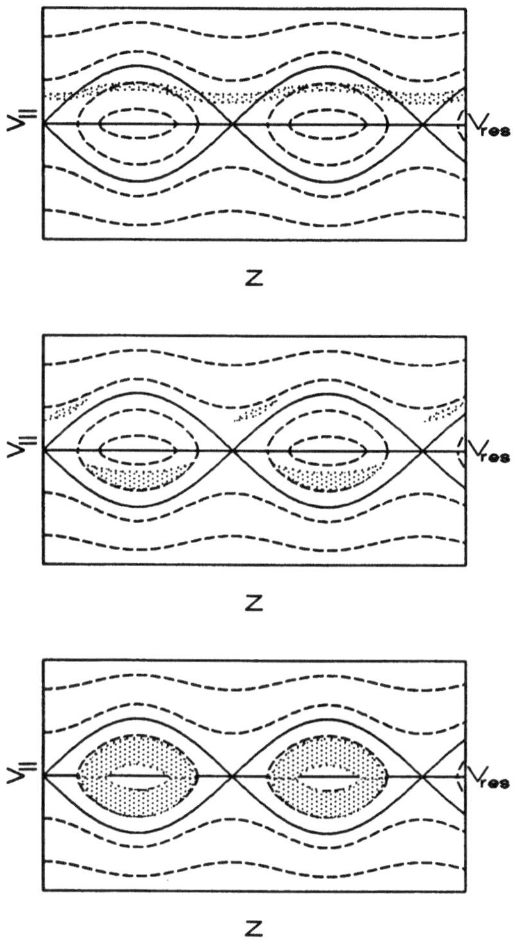

Figure 8.11: Bunching of trapped particles.

ried away and smeared out as shown by the small shaded patches. After several trapping periods the differential angular velocity will smear the trapped particles out as shown in the bottom panel. The non-linear effect is to bunch them to be in phase with the potential troughs in the wave.

An analytical treatment is not easy. Drummond *et al.* [34] treat the case of a cold background and a monoenergetic beam in a semi-quantitative way. Computer simulations confirm the bunching effect.

8.5.2 Phase Bunching in Cyclotron Resonance

The case of cyclotron resonance is analogous. In the rest frame of the wave, as shown in Figure 8.6, the particle rotates with angular velocity $\Omega - k_{||}v_{\mathrm{res}}$. In the wave rest frame $\mathbf{E}_R = 0$. The angular velocity of the field vector $\mathbf{B}_R$ is $\Omega - k_{||}v_{||}$. The rate of change in the phase difference χ is the difference of these

$$\frac{d\chi}{dt} = k_{||}(v_{\mathrm{res}} - v_{||}) = \Delta v_{||}$$

Also, from (8.13),

$$\frac{dv_{||}}{dt} \propto \sin\chi$$

Thus exactly the same discussion can be carried through with the coordinate z in Figure 8.10 replaced by χ. The bunched particles are now not trapped in bunches along the z-axis but in phase angle. At a point in space, as the wave advances, the bunched particles constitute a circularly polarized transverse current density.

8.6 Waves Resonant with a Test Particle

In Section 8.4 we have assumed the existence of a wave which is resonant with a test particle in velocity space. This implies the assumption that there exists a wave with an appropriate frequency and direction such that the resonance condition $\omega - k_{||}v_{||} = l\Omega$ can be met. This is not always the case. The real part of the dispersion relation may allow no waves which are propagated with the appropriate velocity. More than one wave may resonate with the particle. Waves with different frequencies and direction may satisfy the resonance condition for different values of n. In this section we illustrate the nature of the interaction for some specific examples.

8.6.1 Resonance with Electron Plasma Waves

Again consider the case of a low temperature Maxwellian distribution on which is superimposed a weak particle beam such as that shown in the right-hand panel of Figure 6.5. Because the beam is assumed weak growth due to the beam is small. The real part of the dispersion relation is well represented by the warm plasma approximation. We consider the case where the electron plasma wave is coupled to the Z-mode at the resonance cone as illustrated in Figures 5.1 and 5.2. It is convenient to normalize frequencies in terms of the gyrofrequency Ω rather than using the parameters X and Y of cold plasma theory. The resonance condition may then be written

$$\frac{\omega}{\Omega}\left[1 - n_{\|}u_{\|}\right] = l$$

This can be solved for $n_{\|}$. Just as in the case of the Booker quartic (4.1) this can be substituted in the dispersion relation (5.32) using $\cos\theta = n_z/n$, $\sin\theta = n_x/n$ and the result solved for n_x. The result is shown in Figure 8.12. The thermal speed has been chosen to be small, about one hundredth of the speed of light. The refractive index is, at its largest, about 30, implying a phase speed about three times the thermal speed and justifying the warm plasma assumption. The test particle is an electron assumed to be associated with the beam and has $v_{\|}$ about seven times the thermal speed. With these parameters the three resonances corresponding to $l = +1,\ 0,\ -1$ occur at values of $n_{\|}$ denoted by the vertical lines at A, B and C. There are further resonances corresponding to $l > 1$ to the left of A. Resonances corresponding to $l < -1$ lie to the right of C and do not intersect the refractive index curve. In order to determine the relative importance of these resonances it is necessary to take account of several effects.

First it should be remembered that the resonance depends on the polarization. In this case over much of the range shown the electrostatic approximation applies so that the wave is approximately linearly polarized parallel to the wave normal. For most of the range of interest the wave normal makes a small angle with the magnetic field and the strongest wave component is $E_{\|}$. The small component of polarization perpendicular to the magnetic field can be resolved into two equal and opposite circular polarizations B_R and B_L.

Next we need to consider the nature of the Bessel functions in (8.20). The argument is the ratio of the gyroradius $mv_{\perp}/qB_0$ of the resonant particle to the reciprocal of the wave number $k_{\perp}^{-1} = \lambda_{\perp}/2\pi$. If this is small then $J_0 \simeq 1$ and J_l, $l \neq 0$ is small. In this case a simple calculation shows that these conditions are met at B and C for reasonable values of $v_{\perp}$ of the same order of magnitude as $v_{\|}$. Thus only the $l = +1$ gyroresonance with B_R, the $l = -1$ resonance with B_L and the $l = 0$ Čerenkov resonance with $E_{\|}$ will be significant. Of these the Čerenkov resonance is the most important because of the nature of the polarization. Because of the relatively large wave normal angle the resonance at A may be more important for a test particle with a fairly large value of $v_{\perp}$.

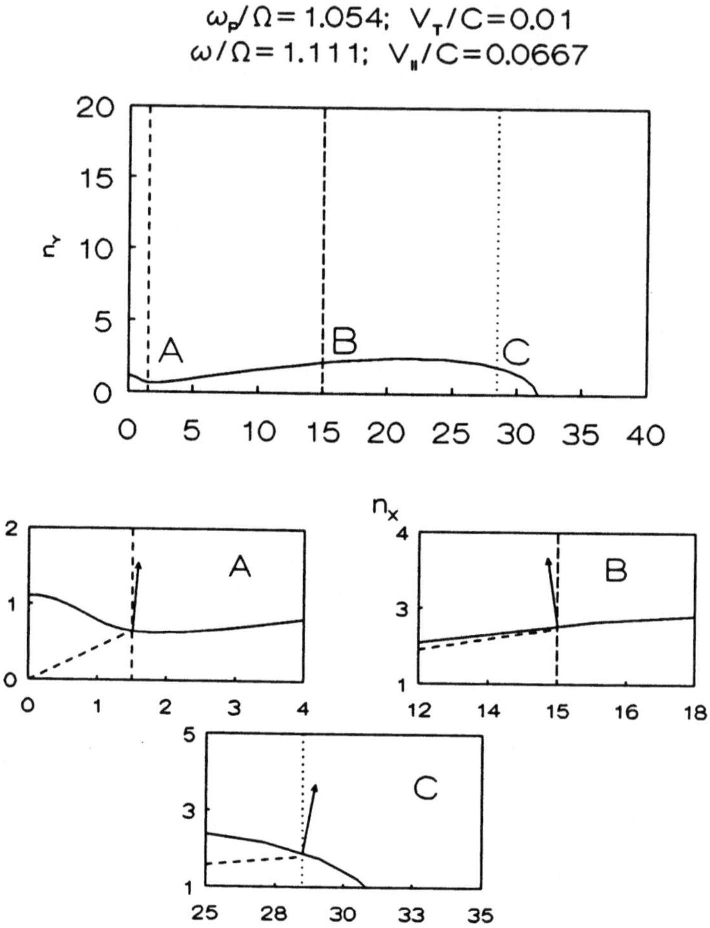

Figure 8.12: Resonance with the modified Z-mode.

The small panels in Figure 8.12 show the directions of wave normal and ray at each resonance. It is of interest to note that the ray is nearly transverse to the magnetic field in the cases shown.

This example, while intrinsically interesting, is not of much practical importance in the magnetosphere. In the next subsection we consider a more relevant case.

8.6.2 Gyroresonance and the Whistler Mode

In order for gyroresonance to occur the particle must interact with a wave for which the circularly polarized component, B_R or B_L in (8.20), is strong. An electron gyrates in the right-handed sense and interacts strongly with the right-handed B_R component of a wave. The opposite is true of a positive ion.

Whistlers propagated along the field are right hand circularly polarized[1] and therefore resonate with electrons. Whistlers propagated at other angles have a more complicated polarization. When the wave normal makes a small angle with the magnetic field the polarization is approximately circular and right handed; almost all the wave-particle interaction is with B_R. Near the resonance cone the wave is approximately electrostatic and is linearly polarized along the wave normal; wave-particle interaction is with two equal and opposite circularly polarized components and a component linearly polarized along the field. Between these extremes the wave is elliptically polarized. It can be seen that gyroresonance occurs mainly for the nearly longitudinal direction of propagation. As the wave normal angle increases the Čerenkov interaction becomes more important.

The condition for gyroresonance between electrons and right handed waves may be written

$$\frac{\omega}{k_\parallel} = v_\parallel + \frac{\Omega}{k_\parallel}$$

For whistlers $\omega < \Omega$ and thus this condition can only be met for negative values of $v_\parallel$. The particle travels in the opposite direction to the wave and therefore sees the wave frequency Doppler shifted upwards to the gyrofrequency. Consider Figure 8.13. This represents the resonance between whistlers of different frequencies and a test particle with a given negative parallel velocity. Each of the panels shows a refractive index surface for a different frequency in the whistler mode. The vertical line represents the parallel phase speed corresponding to the gyroresonance

[1] Our definition of polarization in terms of R and L components is in terms of the positive magnetic field direction and this statement is true in general. If we define polarization in terms of the wave normal direction as in Chapter 2 electrons propagated in the negative magnetic field direction gyrate in a left-handed sense about the wave normal direction and interact with whistler waves which have left handed polarization.

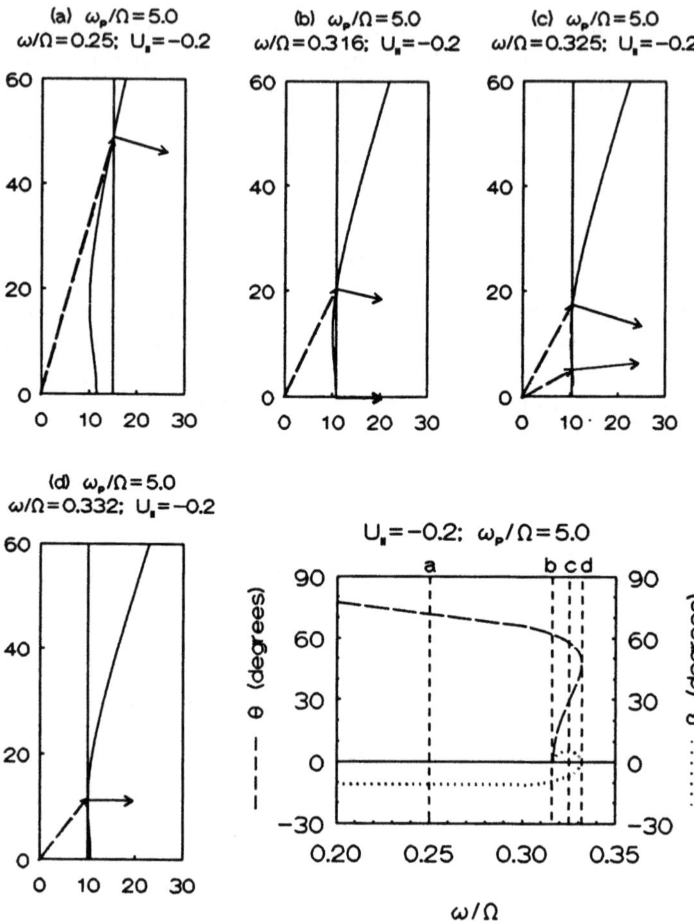

Figure 8.13: Gyroresonance condition for a test particle with $v_{\|} = 0.2c$ with the whistler mode. Panels (a) to (d) show the whistler refractive index surface for different frequencies. The vertical line represents the gyroresonance condition. The dashed arrow is the refractive index vector and the solid arrow shows the corresponding ray direction. The last panel shows the wave normal direction θ and the ray direction β of the resonant wave as a function of frequency.

condition. The intersection of this line and the refractive index surface defines the wave normal angle of the wave resonating at that frequency. The ray is normal to the refractive index surface at this point as shown by the solid arrow. The last panel shows how the wave normal direction and ray direction of the resonant wave vary with frequency.

The picture is quite complicated. Because of the points of inflection in the refractive index surface there are regions where whistlers of the same frequency and different wave normal angles resonate with the same particle. Other particles may resonate with the whistler through different resonance processes. For a full picture the behavior must be summed over all particles and frequencies. Nevertheless, when it is borne in mind that the strongest whistlers may have wave normals making small angles with the field, and that the dominant polarization is right-hand circular, such a picture can be very helpful in visualizing the physics of whistler growth and decay.

8.7 Summary

- The locus of non-relativistic resonant particles in velocity space is a straight line parallel to the $v_\perp$ axis. Particles resonate either through the Čerenkov resonance, in which $v_\parallel$ matches the phase speed along the magnetic field, or through gyroresonances in which the particle sees a Doppler shifted wave frequency which is equal to its gyrofrequency or one of its harmonics.

- The locus of relativistic resonant particles is modified to an ellipse.

- A resonant particle experiences forces due to the wave fields and has its energy and parallel momentum changed by the wave. No matter what the nature of the resonance $dW/dp_\parallel = \omega/k_\parallel$. This condition defines a locus in velocity space, called the single wave characteristic, along which the particle moves under the action of the wave. For Čerenkov resonance the single wave characteristics are parallel to the $v_\parallel$-axis. For gyroresonances they are circles centered on $v_\parallel = \omega/k_\parallel$. For relativistic particles they are ellipses.

- The net flux of resonant particles along single wave charcteristics is in the opposite direction to the component of the gradient of the distribution function along the characteristic. The resonant particles exhibit a diffusion in velocity space. The analogue of the r. m. s. speed is the mean acceleration $\langle m^{-1}dp_\parallel/dt\rangle$. The analogue of the collision time is the wave-particle interaction time τ. The diffusion coefficient is of the form $D = \frac{1}{2}\tau\Theta^2$. If the direction of the diffusion is in the direction of decreasing particle energy then the wave gains energy at the expense of the resonant particles and *vice versa.*

- Non-linear effects may lead to particle bunching. In the case of Čerenkov radiation particles are bunched in the potential troughs relative to the wave. In the case of cyclotron resonance they are bunched at particular values of χ the angle between $\mathbf{v}_\perp$ and the wave vector.

- The precise nature of the resonant interaction requires an understanding of the wave polarization and its refractive index properties.

Part II

Applications of the Theory to Plasma Wave Observations

Chapter 9

Magnetospheric Plasmas

9.1 Introduction

Above an altitude of a few hundred kilometers all the matter in the geospace environment is in the plasma state. Its dynamics are governed by the interaction between the earth's magnetic field and the solar wind. The region over which the geomagnetic field is a dominant influence is called the magnetosphere. It was the subject of an intensive collaborative effort during the International Magnetospheric Study (IMS) and many results of that program are summarized in [121]. The waves which are the subject of this book appear throughout the region. We set the stage with a brief descriptive account of the formation and structure of the magnetosphere and the types of wave which can exist in each region. In such a brief account it is only possible to outline the chief features. It must be recognized that the region is still the subject of intensive study and controversy; it is not possible to be dogmatic about many of the processes that are so confidently outlined below! Much more detail is given in, for example, [47, 37, 76].

9.2 Structure of the Earth's Magnetosphere

9.2.1 Formation of the Magnetosphere

The magnetosphere is formed by the interaction of the solar wind and the earth's magnetic field. The solar wind consists of protons and electrons which are accelerated away from the sun. It "blows" towards the earth at a speed of a few hundred kilometers per second with an energy density of about $4 \times 10^{-10}\,\mathrm{J\,m^{-3}}$.

If the earth were isolated in space its magnetic field would be approximately dipolar. The properties of such a field are outlined in Appendix C. At about 10 earth radii its energy density is of the same order as the solar wind energy density. Here the geomagnetic field has a significant effect on the solar wind. A balance is set up between the solar wind pressure and plasma pressure so that the plasma pressure decreases rapidly with increasing magnetic field. The gradient drift motions of the protons and electrons are associated with a current sheet (the *Chapman-Ferraro currents*) whose magnetic field is such as to cancel the geomagnetic field everywhere on the sunward side of the current. On the earthward side the magnetic field is increased by this current sheet; immediately earthward of the sheet the field is doubled. The net effect is that the earth's magnetic field acts as an obstacle to the solar wind which is deflected round it, carving out a cavity known as the *magnetosphere*. The boundary be-

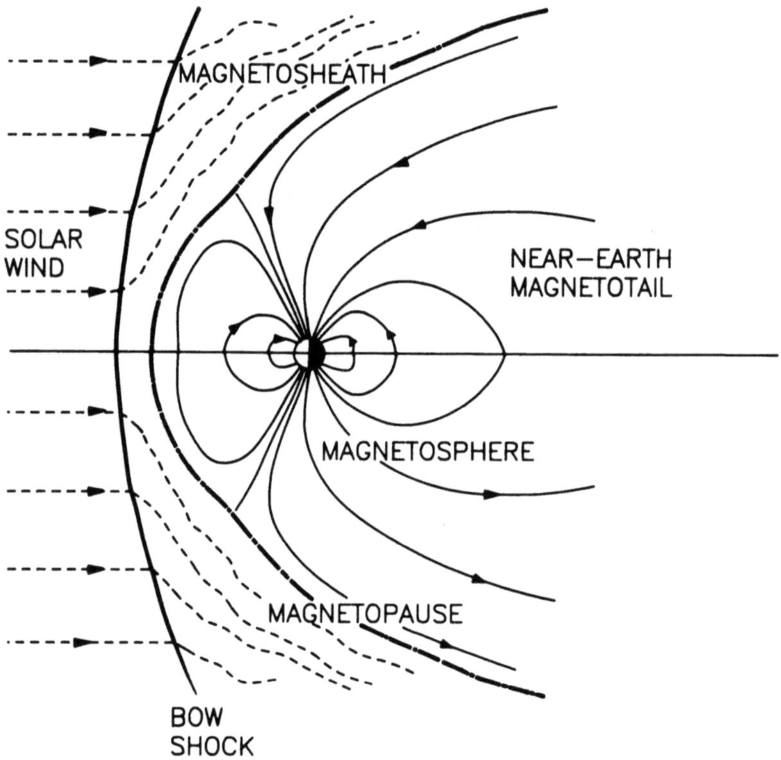

Figure 9.1: Closed magnetosphere model with bow shock.

tween the solar wind and magnetosphere, on which the Chapman-Ferraro current flows is called the *magnetopause*. Its shape is determined by the equilibrium between the solar wind pressure, including the kinetic energy density, on the one hand, and the magnetic pressure on the other. In this simple picture the magnetic field is confined to the inside of the magnetosphere and the plasma to the outside. Such a model is called a *closed* magnetosphere.

This structure is complicated by the fact that the earth moves relative to the solar wind at a speed greater than the speed of wave propagation in the plasma. The result is the formation of a shock wave known as the *bow shock*. The region between this shock and the magnetopause is turbulent and irregular and is known as the *magnetosheath.*

A sketch of the closed magnetosphere model is shown in Figure 9.1. The model is somewhat oversimplified but can be regarded as a first approximation to the magnetospheric behavior. Elaborations of this picture are outlined below.

9.2.2 Plasma Convection

In the magnetohydrodynamic approximation, which applies to the large scale motion of the magnetospheric plasma, it can be shown that magnetic field lines are effectively "frozen in" to the plasma. This means that they appear to move with the plasma so that all the particles which at one instant lie on a field line will, at a later instant, still lie on a single field line. If the particle pressure is large compared with the magnetic pressure the plasma flow is described by fluid dynamic equations and the magnetic field is carried along with it. This is because currents are induced in the plasma which have the effect of changing the magnetic field in such a way that this occurs. On the other hand, if the magnetic pressure is much larger than the plasma pressure, the magnetic field is undistorted and the velocity of the plasma perpendicular to the magnetic field is constrained to maintain the frozen-in field line picture. There is another way of describing the situation which is useful in this case. The component of the plasma velocity perpendicular to the magnetic field has the value $\mathbf{E} \times \mathbf{B}/B^2$. Since this is perpendicular to both $\mathbf{B}$ and $\mathbf{E}$ the plasma drifts along equipotentials perpendicular to the field lines.

The sun itself has a magnetic field with energy density much less than the solar wind energy density. By the time the solar wind reaches the earth, the field lines are stretched out to be more nearly parallel to the ecliptic plane and make an angle of about 45^0 with the sun-earth line. The field may, however, have a component perpendicular to the ecliptic whose sense, northward or southward, has substantial influence on the structure of the magnetosphere.

In Figure 9.2 the magnetic field configuration in the noon-midnight meridian is shown for a southward solar wind field. At the subsolar point A the solar wind and earth magnetic fields are oppositely directed. This leads to the formation of a neutral point at the subsolar point. In these circumstances *magnetic merging* or *reconnection* can take place. The frozen-in field line picture breaks down at the neutral point; the particles which were originally all located on field line numbers 1 and $1'$ lie on field line number $2'$. At the neutral point A the field is zero and field line merging takes place; solar and terrestrial magnetic fields become connected. The plasma is now convected onto two different field lines, 3 and $3'$. It is as if the field lines had broken and been reconnected in a different way. Successive positions of the field line are shown in Figure 9.2

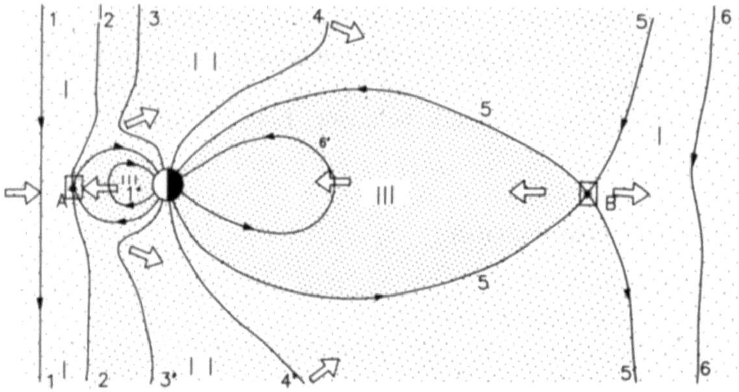

Figure 9.2: Reconnection as seen in the noon midnight meridian.

labeled 4, 4′ and 5. The magnetosphere is drawn out into a long tail, the *magnetotail*, which extends for distances many hundreds of earth radii behind the earth. It can be seen from the figure that another neutral point B forms in the tail on field line number 5. Merging takes place here as well; plasma is convected to the front of the magnetosphere on field line 6′ and down the tail to the solar wind on field line number 6.

Because the earth's magnetic field now extends out into the magnetosheath region, such a model is called an *open* magnetosphere. Of course the three-dimensional picture is more complicated than this. Far from the earth convection is dominated by electric fields of magnetospheric origin; near the earth the plasma is convected by electric fields of terrestrial origin which cause it to corotate with the earth. Figure 9.3 shows the resultant equipotentials of these two fields or equivalently the convection pattern of the cold plasma in the equatorial plane. The path followed by the cold plasma is the path followed by the frozen-in field lines. Energetic particles follow a different path because of energy-dependent drifts arising from the gradient and curvature of the magnetic field. Near the earth there is a stagnation point C where the electric field is zero. The equipotential which passes through this point separates plasma which corotates from plasma which does not.

Parts of the diagram are quite speculative. Two regions with different shading are shown. One shaded region shows field lines which are convected to the front of the magnetosphere where they merge with solar wind field lines along the reconnection line AB. This indicates what happens in the equatorial plane near the noon-midnight meridian shown in Figure 9.2. The associated plasma is swept over the polar caps. The

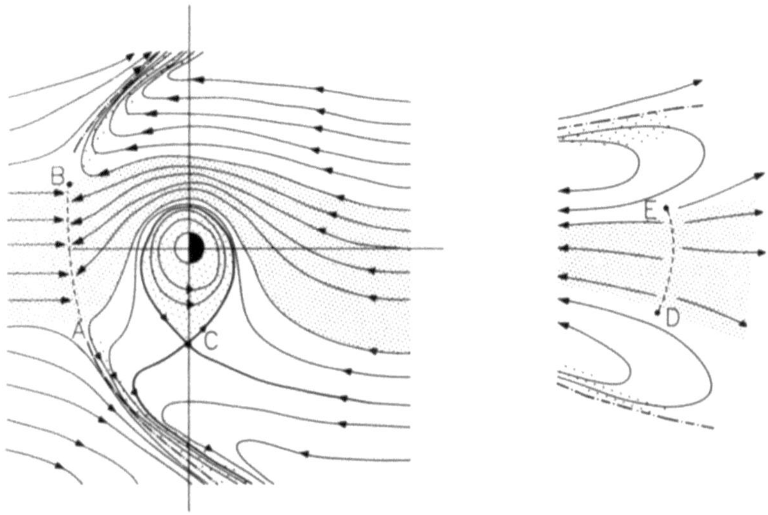

Figure 9.3: Plasma convection in the equatorial plane. AB and DE are lines along which merging is taking place. C is a stagnation point. The contour through C separates corotating plasma from convected plasma.

other shaded region shows the convection of plasma on field lines which do not merge but remain inside the magnetosphere and are swept back in a boundary region (the *low latitude boundary layer*) through some kind of viscous interaction. The relative importance of merging and viscous interaction is a subject of controversy. The weight of opinion is that merging is more important.

Far down the tail merging may take place along the line DE. Plasma on parts of the field lines is convected back towards the sun. The rest escapes to the solar wind.

This description gives the impression of a steady and uniform process. This is by no means the case. The magnetosphere is very dynamic. The merging at the front of the magnetosphere varies with conditions. It can be a steady process but may also happen sporadically through *flux transfer events.* The relative importance of steady merging and flux transfer events is not certain. The merging in the tail is also a fluctuating process. It appears to be related to the rate of merging at the front of the magnetosphere and hence to the direction of the magnetic field in the solar wind. A northward turning of this field is associated with the occurrence of *magnetospheric substorms* in which there is a dramatic reconfiguration of the magnetosphere (e.g. [92]).

9.2.3 Magnetic Structure of the Magnetosphere

The processes described above result in a division of the magnetospheric environment into three primary regions depending on the magnetic field topology. In Figure 9.2 they are labelled I, II, and III. In region I the field line nowhere meets the earth's surface (the field lines "have no feet on the ground"), in region II one end of the field line meets the earth's surface and the other connects with the solar wind magnetic field to end at the sun ("one foot on the ground"), in region III both ends of the field lines meet the earth's surface ("both feet on the ground"). We consider each of these in turn. It must be emphasized that what we describe is a steady state situation, different from the actual dynamic behavior of the magnetosphere which is highly variable both spatially and temporally. It nevertheless gives an average picture which is adequate for the purposes of providing a framework for locating the plasma populations which support the waves of interest in this book.

Region I: The Region I field lines essentially follow the solar wind as it blows past the magnetosphere except where, as described below, they participate in reconnection processes at the subsolar point and in the magnetotail. They may be connected with the sun or form closed loops on a scale that is large compared with the size of the magnetosphere.

Region II: The history of a Region II field line is seen in Figure 9.2. When the solar wind impinges on the subsolar point reconnection takes place. A Region I field line in the solar wind connects with a Region III field line, which has been convected to the day side of the magnetosphere, to form two Region II field lines, one convected over the northern and one over the southern polar cap. The polar caps can, in fact, be regarded as the regions defined by the intersection of the Region II field lines with the ionosphere. As the solar wind drags the field line backwards it stretches it out into the magnetotail. Far down the tail a southern and a northern Region II field line merge forming a Region III field line, which is again convected sunwards, and a Region I field line, which is swept away by the solar wind to the distant regions of the solar system.

Region III: The field lines in Region III are often labeled by a coordinate L. This represents the approximate radius, measured in terms of the earth's radius R_E, at which the field line crosses the equatorial plane. In this region the field lines fall into two classes. For smaller values of L the magnetospheric field lines approximately corotate with the earth. Their equatorial crossing point follows a path inside the separatrix defined by stagnation point C in Figure 9.3. The electric field associated with their convection maps down into the ionosphere where the frozen-in picture fails because of the finite conductivity. Here the electric fields are associated with currents rather than convection. It is possible to think of the feet of the field line as being dragged along by the ionosphere as the field line corotates. For larger values of L field lines do not corotate. Instead their equatorial crossing points follow paths which take them to the magnetopause. The plasma is either convected in from the tail to the sunward side of the magnetosphere and back along the flanks in the boundary layer to the tail or it merges with solar wind field lines to form part of a field line in Region II. The feet of such field lines follow closed paths in the ionosphere.

The sources of the magnetic field in Region III are the earth's main field due to dynamo processes in the core and the fields due to external current sources such as the ring current, the Chapman-Ferraro current, cross-tail currents, Birkeland currents and ionospheric current systems. For our purposes it is only necessary to have a crude idea of the magnitude and direction of the field. The main field is approximately dipolar with values at the earth's surface of about 32 000 nT at the equator to 64 000 nT at the poles. It drops off radially as the inverse cube of the radius.

9.3 Cold Plasma Populations in the Magnetosphere

Within the magnetosphere there are a number of different particle populations with different energies and densities. These arise from two sources, the solar wind and the ionospheric plasma. The location and energy of these sources is determined by complicated dynamic processes. There is strong ordering produced by the structure of the geomagnetic field. The exact composition of magnetospheric plasma is still being intensively studied.

Some of the plasma populations are of relatively low energy (a few electron volts or less) and approximately in thermal equilibrium (e.g. Chappell and Gringauz [24]). Some of these are described briefly below.

9.3.1 The Ionosphere

The ionosphere is the ionized region of the upper atmosphere lying between about 90 km and 500 km. It forms the inner boundary of the magnetosphere. It arises because the solar radiation is absorbed by the upper atmosphere, dissociating the molecules, and liberating free electrons. At great heights there are few molecules; at low heights most of the ionizing radiation has been absorbed. The result is a peak in the production of ionization corresponding to a particular absorption wavelength. Another important source of ionization at high latitudes is the precipitation of energetic charged particles. An equilibrium is reached in the upper atmosphere between the production, loss and diffusion of ionization. The result is the formation of a series of ionized layers in the upper atmosphere. For historical reasons they are labeled the D-, E- and F-layers. The F-layer is split into the F1- and F2-layers.

The plasma composition of the ionospheric layers is essentially determined by the local composition of the neutral atmosphere. Nitrogen and oxygen ions are important in the lower regions; helium becomes important above the F2 maximum; hydrogen dominates in the so-called protonosphere above about 900 – 1 000 km. Most of the ionospheric wave phenomena described in this book occur at frequencies very much greater than the ion gyrofrequencies. Thus only the electrons are of importance. Their density and temperature typically vary as shown in Figure 9.4, peaking at values of about 10^{12} m^{-3}. Their temperature is of the order

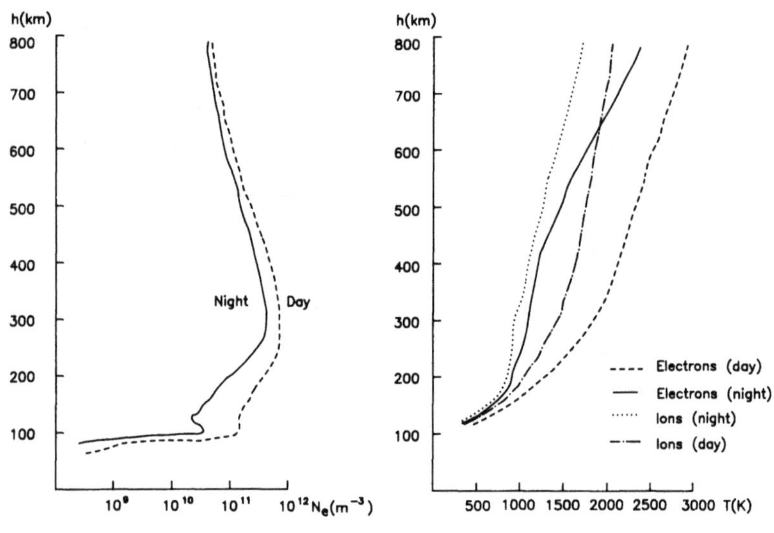

Figure 9.4: The ionosphere for typical mid-latitude conditions. (a) Electron density. (b) Electron and ion temperatures.

of 1000 K corresponding to energies of about 0.1 eV.

9.3.2 The Plasmasphere

The plasmasphere is formed in the region defined by those field lines which corotate with the earth. Its outer boundary is called the plasmapause. In Figure 9.3 the line separating the region of corotation from the outer convection region is the intersection of the plasmapause with the equatorial plane. Within this region cold positive ions, largely protons, and electrons are trapped on the corotating field lines. Their gravitational potential energy is comparable with their thermal energy. Under the influence of gravity and the geomagnetic field they can reach a state of diffusive equilibrium with the ionosphere so that the density in the plasmasphere is relatively high. Near the equatorial plane in the outer parts of the plasmasphere the electron density is of the order of 10^9 m^{-3}. It increases to ionospheric densities as the altitude decreases. Ion and electron temperatures are typically 1 000 – 2 000 K, corresponding to energies of 0.1–0.2 eV.

The mechanism setting up the equilibrium of the plasma along the field lines is called *ambipolar diffusion* [4]. The production of the ions is governed by processes in the ionosphere. They can move freely along the field lines but not across them. The lightest particles would diffuse to the greatest altitude if they were neutral; because of the charge any separation between electrons and ions sets up an electric field. The result is to tie electrons and ions together so that they behave like particles having the mean mass (essentially half the ion mass), hence the term *ambipolar*. The particles then set up an equilibrium under gravity while constrained to move along the field line. The net effect is that, in the upper ionosphere, the dominant ions are oxygen and hydrogen. At several hundred kilometers altitude there is a layer of helium. Above about 1000 km protons dominate.

Of course the actual behavior of the plasmasphere is dynamic; disturbed magnetospheric conditions change the convection pattern and field lines which formed part of the plasmasphere may be convected to the magnetospheric boundary where plasma is lost. Alternatively, as magnetic conditions quieten, field lines which formerly formed part of the outer convection pattern may move inside the corotation boundary. They then contain little plasma; plasma diffuses up from the ionosphere to fill the flux tubes on a time scale of many hours or days.

9.3.3 Cold Plasma Outside the Plasmapause

Outside the plasmapause the cold plasma does not corotate; it is convected to the magnetopause. The flux tubes expand to a very large volume and may, after reconnection, be connected to the solar wind plasma. In such circumstances a diffusive equilibrium is never reached and the cold plasma density is always much less than within the plasmasphere. In this region the ion density is of the order 10^6 m^{-3}. The history and source of these ions are still a matter of controversy and many more observations are necessary. Nevertheless it is now well-known that there are not only protons but also a significant population of heavy ions. These may be energized but details are beyond the scope of this book.

9.3.4 The Polar Wind

In the polar cap the cold plasma leaves the ionosphere in the form of an upward "polar wind". This consists chiefly of H^+ and He^+ ions. They may be significantly accelerated by the electric field associated with the charge separation under gravity of heavy oxygen and nitrogen ions and electrons.

9.4 Hot Plasma Populations

There are many energetic particle populations inside and outside the magnetosphere which are far from thermal equilibrium, although the Coulomb force still ensures that they are electrically neutral on scales longer than the Debye length. Because their mean free paths are extremely large, the times required to attain thermal equilibrium are long compared with time scales of interest in wave propagation. They have free energy associated with their non-equilibrium state and this is available for the excitation of waves. They are an important source of plasma waves and the wave-particle interactions are the main route by which such plasma populations can relax to equilibrium. Examples of such hot plasma populations are listed below. The locations of the various plasma regions of the magnetosphere are illustrated in Figure 9.5.

9.4.1 Solar Wind and Magnetosheath

The solar wind consists largely of protons and electrons flowing away from the sun with velocities of a few hundred kilometers per second. Their density is of the order of 10^6 m^{-3} and their energies are typically a few eV. The magnetic field is a few nanotesla. The energy density of the bulk flow is thus about one hundred times greater than the thermal energy and fifty times greater than the magnetic energy density. We see that the flow is both supersonic and super-Alfvénic and this results in the formation of the bow shock, a detached collisionless shock a few earth radii ahead of the subsolar point on the magnetopause. The plasma on passing through the shock into the magnetosheath is is reduced to subsonic, sub-Alfvénic flow. Its bulk velocity is reduced although it may still be of the order of several hundred km s^{-1}. The energy is increased to some hundreds of electron volts and the magnetic field is increased to some tens of nanotesla. The plasma is turbulent as a result of the passage through the shock. The feet of region II field lines define the extent of the polar cap. These field lines map to all parts of the magnetosheath.

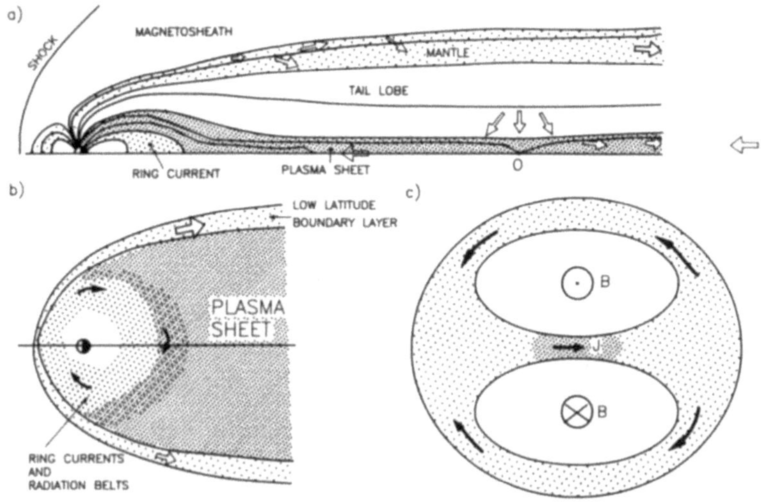

Figure 9.5: Hot plasma regions in the magnetosphere. Open arrows denote flow velocities. Solid arrows denote current densities. (a) Noon-midnight meridian. (b) Equatorial plane. (c) Cross section of tail as seen from the earth.

9.4.2 Boundary Layers

The magnetopause is not the sharp boundary shown in Figure 9.1. Even in this simple picture of a closed magnetosphere we expect diffusion of hot plasma across the boundary between the region of magnetic field and the region of field free plasma. The result is a boundary layer varying in thickness from a few hundred kilometers to about 1 earth radius, over which conditions change from magnetosheath to magnetosphere behavior. In the case of an open magnetosphere the magnetopause is no longer the delineation between the region of magnetic field and the region of solar wind plasma. Indeed, near the boundary the magnitude of the magnetosheath magnetic field may be comparable with the magnetospheric magnetic field. Nevertheless, across the boundary, well defined changes in magnetic field direction and plasma properties occur.

Observationally [83] the magnetopause can be defined on an outbound orbit as the first sharp turn of the magnetic field to a consistently different direction. The boundary layers are then, in general, observed to lie inside this. It should be noted that such a definition does not exclude the possibility that open field lines can lie inside the magnetopause. A region II field line originating at high latitude can follow the general shape of a region III field line for much of its length before abruptly changing direction and penetrating the magnetopause. Such a field line lies inside the magnetopause in a region of open field lines.

The boundary layer can be divided into a number of different plasma regions as illustrated in Figure 9.6 [83, 46, 84]. The properties of the

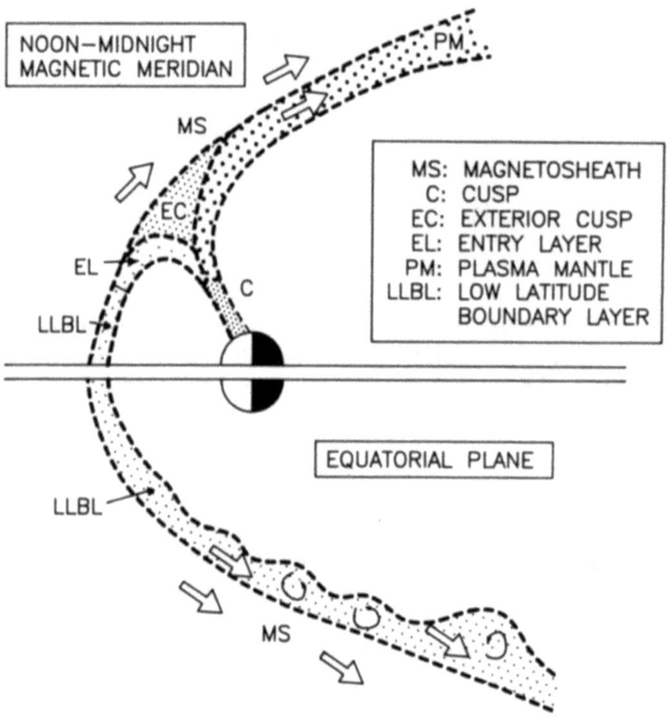

Figure 9.6: Plasma in the boundary layers of the magnetosphere.

plasma in these regions are in general similar to those of the magnetosheath plasma. The *polar cusp* or *cleft* is a region on the sunward side of the polar cap where plasma can easily enter the magnetosphere. The plasma in the exterior cusp is indistinguishable from the magnetosheath plasma and, indeed, lies outside the magnetopause as defined above.

The *entry layer* lies on the sunward side of the cusp. It occupies a limited region at latitudes near 60^0 and local times between 0800 and 1600. It contains plasma with proton and electron energy spectra very similar to magnetosheath plasma. It is distinguished from the magnetosheath plasma by a smaller and fluctuating flow velocity.

The *plasma mantle* is a high latitude boundary layer, extending backwards from the cusp to the distant tail. It shows ordered flow along the magnetic field with velocities comparable to those in the magnetosheath. As one moves outwards through the mantle towards the magnetopause proton energies increase from a few hundred eV to about 1keV. The distribution is anisotropic with $T_\perp > T_\parallel$. Electron spectra are similar to those in the magnetosheath.

The *low latitude boundary layer* extends from near the subsolar point backwards to the tail. It contains magnetosheath-like plasma. The flow velocity is roughly at right angles to the magnetic field. Eddies have been observed in the tail region.

9.4.3 Tail Lobes and Plasma Sheet

To a first approximation the magnetotail consists of two *lobes* containing tenuous plasma separated by a denser hotter *plasma sheet.* In the upper lobe the magnetic field is directed towards the earth and in the lower lobe away from it. Its magnitude in the lobes is about 20 nT. Such a field configuration is associated with a cross-tail current as shown in Figure 9.5c. This current is carried by the relatively hot plasma in the plasma sheet. For self consistency the sum of the magnetic pressure $B^2/2\mu_0$ and the plasma pressure must be constant which implies the existence of the sheet. The source of the plasma is the solar wind and it is convected from the mantle through the lobes by processes associated with the dynamo action of the solar wind. Observations of the sheet show that it is thin – only a few earth radii across. The plasma in it is fairly isotropic with an effective temperature of some 10^7 K. Proton energies are of the order of a few keV and electron energies some hundreds of eV. The particle density is about 10^6 m^{-3}.

Again this description is greatly oversimplified. In the first place there is a small component of magnetic field perpendicular to the sheet as shown in Figure 9.5a. Magnetic merging can take place at a neutral line. The sheet is very dynamic; its formation is associated with the storage of large amounts of energy in the tail. This energy is released during substorms when the cross-tail current is diverted through the ionosphere, the tail collapses to a more nearly dipolar configuration. Particles are energized and precipitated into the upper atmosphere at high latitudes, producing dramatic auroral displays. Energized plasma is injected into the inner magnetophere as described below.

9.4.4 Radiation Belts and Ring Current

We have seen how the dawn-dusk and corotation electric fields combine to define a region within which cold plasma is essentially trapped so that it effectively corotates with the earth. Hot plasma is subject to the same fields but also experiences other drifts due to the gradient and curvature of the magnetic field which modify the trapping region. Nevertheless the net effect is that there is an energy dependent trapping boundary within which energetic particles can be stably trapped to form the radiation belts. Within these belts injection and loss processes are on average in balance. The energies of protons in the inner regions extend to hundreds of MeV. The bulk of the particles have energies of the order of tens of keV.

At times of substorms particles from the magnetotail are injected into the radiation belt region. They lead to a large increase in plasma pressure and a resulting *ring current* which substantially modifies the geomagnetic field on the ground.

9.5 Waves in the Magnetosphere

9.5.1 Cold Plasma Waves

Many of the waves observed in the magnetosphere are cold plasma modes. It is useful to understand which of these modes can be propagated in the various regions of the magnetosphere. This can be done in terms of the classification of the CMA diagram.

Consider a schematic model in which the electron and ion densities and the magnetic intensity are functions of altitude. The conditions for cut-offs and resonances can be used to calculate f as a function of altitude for each CMA boundary. Let $f_P = \omega_P/2\pi$ be the plasma frequency, $f_e = \Omega_e/2\pi$ the electron gyrofrequency and $f_1 = \Omega_1/2\pi$, $f_2 = \Omega_2/2\pi$ the ion gyrofrequencies of ion species 1 and 2. Then the conditions for CMA boundaries in Tables 3.1 and 3.2 can be expressed in terms of frequency as shown in Table 9.1

If frequency is plotted against altitude, we find which CMA regions correspond to each frequency at each altitude. There are two essentially different cases. At low to medium latitudes the ion and electron densities fall off rather slowly with altitude above the $F2$-maximum of the ionosphere because of the presence of the plasmapause, with the consequence that the plasma frequency in the magnetosphere is everywhere greater than the electron gyrofrequency. At high latitudes the ion and electron densities fall off much more rapidly so that there is an extensive region above the ionospheric $F2$-maximum in which the plasma frequency is less than the electron gyrofrequency. The CMA regions are shown schematically for these two cases in Figures 9.7 and 9.8. They are drawn roughly to scale for typical conditions but there is no attempt to provide an accurate model. In Figure 9.7 which represents low latitudes and includes the plasmapause f_P is greater than f_e everywhere.

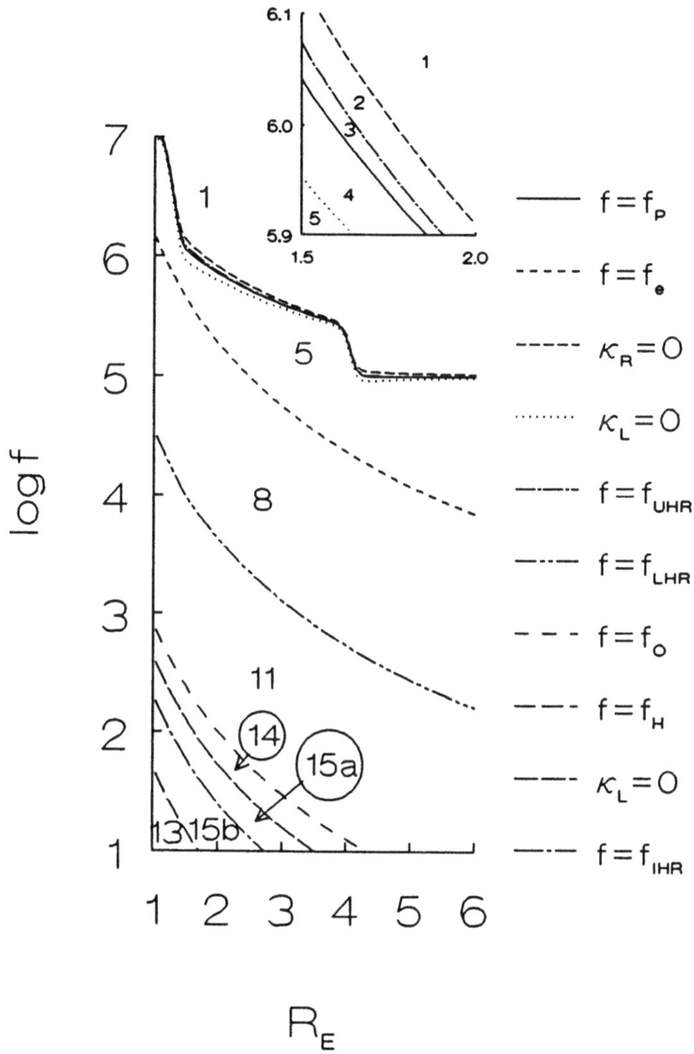

Figure 9.7: CMA regions at low latitudes.

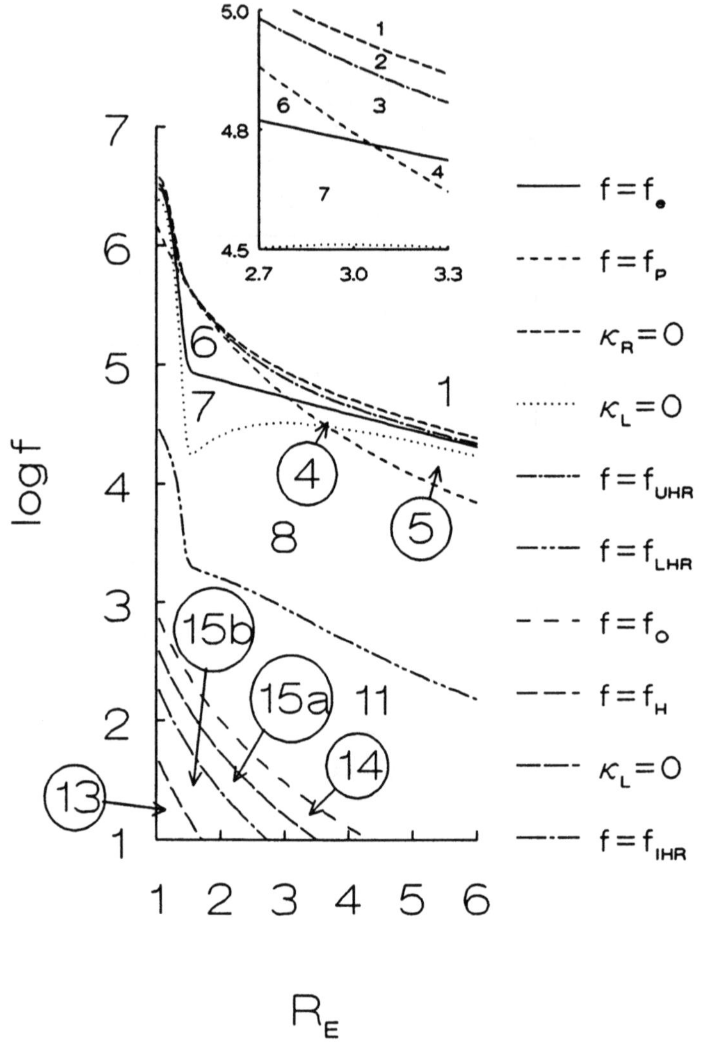

Figure 9.8: CMA regions at high latitudes.

Condition	Formula for Frequency
$\kappa_3 = 0$	$f = f_P$
$\kappa_R = \infty$	$f = f_e$
$\kappa_R = 0$	$f = \frac{1}{2}f_e + \sqrt{\frac{1}{4}f_e^2 + f_P^2}$
$\kappa_L = 0$	$f = \frac{1}{2}(f_1 - f_e) + \sqrt{\frac{1}{4}(f_1 + f_e)^2 + f_P^2}$ There are two useful approximations $f = -\frac{1}{2}f_e + \sqrt{\frac{1}{4}f_e^2 + f_P^2}$; $f_P^2 > f_e f_1$ and $f = f_1 + f_P^2/f_e$; $f_P^2 < f_e f_1$
$\kappa_1 = 0$	$f = \sqrt{f_e^2 + f_P^2}$ (Upper hybrid resonance) $f = \sqrt{f_e f_1 (f_P^2 + f_e f_1)/(f_e^2 + f_P^2)}$ (Lower hybrid resonance) $f = \sqrt{f_1 f_2}$ (Ion hybrid resonance)
$\kappa_L = \infty$	$f = f_1$; $f = f_2$

Table 9.1: Cut-off and resonance frequencies for a plasma with two species of positive ion.

The Free Space Modes

Two modes connect with free space. The ordinary free space mode can be propagated in regions 1, 2, 3 and 6. It is cut off at the plasma frequency f_P. Its polarization labeling is OL. If the frequency is high enough it can be propagated to the ground. At lower frequencies the ionosphere forms a barrier between it and the ground. The extraordinary free space mode can be propagated only in region 1. It is cut off at $f = \frac{1}{2}f_e + \sqrt{\frac{1}{4}f_e^2 + f_P^2}$ which is always greater than the ordinary mode cut-off. Its polarization labeling is RX.

The Whistler, Ion Cyclotron Whistler and Alfvén Modes

The classical whistler mode described by Storey [101] occurs in regions 7 and 8 of the CMA diagram. Its labeling is OR in regions 7, 8a and 8b. It changes to XR in region 8c. In regions 7 and 8 it has a resonance cone. In region 11 the whistler dispersion relation is modified by the effect of heavy positive ions.The whistler mode is unaffected by the ion gyroresonance. In a single ion plasma the refractive index surface is continuous across the ion gyroresonance into region 13. For frequencies very much less than the ion gyrofrequency it becomes the fast isotropic Alfvén wave. It can be seen that propagation in the whistler mode is possible over a wide band of frequencies in much of the magnetosphere.

In region 13 another mode appears. It is an open mode with a resonance cone known as the ion whistler (e. g. proton whistler or helium whistler depending on the ion species) or ion cyclotron wave. In a multi-ion plasma the situation is complicated and there is ambiguity in labeling the surfaces. This situation will be considered in the discussion of ion whistlers later.

The whistler mode is, in principle, cut off by the plasma resonance before it reaches the ground. In practice, for most frequencies of interest, the level at which this occurs is in the D-region of the ionosphere. Here collisions with neutral particles dominate and collisionless theory fails. Then the region between the bottom of the ionosphere and the level where collisions can be neglected and f is much smaller than f_P has a thickness which is small compared with typical whistler wavelengths. Whistlers penetrate it easily.

The Z-Mode

The Z-mode exists in regions 3, 4, 6 and 7. At low latitudes we see from Figure 9.7 that regions 3 and 4 are confined to a narrow band of frequencies near the plasma frequency while regions 6 and 7 are absent. At high latitudes the situation is different. In the region where the plasma frequency is less than the gyrofrequency it can be seen from Figure 9.8 that these regions occupy a relatively wide band of frequencies.

The Z-mode cannot reach the ground. If it is generated in the magnetosphere, as it encounters increasing electron density in the ionosphere, it encounters the $\kappa_L = 0$ cut-off. On the other hand, if it is generated below the $F2$ maximum, as electron density decreases with decreasing altitude, it encounters the upper hybrid resonance.

9.5.2 Hot Plasma Waves

Many of the hot plasma populations in the magnetosphere coexist with cold plasmas. They have unstable velocity distributions but are close to stability because of quasi-linear processes. In many cases these processes involve wave-particle interactions between cold plasma modes and resonant particles in the hot plasma distribution. Thus, even in the presence of hot plasmas, an important part of the problem is understanding the cold plasma waves in various regions of the CMA diagram.

Some of the waves occurring are intrinsically hot plasma waves arising from the properties of the anisotropic distributions discussed in Chapter 7. The classification of such waves is difficult because of the great richness of possible modes. For this reason study of them usually involves an *ad hoc* approach. A number of such waves are introduced in the following chapters.

9.6 Summary

- The magnetosphere is formed by the interaction of the earth's magnetic field and the solar wind. As a result of this interaction a number of different regions containing plasma with different characteristics are formed.
- Important cold plasma populations with different characteristics occur in the ionosphere, the plasmasphere, the high latitude regions and the polar wind.
- Important hot plasma populations occur in the solar wind, the boundary regions, the magnetotail lobes, the plasma sheet, the ring current and the radiation belts.
- The types of cold plasma wave which can exist at a particular frequency in a particular part of the magnetosphere can be classified in terms of the scheme of the CMA diagram.
- Many of the effects of hot plasma arise as a result of wave–particle interactions with cold plasma modes. Intrinsic hot plasma modes such as the Harris instabilities may also be important in hot plasma regions.

Chapter 10

Waves in the Plasmasphere – 1. Whistler Observations and Basic Theory

10.1 Introduction

The earliest examples of naturally generated magnetospheric plasma waves to be extensively studied were whistlers. They were first described in detail by Barkhausen [8]. Full understanding of their nature was not reached until the work of Storey [101]. The classic book by Helliwell [53] describes their properties, as understood in 1965, in great detail, and is still an excellent reference. It also contains an interesting description of the history of work in the very low frequency (v. l. f.) band until the beginning of the International Geophysical Year (IGY) in 1957.

Whistlers are phenomena in the v. l. f. band. Their source is a lightning stroke which emits a v. l. f. impulse. This impulse is strongly modified by propagation effects as it is propagated through the magnetosphere in the whistler mode. In this chapter we first consider some properties of the whistler mode which are relevant to understanding the observations. We also study related modes which are important in understanding the ion cyclotron whistlers discussed in Chapter 12. Observations of whistlers and other lightning induced phenomena are then described and a general outline is given of what is now understood about their propagation. An elementary theory is then presented which explains the major features of the observations. A more detailed treatment of each of the several stages in their propagation is given in Chapter 11.

10.2 Properties of the Whistler Mode

In this section we make extensive use of the nomenclature of the CMA diagram described in Chapter 3. We discuss the nature of propagation in the whistler mode above and below the ion gyrofrequency and derive appropriate formulae for the analysis of whistlers.

10.2.1 Refractive Index Surface and Ray Direction

For frequencies above the lower hybrid resonance the whistler mode occurs in regions 7 and 8 of the CMA diagram. In both these regions there is a resonance cone so that the refractive index surface is open; propagation perpendicular to the magnetic field does not occur. In region 7 the whistler mode coexists with the Z-mode. In region 8 it is the only propagated mode. Below the lower hybrid resonance frequency, in region 11, the shape of the surface is modified by the effect of positive ions. The result of this modification is that the resonance cone disappears and the refractive index surface is closed so that propagation perpendicular to the magnetic field becomes possible. Examples of the refractive index surface are shown in Figure 10.1. Other examples as well as the ray and group

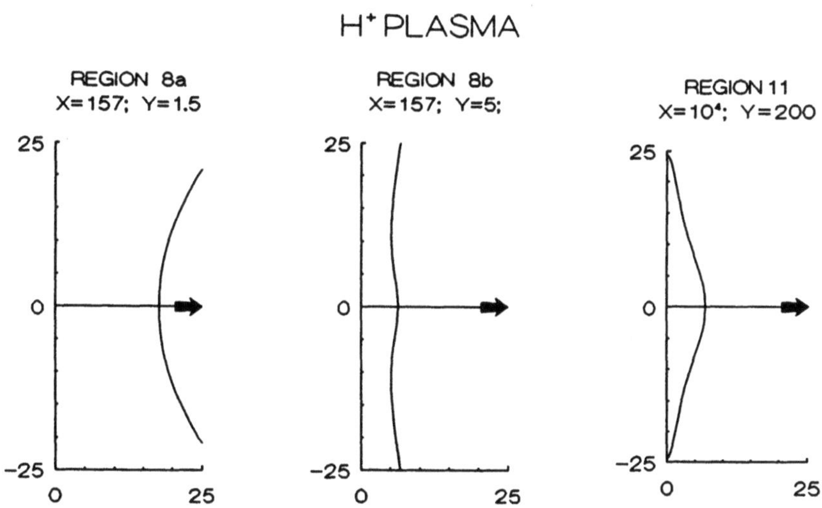

Figure 10.1: Whistler mode refractive index surface for an H^+ plasma in various regions of the CMA diagram. The heavy arrows show the magnetic field direction

velocity surfaces are shown in Figure 3.17. In much of region 7 and all of 8, the angle of the resonance cone is quite near 90^0. The result is that the ray direction, which is normal to the refractive index surface, makes a small angle with the magnetic field, no matter what the wave normal direction. In region 11 the refractive index surface is closed so that, for wave normal angles near 90^0, the ray may also be at right angles to the magnetic field.

In region 7 the polarization labeling of the whistler mode is OR. When propagated along the magnetic field the wave is right hand polarized.

Although it cannot be propagated at right angles to the field, because the coexisting Z-mode is an extraordinary mode for propagation at right angles to $\mathbf{B}_0$, the whistler mode is labeled O. By continuity, in region 8a, we still label it OR, although there is no way of establishing the O-polarization by considering the wave itself. In region 11, where the surface is closed, it turns out that the labeling is XR. It is convenient to regard the switch from O to X-polarization as occurring at the cross-over frequency between regions 8b and 8c.

10.2.2 Quasi-Longitudinal Propagation for an Electron Plasma

The highest ion gyrofrequency occurring in the magnetosphere is the proton gyrofrequency near the earth's surface. At 1000 km, where H^+ ions begin to dominate, this frequency is a few hundred hertz and drops off rapidly as the cube of the radius. Thus, in treating the propagation of waves in the v. l. f. range of a few kilohertz and above, the effect of positive ions can be ignored. In addition, over most of the magnetosphere, and especially within the plasmasphere, the plasma frequency is much greater than the gyrofrequency. The result is that we can always make the approximation $\omega_p^2 \gg \omega\Omega$ which implies $X \gg Y \sim 1$. The approximate expression for the refractive index 2.78 is then

$$n^2 \simeq -\frac{X}{1 - Y\cos\theta} \tag{10.1}$$

At lower L-values we can make an additional approximation $\Omega \gg \omega$ or $Y \gg 1$. In this case, over a wide range of angles such that $|\theta| \ll 1/Y$, we can use the *Storey approximation* [101]

$$n^2 = X/Y\cos\theta \tag{10.2}$$

In many cases the wave normal makes a relatively small angle with the magnetic field. If θ is less than about 15^0 we can set $\cos\theta = 1$ and use the strictly longitudinal form of these formulae.

The group refractive index in the longitudinal case can easily be found from

$$n' = n + \omega\frac{\partial n}{\partial\omega}$$

The result is

$$n' = \frac{1}{2}n\frac{Y\cos\theta}{Y\cos\theta - 1} \tag{10.3}$$

or, in the Storey approximation,

$$n' = \frac{1}{2}n \tag{10.4}$$

10.2.3 The Refractive Index Surfaces Below the Ion Gyrofrequency

In a single ion plasma the whistler mode is unaffacted by the ion gyroresonance; it has R-polarization and the ions resonate with waves with L-polarization. Thus in region 13 the whistler refractive index surface is similar to that in region 11. In region 13 another wave, the ion cyclotron whistler, appears. The refractive index surface for this wave is similar to that for the whistler mode in region 8 but it has L-polarization. At frequencies very much lower than the ion gyrofrequency the whistler mode becomes isotropic and non-dispersive and propagates with the Alfvén speed; at these frequencies it is simply the fast Alfvén wave. The ion cyclotron whistler, on the other hand, becomes the slow Alfvén wave at these ultra low frequencies.

The situation in a plasma with more than one species of positive ion is much more complicated. As an example consider the two-ion plasma of Figure 10.2 This shows the refractive index surfaces in successive regions of the CMA diagram as the frequency changes from a value above the upper to below the lower ion gyrofrequencies. The positive ions are 50% protons and 50% HE^+. The additional CMA regions 14, 15a and 15b occur between the ion gyrofrequencies. In region 11 only the XR whistler mode exists. The refractive index surface is closed. As the proton gyroresonance is crossed the whistler mode is unaffected and another surface, the proton whistler surface, appears. This slow wave has OL polarization and the refractive index surface is open. The boundary between regions 14a and 14b is the cross-over boundary $\kappa_2 = 0$. On this boundary the two surfaces touch at $\theta = 0$; for longitudinal propagation the polarization and speed of both modes are the same and there is coupling between them. The polarization is linear and perpendicular to the plane containing the wave normal and magnetic field direction; it can be synthesized by superposing oppositely circularly polarized waves with the same speed. On the other side of this boundary, in region 14b, the polarization label of the fast X wave with a closed refractive index surface has changed from R to L and that of the slow O wave with an open surface from L to R. There is an ambiguity in how to identify the whistler mode. It is conventional to label the R-surface as the whistler mode and the L-surface as the proton whistler. The reason for this will be discussed later. The proton whistler is cut off at the boundary between regions 14 and 15a. The whistler is the only mode in regions 15a and 15b; it is open in 15a and closed in 15b. The behavior in these regions is analogous to the behavior in regions 8 and 11. Again across the helium gyrofrequency the whistler is unaffected and the ion cyclotron whistler which appears in region 13 is now the He^+ whistler. The behavior in this region is the same as for a one-ion plasma. The generalization to many ion species is

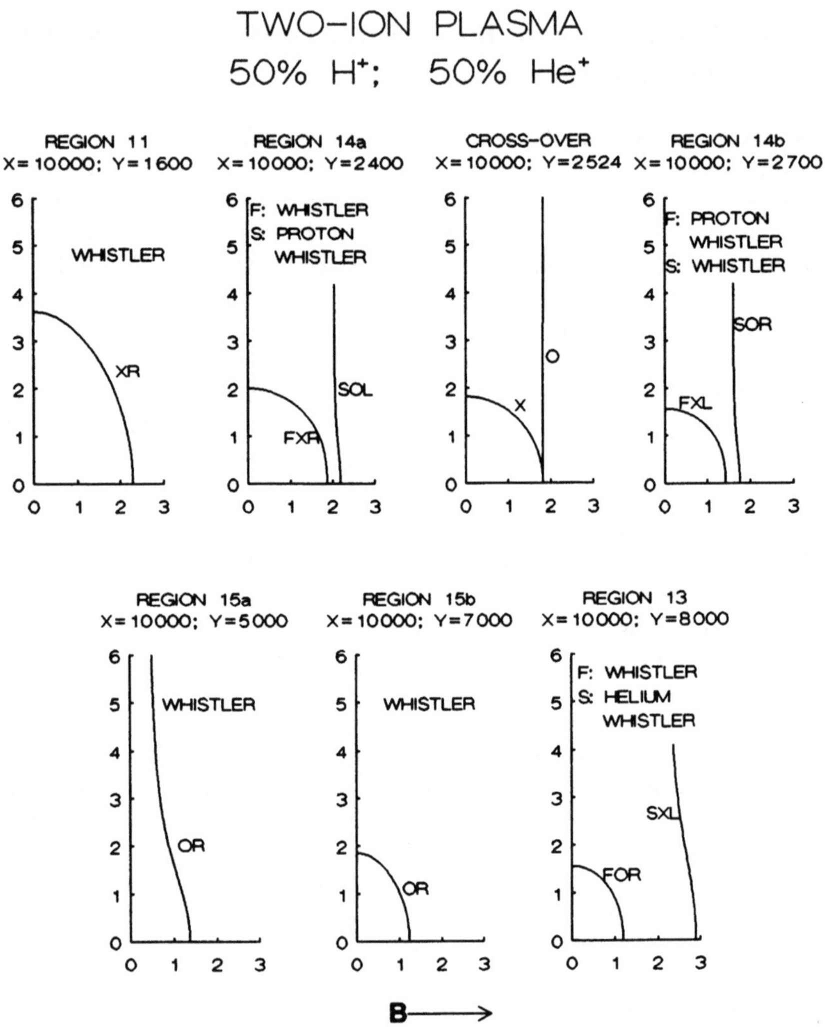

Figure 10.2: Refractive index surfaces in the regions of the CMA diagram near the ion gyrofrequencies in a plasma with positive ions consisting of 50% protons and 50% helium ions.

obvious. There is a set of regions like regions 14, 15a and 15b between each pair of ion gyrofrequencies.

At the cross-over frequency $\kappa_2 = 0$. Thus $\kappa_L = \kappa_R = \kappa_1$. The equation for one of the refractive index surfaces then becomes $n^2 = \kappa_1$; it is a sphere of radius $\kappa_1^{1/2}$. The other surface is an open surface given by $n^2 = \kappa_1\kappa_3/(\kappa_1 \sin^2\theta + \kappa_3 \cos^2\theta)$. It touches the first surface at $\theta = 0$.

10.3 Observations of Whistlers

10.3.1 Ground Based Observations

At medium latitudes a sensitive broad band (2 – 20 kHz) audio amplifier connected to an appropriate antenna receives a variety of signals. The whistler is one of the most distinctive types. It is thus named because, if a loud speaker is attached to the amplifier, it sounds like a descending whistle lasting for about a second or less. A dynamic spectrum of such a signal is shown in Figure 10.3. The lightning strokes which are the

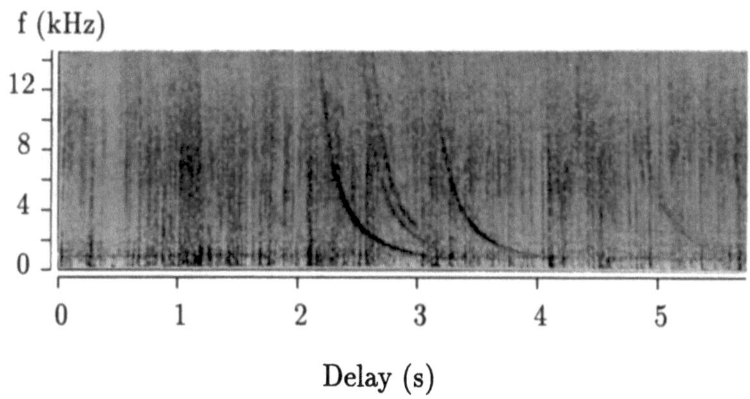

Figure 10.3: Dynamic spectrum of a low latitude whistler observed at Durban (Geomagnetic Latitude 31^0 S, $L = 1.36$). A Fourier analysis of the broad band signal is performed every 10 ms and the spectrum plotted vertically as a function of frequency. Successive spectra are plotted as a function of time. The grey scale represents the magnitude of the power spectrum.

sources of such signals provide a broad band impulse seen in the figure as a vertical trace. The impulse is propagated along the magnetic field through the plasmasphere to the opposite hemisphere. Because of the highly dispersive nature of the medium different frequencies have different delays and as a result the impulse is smeared out into a whistling tone. A detailed picture of the process as now understood is given in Figure 10.4. The following is a brief description of each stage in the figure. Greater

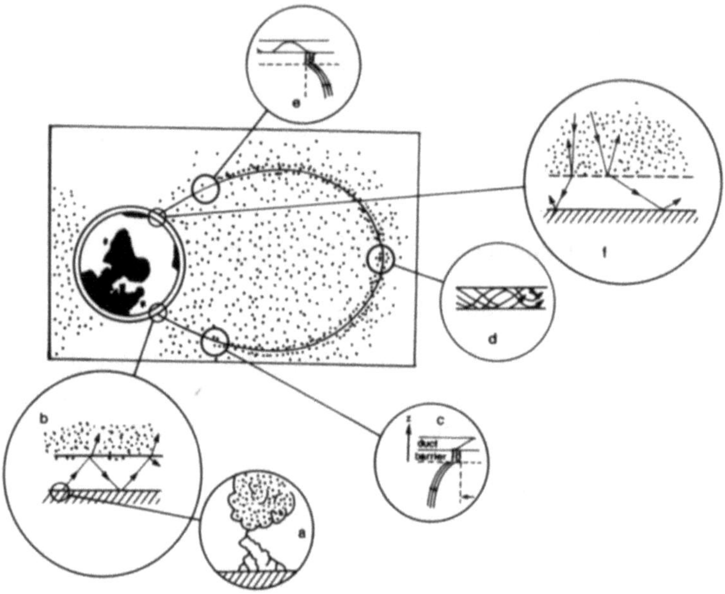

Figure 10.4: Schematic representation of key processes in the generation and propagation of ducted whistlers [114]: (a) radiation from lightning stroke; (b) excitation of earth-ionosphere cavity and lower ionosphere; (c) excitation of duct; (d) guiding in duct; (e) leakage from duct; (f) propagation through lower ionosphere into earth–ionosphere cavity.

detail is given in the remaining sections of the chapter.

1. The whistler is initiated by the electromagnetic radiation generated by a lightning stroke (11.2). This has a peak in the v. l. f. range (1–40 kHz). The duration of the impulse is a few milliseconds. Other higher frequencies radiated by the source are of no consequence to whistler propagation.

2. The signal excites waveguide modes in the earth–ionosphere cavity (§11.2). These may be propagated large distances almost undispersed except near the cut-off frequency where there is some additional delay. The ionosphere is not a perfect reflector. Some of the energy in the waveguide leaks into the ionosphere and is propagated upwards (§11.4). As discussed below, the whistler mode is a slow wave and highly dispersed. Whatever the wave normal direction the ray makes a small angle ($< 20^0$) with the magnetic field. The signal thus roughly follows the magnetic field direction. If there

are no ducts the ray is refracted through the magnetosphere as an unducted whistler (§11.5).

3. If field aligned ducts of enhanced ionization are present they may be excited by the whistler mode signal; part of the signal energy is trapped and guided along the field line (§11.6).

4. The energy trapped in the duct is accurately guided to the opposite hemisphere. The mechanism is similar to that involved in light propagation in optical fibers (§11.6).

5. Energy leaks from the duct with wave normal making a small angle with the geomagnetic field. The wave is thus propagated downwards.

6. The earth–ionosphere waveguide is excited by the downgoing whistler. A substantial fraction of the energy is reflected. The signal below the ionosphere may be propagated some distance and received by a station well removed from the foot of the duct.

At higher latitudes the variety of signals is much richer. Whistlers are still very prominent. Their behavior is more complicated, often consisting of multiple traces with both rising and falling tones, as shown in Figure 10.5. The multiple traces arise from multiple ducts, giving many propagation paths with different delays. The "nose" structure of each trace arises from the nature of the dispersion relation (§10.4).

Whistlers may undergo more than one hop as energy reflected at the ionosphere is propagated back along the duct. A whistler which has been propagated from one hemisphere to the other is a *single hop* whistler. A whistler which is reflected at the ionosphere and received in the same hemisphere in which it is generated is a *double hop* whistler. Multihop whistlers (see Chapter 13) are also observed.

10.3.2 Satellite Observations

Whistlers can also be observed with broad band receivers on spacecraft. Generally spectrograms of the data from such instruments show a very rich variety of phenomena, including, on occasion, whistler signals which appear very similar to those observed on the ground. An example is shown in Figure 10.6. The satellite was on a field line of low L-value. The data feature a number of unducted whistlers with quite small dispersion extending over the whole frequency band. These are unducted whistlers which have reached the satellite after a fractional hop. Near the beginning of the trace is a whistler of larger dispersion. This is an unducted whistler from the opposite hemisphere which has traveled over a longer path. An interesting feature occurs below 0.5 kHz. This is a set of traces which

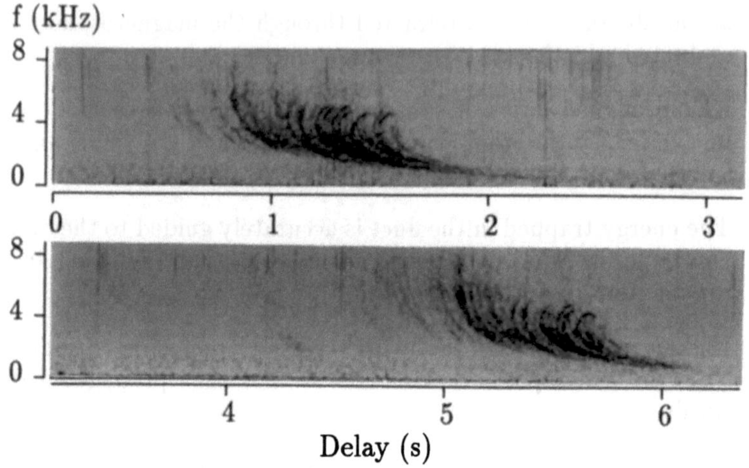

Figure 10.5: Dynamic spectrum of a high latitude whistler from Sanae, Antarctica showing many components.

show a nose-like feature and which appear to grow out of the whistler trace. The delay becomes infinite at the proton gyrofrequency (277 Hz). These are examples of *proton whistlers* which are produced by positive ion effects on propagation. There is also a band of noise *proton hiss* near 1 kHz. Note that the intensity modulation is due to the satellite spin. Proton whistlers were first observed by Smith *et al.* [96]. Analogous helium whistlers were first reported by Barrington *et al.* [9]. Ion cyclotron whistlers of this type are discussed later in Chapter 12.

10.4 Elementary Theory of Whistlers

As we shall see (§11.2), the signal emitted by a lightning stroke in the v. l. f. range is a short broad-band pulse with a duration of about 10 ms. That part of it which penetrates the ionosphere travels with the group velocity, which depends strongly on frequency. Different frequencies have different speeds so that the signal is highly dispersed. In many cases ducts guide the signal along the magnetic field so that the wave normal is always approximately field-aligned. It is thus useful to present an analysis of the propagation of pulses of whistler mode waves which are strictly field aligned.

Because the wave normal is parallel to the magnetic field the group velocity is in the same direction as the phase velocity. The time interval dt' taken by a wave packet to travel a distance ds is $dt' = ds/V_G = \frac{1}{c}n'\,ds$.

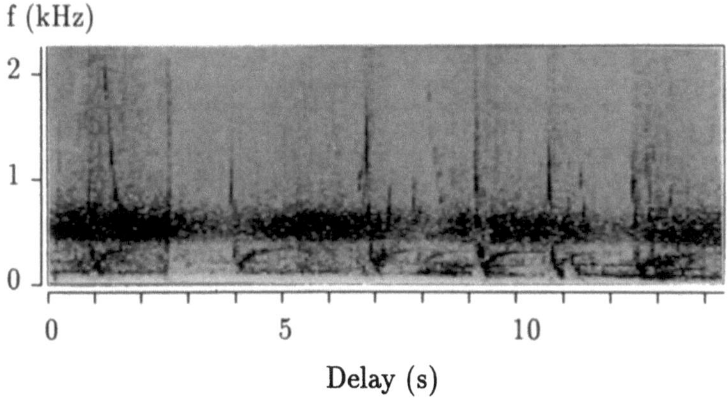

Figure 10.6: Proton whistler data from ISIS2, 13:21 UT, Oct. 13 1976. Altitude 2178 km. Latitude 2.8^0. Longitude -71.6^0. Magnetic Latitude 14.5^0. Magnetic dip 28^0. $L = 1.36$. Electron gyrofrequency is 511 kHz. Proton gyrofrequency 277 Hz.

Thus the time delay over the whole path is

$$t' = \frac{1}{c} \int_{\text{path}} n' \, ds \tag{10.5}$$

10.4.1 Low and Mid-Latitude Whistlers – Dispersion

Whistlers observed at low latitudes have been propagated along field lines on which both plasma frequency and gyrofrequency are large compared with wave frequency. The Storey approximation for the group refractive index is applicable. The delay (10.5) can then be written

$$t' = D/\sqrt{f} \tag{10.6}$$

where D is called the *dispersion*. In the Storey approximation it is a constant for a particular whistler path and is given by

$$D = \frac{1}{\sqrt{8\pi\mu_0 e}} \int_{\text{path}} \sqrt{N(s)/B_0(s)} \, ds \tag{10.7}$$

In Figure 10.7 we show a plot of delay t' against the square root of the frequency for the low latitude whistler shown in Figure 10.3. As expected

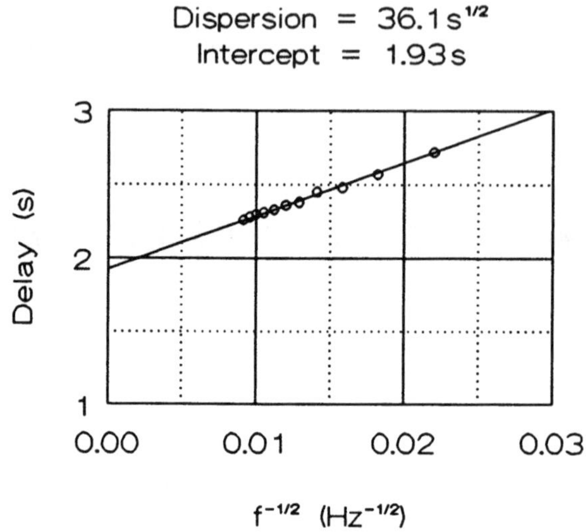

Figure 10.7: Dispersion analysis of the whistler shown in Figure 10.3

it fits a straight line very accurately, showing that Storey's assumption of longitudinal propagation is a close approximation. The spherics which initiated the various whistlers can be seen as vertical traces on the dynamic spectrum. The intercept of the straight line cuts the t' axis at the position of the initiating spheric for the whistler which has been analysed. The slope of the straight line is the dispersion. This is proportional to $\int_{\text{path}} \sqrt{N(s)/B_0(s)}\, ds$. If we assume that the duct along which the whistler was propagated coincides with the field line which intersects the ground at the receiving station then the magnetic field along the path is known. We can regard $1/\sqrt{B_0}$ as a weighting factor. Then the weighted mean of the electron density along the path is

$$\langle N \rangle = \frac{\int_{\text{path}} \sqrt{N(s)/B_0(s)}\, ds}{\int_{\text{path}} 1/\sqrt{B_0(s)}\, ds} \tag{10.8}$$

It was this which allowed Storey to estimate the electron density along the path of mid-latitude whistlers in England and discover that it was very much greater than expected. This was the first observation of the enhanced plasma density in the plasmasphere.

10.4.2 High Latitude Whistlers – The Nose Frequency

At higher latitudes near the plasmapause the magnetic field near the equatorial plane is small enough so that the gyrofrequency is comparable with the wavelength. In this case $X \gg Y^2 \sim 1$ and we get

$$t' = \frac{1}{2\omega^{1/2}} \int_{\text{path}} \frac{\Omega \omega_P}{(\Omega - \omega)^{3/2}} \, ds \tag{10.9}$$

In a uniform medium it is easy to see that the group velocity is maximum at $\omega = \frac{1}{4}\Omega$. Because Ω varies substantially along the path the frequency component having maximum speed varies along the path. The net effect, when integrated along the path, shows that there is a particular frequency which arrives first. In Figure 10.8 we show a number of synthetic whistlers computed in a simple model of the plasmasphere using this formula. The electron density distribution assumes diffusive equilibrium [4] with $\omega_P = \omega_{P,\,eq.} \exp(-2.5/L\cos^2\theta)$ and dipole field with $\Omega = \Omega_{eq.}\sqrt{4 - 3\cos^2\theta}/L^3\cos^6\theta$. It can be seen that there is a well defined *nose frequency*, the frequency of minimum time delay. The upper cut-off frequency of the computed whistlers is half the minimum gyrofrequency on the path. This is because propagation in enhancements of ionization is only possible below half the gyrofrequency (§11.6). The lower panel of Figure 10.8 shows the nose frequency and the upper cut-off frequency as functions of L.

The analysis of such whistlers is not as easy as the low latitude case. Various fitting formulae can be used . On empirical grounds Dowden and Allcock [33] have shown that the reciprocal of the dispersion of a nose whistler is linear in f. This leads to an expression for the delay of the form

$$t' = t_0 + \frac{1}{\sqrt{f}} \frac{D_0}{1 - f/af_n} \tag{10.10}$$

where D_0 is the zero order dispersion, t_0 the position of the initiating spheric and f_n the nose frequency. This formula gives a minimum delay where $d\tau/df = 0$ leading to $a = 3$. Walker [113] has discussed the derivation of this formula. The Storey formula is an approximation for small f. We therefore expand (10.9) in powers of f/F, the ratio of frequency to gyrofrequency, for a better approximation, getting

$$D = \frac{1}{2c}\left\{I_0 - \frac{3I_1}{2}\left(\frac{f}{F}\right) + \frac{15I_2}{8}\left(\frac{f}{F}\right)^2 + \ldots\right\} \tag{10.11}$$

where I_0, I_1, I_2 are proportional to f_p and are various integrals of moments of the density distribution about the field line. In a dipole magnetic field (Appendix C) with a plasma in diffusive equilibrium they are

$$\begin{aligned}
I_0 &= R_E f_{p,\text{eq}} \int_0^\Lambda \cos^4\theta(4 - 3\cos^2\theta)^{1/4} e^{5/2L\cos^2\theta} \, d\theta \\
I_1 &= R_E f_{p,\text{eq}} \int_0^\Lambda \cos^{10}\theta(4 - 3\cos^2\theta)^{-1/4} e^{5/2L\cos^2\theta} \, d\theta
\end{aligned}$$

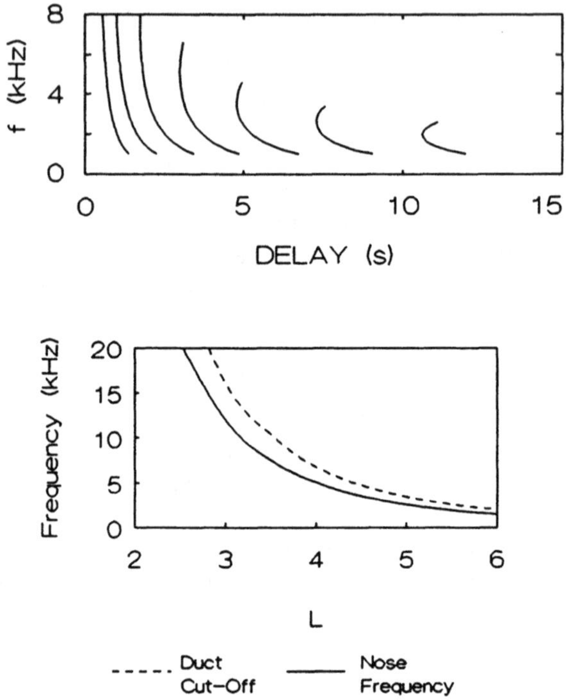

Figure 10.8: Synthetic dynamic spectra computed for whistlers propagated along different field lines in a dipole field model with plasma in diffusive equilibrium. The delays shown are for propagation along the field lines L = 2.5, 3.0, 3.5, 4.0, 4.5, 5.0 and 5.5. The lower diagram shows the nose frequency as a function of L as well as the cut-off frequency.

$$I_2 = R_E f_{p,\text{eq}} \int_0^{\Lambda} \cos^{16}\theta (4 - 3\cos^2\theta)^{-5/4} e^{5/2L\cos^2\theta}\, d\theta$$

where $f_{p,\text{eq}}$ is the equatorial plasma frequency and Λ the latitude at which the field line reaches the ground. When $f \sim f_n$ many terms are needed for sufficient accuracy. The Dowden and Allcock expression is an excellent approximation over a wide band of frequencies. To see why this is so we can regard $f_p/\sqrt{F}$ as a weighting function and we can then express I_1 and I_2 in terms of I_0 and weighted values of F^{-1} and F^{-2}.:

$$I_1 = \left\langle \frac{1}{F} \right\rangle I_0 ; \quad I_2 = \left\langle \frac{1}{F^2} \right\rangle I_0 \tag{10.12}$$

If (10.11) is inverted we get

$$\frac{1}{D} = \frac{2c}{I_0}\left\{1 + \frac{3}{2}\left\langle\frac{1}{F}\right\rangle\frac{f}{F} + \frac{3}{8}\left[6\left\langle\frac{1}{F}\right\rangle^2 - 5\left\langle\frac{1}{F^2}\right\rangle\right]\left(\frac{f}{F}\right)^2 + \dots\right\} \tag{10.13}$$

The quantities $\langle 1/F\rangle^2$ and $\langle 1/F^2\rangle$ are positive and do not differ much in magnitude for any reasonable plasma distribution. Thus the second order term is negligible even when $f/F \sim 0.5$. The assumption that D^{-1} is linear in f is thus well justified.

Another fitting formula is due to Bernard [10] who deduced it with a theoretical justification. It is

$$t' = t_0 + \frac{D_0}{\sqrt{f}}\frac{F_{eq} - Af}{F_{eq} - f} \tag{10.14}$$

where

$$A = \frac{(3f_n - F_{eq})F_{eq}}{f_n(f_n + F_{eq})} \tag{10.15}$$

In either of these formulae there are unknown parameters D_0 and t_0. In the Bernard formula there is also a parameter F_{eq}. However a in the Dowden Allcock formula is essentially a measure of the same quantity and can be regarded as determined by the data. They found it to be 3.1 ± 0.04. There are a number of methods by which the formulae can be fitted to the data to find the unknown parameters and hence, from a model, the field line on which the wave has been ducted and the equatorial electron density.

10.5 Use of Whistlers as a Magnetospheric Probe

The earliest use of whistlers as a probe was Storey's [101] early estimate of the electron density at a few earth radii described in §10.4. At the time it was assumed that the electron density fell off at the same rate as the neutral atmospheric density and was therefore negligible above a few hundred kilometers. This was the first observation of a significant cold plasma density in what is now known to be the magnetosphere.

A second dramatic discovery was Carpenter's [21] observation of a *knee whistler* in data from Siple Antarctica. In Figure 10.5 the delay of the nose increases with decreasing nose frequency. In Carpenter's data the trace with the lowest nose frequency had a much smaller delay. He attributed this to the fact that this part of the signal had been propagated in a region where the background plasma density was reduced by nearly two orders of magnitude. He thus had discovered the existence of the plasmapause, soon confirmed by satellite observations.

Since then the motion of ducts has been tracked by observing the changing nose frequencies in successive whistler groups [22, 89]. This has allowed the large-scale electric field to be deduced since the observed velocity $\mathbf{E} = -\mathbf{v} \times \mathbf{B}_0$ and the magnetic field is known. Further advances in this technique have been made with observations of the point at which the signal emerges from the ionosphere, using direction finding equipment.

10.6 Summary

- Propagation in the whistler mode in the magnetosphere is possible over a wide band of frequencies in the v. l. f. range and lower.
- Observations of v. l. f. signals on the ground and in satellites show phenomena called whistlers which are discrete signals with a characteristic dependence of delay time on frequency.
- The source of a whistler signal can be identified as the radiation emitted by a lightning stroke.
- The features of whistlers can be well explained by assuming an impulsive broad band source which emits v. l. f. waves. These are guided precisely along the earth's magnetic field. Cold plasma propagation theory can then be used to predict their delay with great accuracy.
- Additional whistler-like phenomena observed by satellite receivers are associated with the effects of heavy positive ions. They are called ion cyclotron whistlers.
- Theoretical fits can be made to whistler observations to deduce the path on which the signal has traveled and the electron density along the path. Whistler observations were the first to show enhanced plasma density in the plasmasphere and the existence of the plasmapause.

Chapter 11

Waves in the Plasmasphere – 2. Details of Whistler Propagation

11.1 Introduction

In this chapter we consider in more detail the various stages in the generation and propagation of a whistler which are outlined in Chapter 10. We study the nature of radiation from lightning, the propagation of spherics below the ionosphere and the propagation of ducted and unducted whistlers in the magnetosphere.

11.2 Lightning as a Source of Electromagnetic Radiation

Lightning strokes are strong sources of electromagnetic radiation, particularly in the v. l. f. frequency band. The properties of this radiation are summarized in detail by Volland [108] and Lewis [75]. Figure 11.1a is an idealization of data presented by these authors. It shows the spectral density of the electric field of the wave emitted by the lightning stroke normalized to a distance of 10 km. The impulse is typically about 10 ms long. Part of it penetrates the ionosphere and is propagated upwards. The matching is relatively poor. Figure 11.1b [R. Friedel, private communication] shows an estimate of the intensity above the ionosphere expressed as attenuation relative to the intensity of the 1 kHz component at the ground. The remainder of the signal may travel large distances in the earth–ionosphere waveguide, being reflected back and forth between the conducting ionosphere and the ground.

The radiation is strong in the v. l. f. band between about 1 and 10 kHz. It is also this band which is least attenuated in passing through the ionosphere. Thus there is a substantial source of energy of broad band v. l. f. radiation which can be propagated into the magnetosphere.

11.3 Propagation in the Earth-Ionosphere Waveguide

Most of the radiation emitted by a lightning stroke is propagated by being reflected back and forth between the lower ionosphere and the

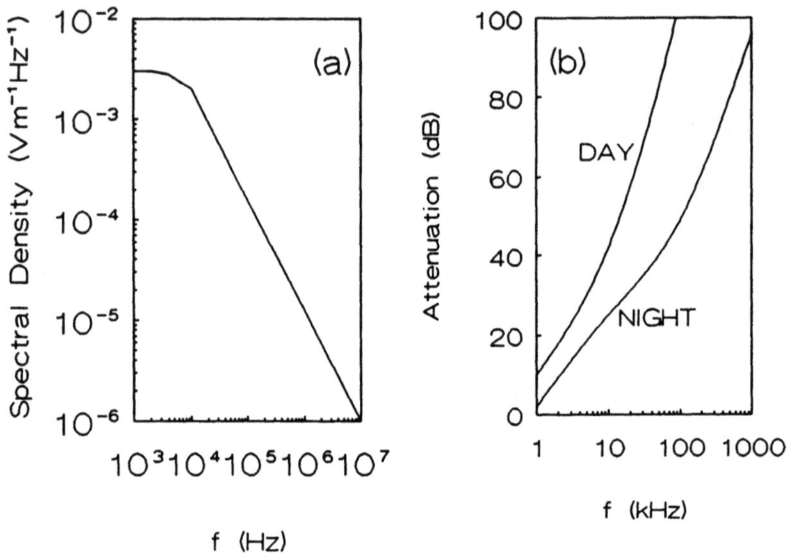

Figure 11.1: Properties of radiation from lightning stroke. (a) Idealized spectral density of electric field. (b) Relative intensity of radiation penetrating the ionosphere normalized to amplitude of 1 kHz signal at the ground.

ground. Over short distances the propagation can be studied by considering the rays. Over large distances with multiple reflections there are many paths by which a signal can reach a particular point. Furthermore the distance between ground and ionosphere is comparable with a wavelength and interference between the waves which have been propagated along different paths complicates the problem severely. In such circumstances waveguide mode theory is more appropriate. The theory of waveguide mode propagation in the earth–ionosphere cavity is a complex topic to which whole books have been devoted [18, 109]. We shall treat it very briefly in a simple approximation.

The ground and the lower ionosphere both act as imperfect conductors. Ohmic losses occur both in the ground and the ionosphere. In addition some wave energy is transmitted through the ionosphere and therefore lost to the waveguide. If the losses are small we can, to a first approximation treat the earth–ionosphere cavity as a waveguide bounded by two plane perfect conductors.

Let the x-axis be horizontal and the z-axis vertical. Propagation in the free space between the conductors is governed by the dispersion relation

$$\omega = c\sqrt{k_x^2 + k_y^2} \tag{11.1}$$

The horizontal electric field must be zero at the conductors. One possible polarization has the electric field in the y-direction and is called transverse electric or *TE* polarization. The other has the magnetic field in the y-direction and is called transverse magnetic *TM* polarization. Consider the *TE* case. The electric field in the wave must be of the form

$$E_y = E_0 \sin k_z z e^{i(k_x x - \omega t)} \tag{11.2}$$

and the boundary conditions require that

$$k_z = n\pi/h, \quad n = 1,2,3\ldots \tag{11.3}$$

where h is the height of the reflecting layer of the ionosphere. The dispersion relation then requires

$$k_x = \sqrt{\frac{\omega^2}{c^2} - k_z^2} = \sqrt{\frac{\omega^2}{c^2} - \frac{\pi^2 n^2}{h^2}} \tag{11.4}$$

There are thus a number of waveguide modes corresponding to the integers n, each having a different phase speed $V_P = \omega/k_x(n)$. It is trivial to show that the group speed $V_G = d\omega/dk$ is given by

$$V_G = c/V_P \tag{11.5}$$

If the finite conductivity and leakage of the modes are taken into account each mode suffers a different attenuation. After the wave has been propagated a great distance the least attenuated mode will dominate. This mode, under a wide variety of conditions, is the *TE* mode with $n = 1$.

An impulse emitted by a lightning stroke travels with the group speed. Its delay is $\tau = d/V_G$ where d is the distance traveled in the guide. A wave traveling the same distance in free space suffers a delay of $\tau_0 = d/c$. The ratio τ/τ_0, expressed as a function of frequency, is

$$\frac{\tau}{\tau_0} = \frac{1}{\sqrt{1 - (c/2fh)^2}} \tag{11.6}$$

Thus after the signal has traveled a great distance a spectrogram will have the appearance shown in Figure 11.2. There is a waveguide cut-off frequency where $f + c/2h$. Near the cut-off the delay is long. At frequencies well above the cut-off the signal travels almost at the free space speed of light. The result is a signal appearing as a near vertical line on a spectrogram with a small tail near the cut-off frequency. Examples of such *spherics* which appear in spectrograms are called *tweeks*. Often spherics arise from nearby lightning strokes in which case they do not show any waveguide cut-off.

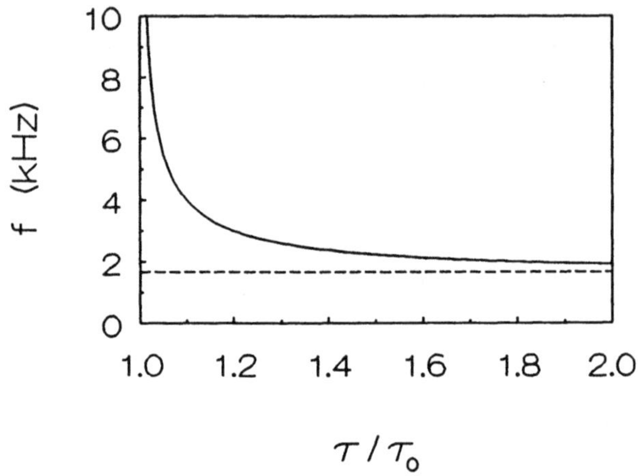

Figure 11.2: Ratio of waveguide propagation time, τ, for lowest order mode to free space propagation time τ_0 in the earth–ionosphere waveguide. $h = 90$ km.

11.4 Transmission of Whistlers Through the Ionosphere

In this section we consider a very crude picture of the penetration of the ionosphere by whistler mode waves, both from above and below. Because the scale lengths of the medium are comparable with a wavelength and collisions may be important many of the approximations we shall use are not very accurate. They are, however, adequate for our purpose since the ionosphere forms only a minute fraction of a whistler's total path. A proper treatment of the penetration of the ionsphere by v. l. f. waves requires a full wave, numerical treatment [87].

Waves in the 2–10 kHz range have free space wavelengths between 150 km and 30 km. Thus the earth-ionosphere cavity has dimensions comparable with wavelength and propagation in this region is often treated using waveguide mode theory. In the ionosphere above the E-region the plasma frequency is typically several megahertz and the gyrofrequency is of the order of 1 MHz. For waves with frequencies of a few kilohertz it can be seen that the refractive index is very large. Typical values are 20–100. The wavelength in this region is much smaller. The length scale of variation of electron density is some tens of kilometers and, insofar as it can be regarded as large compared with the wavelength, ray tracing is appropriate. This approximation is not particularly good but is adequate for our purposes.

In the D- and lower E-regions the collision frequency between electrons and neutral molecules is larger than the gyrofrequency and the collisionless formulae for the refractive index cannot be used. The thickness of these regions is, however, small compared with the free space wavelength and comparable with the wavelength in the ionosphere above. It is possible without severe inaccuracy to regard them as equivalent to a sharp boundary in a simple treatment.

Consider a wave incident from below on a horizontally stratified ionosphere with a large refractive index, n, at an angle of incidence θ_i. The angle of the wave normal in the ionosphere, by Snell's law, is given by

$$\sin\theta = \frac{1}{n}\sin\theta_i$$

and, since n is large, θ is small. This angle θ defines the half angle of the *acceptance cone.* So long as the medium is horizontally stratified the wave normal (but not the ray) is confined to a small cone of angles near the vertical. As altitude increases above the $F2$ maximum the assumption of horizontal stratification is no longer valid and the wave normal may make large angles with the vertical.

A wave incident on the ionosphere from above suffers total internal reflection unless its wave normal lies within a narrow acceptance cone. A relatively small fraction of the total wave energy its transmitted to the earth–ionosphere waveguide.

11.5 Propagation of Unducted Whistlers

After the whistler has penetrated the ionosphere it is refracted upwards. It may get trapped in a duct. Otherwise its path is determined by refraction in the plasmasphere. Its path can be found by using the ray tracing methods of Chapter 4.

Storey [101] in seeking to explain early observations of medium latitude whistlers in the frequency range 2–10 kHz similar to that in Figure 10.3 noted that, in regions 8b and 8c of the CMA diagram, the angle between the ray and the magnetic field was relatively small, even when the wave normal angle was large. In the region of interest the Storey approximation was excellent. In this approximation it can easily be shown that the angle between ray and magnetic field is given by

$$\tan\beta = \frac{\tan\theta}{2+\tan^2\theta}$$

Then the maximum value of β occurs where $\tan^2\theta = 2$. Its value is $\beta_{\max} = \arctan 1/\sqrt{8} = 19.5^0$. Storey assumed that this was sufficient to guide the wave approximately along the magnetic field to the opposite hemisphere. He therefore assumed that whistlers were propagated exactly along the field, and used the formula for strictly longitudinal propagation. He was able to explain their dispersion with a very high degree of success.

Early ray tracing studies [79] showed that this explanation was untenable. The reason for this is illustrated in Figure 11.3. This diagram shows the path of a 3 kHz signal in a simple model of the magnetosphere[1]. It is assumed that the propagation effects of heavy positive ions may be neglected. Sophisticated versions of the ray tracing techniques described in Chapter 4 have been used to obtain the ray. In Figure 11.3a the ray path is shown using a distorting coordinate system. The horizontal coordinate is Φ_M, the magnetic scalar potential in a dipole field and the vertical coordinate is L, the distance measured in earth radii at which the field line crosses the equator. The magnetic field lines are thus mapped parallel to the Φ_M-axis. It is difficult to show the detail in such a diagram in undistorted coordinates because of the enormous differences in scale at different parts of the magnetosphere. It can be seen that the ray remains quite close to a field line for its entire path as suggested by Storey [101].

In Figure 11.3b the angle between the wave normal and magnetic field is shown at points along the ray path. It is immediately apparent that, while the ray remains approximately aligned with the magnetic field, the wave normal is refracted to make a large angle with the magnetic field. When it reaches the ionosphere in the conjugate hemisphere it lies outside the acceptance cone and cannot reach the ground. What happens to the signal in such a case is discussed later when we consider the effect of ions on propagation.

It can thus be seen that Storey's original mechanism is not enough to explain how whistlers reach the ground in the opposite hemisphere. Many of the whistlers observed on satellites, however, have been propagated along paths such as those described here. It is crucial that, if the signal is to be observed on the ground, some additional mechanism must keep the wave normal as well as the ray aligned with the magnetic field.

11.5.1 Nearly Transverse Whistler Propagation

In the ray illustrated in Figure 11.3 it can be seen that, although the ray remains quite closely aligned with the magnetic field, the wave normal does not. The wave normal obeys Snell's law in the generalized form of Chapter 4. The gradients of electron density and magnetic field do not actually produce a large change in the wave normal direction as the ray is propagated. Thus, as the magnetic field direction changes, the angle between magnetic field and wave normal becomes larger and larger. At the same time the ray remains roughly aligned with the magnetic field. This behavior is illustrated by the refractive index surface shown in Figure 11.4. In this case, when the Storey approximation holds, the angle of the resonance cone is approximately 90^0. The figure shows the ray direction

[1]The unpublished model was developed by Rycroft and Alexander. It is described by Strangeways and Rycroft [103]

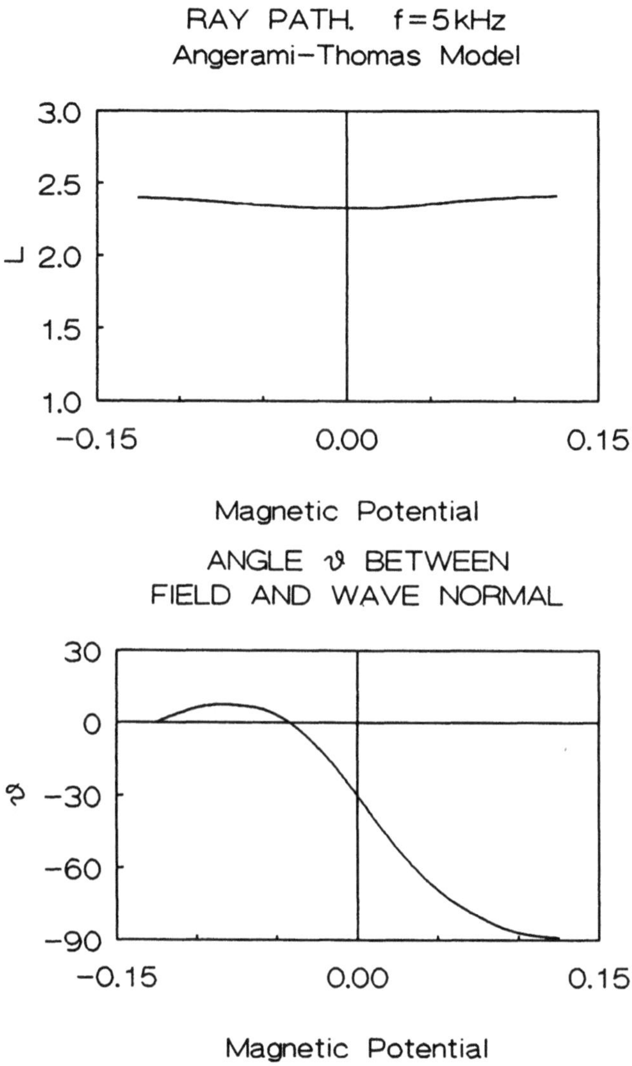

Figure 11.3: Ray tracing for an unducted whistler. (a) Ray path in Φ_M–L coordinate system. (Appendix C) (b) Direction of wave normal [K. J. Ramlall, private communication].

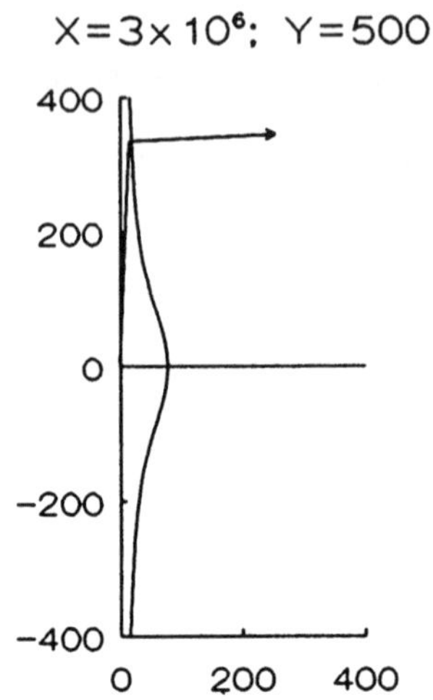

Figure 11.4: Refractive index surface in the Storey approximation.

normal to the surface for a wave normal angle which is close to 90^0. As the wave normal makes larger angles with the magnetic field the ray becomes more and more nearly parallel to the field. The angle between ray and wave normal approaches 90^0. This description is valid in regions 7 and 8 of the CMA diagram where, above the lower hybrid resonance frequency, ion effects are unimportant. We have seen in §5.4.3 how, in a warm plasma near the resonance cone, the polarization is along the wave normal and the whistler wave is coupled to an electron plasma wave. At the same time the group velocity tends to zero. In these circumstances the wave packet travels more and more slowly and, even if the plasma temperature is low, Landau damping ultimately ensures that the wave energy is absorbed by the medium.

11.5.2 Effect of Positive Ions on Propagation

If the ray penetrates to a fairly low altitude in the lower plasmasphere or upper ionosphere then the ion and electron gyrofrequencies may be large enough so that the wave frequency is above the highest ion gyrofrequency but below the lower hybrid frequency. This is region 11 of the CMA diagram. The whistler refractive index surface is then closed as illustrated in Figure 10.1. In this case there is no resonance cone and our discussion of the previous section is not valid. We can understand qualitatively what happens to a ray in these circumstances by considering a Poeverlein construction (§4.2.4) in a horizontally stratified medium as shown in Figure 11.5. The diagram is schematic and not to scale. The

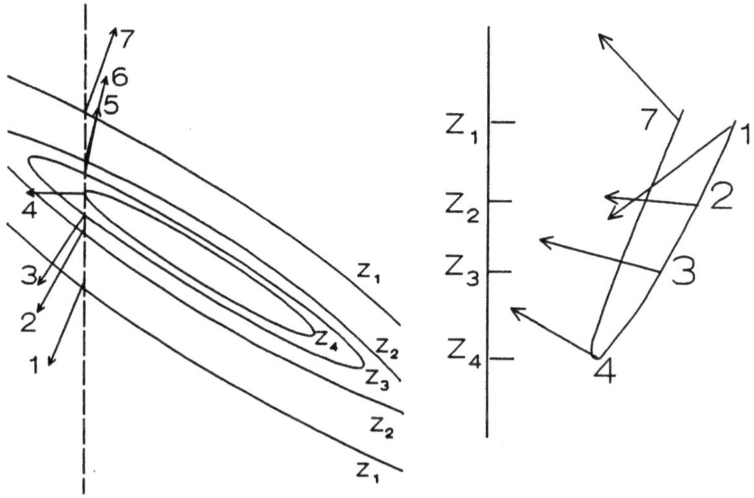

Figure 11.5: Qualitative illustration of Poeverlein construction for reflection of magnetospherically reflected whistlers.

left-hand side shows the refractive index surfaces at four levels, z_1, z_2, z_3 and z_4 above the $F2$-maximum at decreasing altitude. The dotted line represents the condition for Snell's law to be obeyed in the horizontally stratified medium. The successive directions of an initially downcoming ray are labeled 1 to 7. On the right-hand side the shape of the ray is shown with the corresponding wave normal directions. It can be seen that the ray becomes horizontal at level z_4 and the signal is therefore reflected at this level. A whistler reflected in this way is said to be *magnetospherically reflected* and is called an MR whistler [99]. If the surface were open the signal could never be reflected. In Figure 11.6 some rays at various frequencies are shown when the effect of ions on propagation is taken into account. The diagram is a rectangular plot of height against latitude so there is some distortion of the shape of the ray path. The magnetospheric reflection takes place at different altitudes for different frequencies. Only one reflection is shown for clarity. If the trace is continued the ray is reflected back and forth between the northern and southern hemispheres many times, moving further out on each hop. This ray starts at a fairly low latitude so the plasmapause is not included in the model. When the signal eventually reaches the plasmapause it encounters a sharp drop in refractive index so that it is reflected inwards. Whistler mode noise generated by wave-particle interactions can fill the plasmasphere by being reflected in this way.

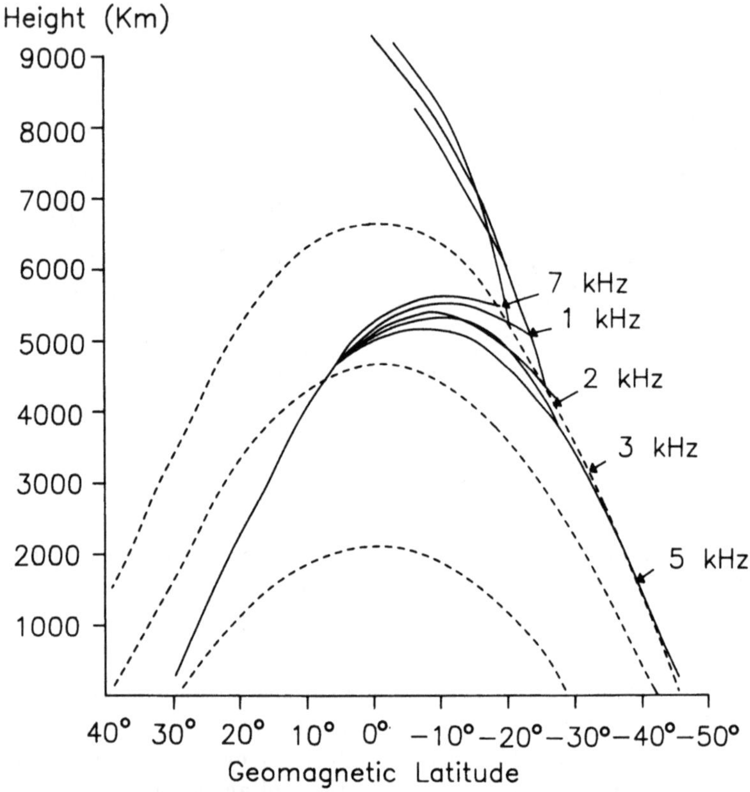

Figure 11.6: Ray paths at frequencies below 5 kHz when the effect of positive ions is included. The wave normal is initially vertical. [P. D. Terry, private communication]

11.6 Propagation in Ducts

The plasma in the plasmasphere is, to a first approximation, in ambipolar diffusive equilibrium along the field lines as described in §9.3.2. The time scale for particles to diffuse across the field lines is very long (several days) compared to motion along the field lines. The result is that any density irregularities are aligned with the magnetic field and, in the absence of disturbing electric fields, may persist for extended periods of time. In particular, the plasma in the outer plasmasphere tends to be very irregular [93]. The result is the formation of irregular ducts of enhanced or depleted ionization along the magnetic field.

The refractive index in these ducts is different from the surrounding medium so that they may act as waveguides rather like optical fibers. When their transverse dimensions are large compared with the wavelength the guiding may be understood by using ray tracing. For smaller ducts it may be necessary to use waveguide theory.

11.6.1 Behavior of Whistler Rays in a Plane Stratified Medium

The electron density in a duct varies rapidly across the duct and only slowly along it. A simple model of a duct is a plane stratified medium with the magnetic field parallel to the stratifications. We consider the propagation of whistler rays in such a medium.

In the upper panels of Figure 11.7 we show sets of refractive index surfaces for various values of X with $Y = 3$. These conditions are typical of regions 7a, 7b, 7c, 8b and 8c of the CMA diagram. The left-hand panel shows surfaces for a wide range of values between $X = 1$ and $X = 300$. The right-hand panel shows details of a number of surfaces near $X = 200$. We can imagine the magnetic field to be in the x-direction and X to be a function of z, measured perpendicularly to the field direction. As a ray is propagated through such a medium Snell's law requires that n_x is constant. The vertical line n_x = constant is thus the locus of the tip of the refractive index vector along the ray. The ray is perpendicular to the refractive index surface. The Poeverlein construction (§4.2.4) can then be used to deduce the ray direction.

Points where the construction line intersects each of the illustrated refractive index surfaces are labeled 1 – 8. The points A, B, C, D are reflection levels, indicated with heavy arrows. In the case of B, C and D the ray is in the x-direction and is thus reflected. The reflection level A is somewhat different. This level occurs where $X = 1$. Here the refractive index surface is degenerate, shrinking to a segment of a line along the magnetic field, as shown in Figure 3.10. If we allow the wave normal direction to vary continuously from making a small positive angle with

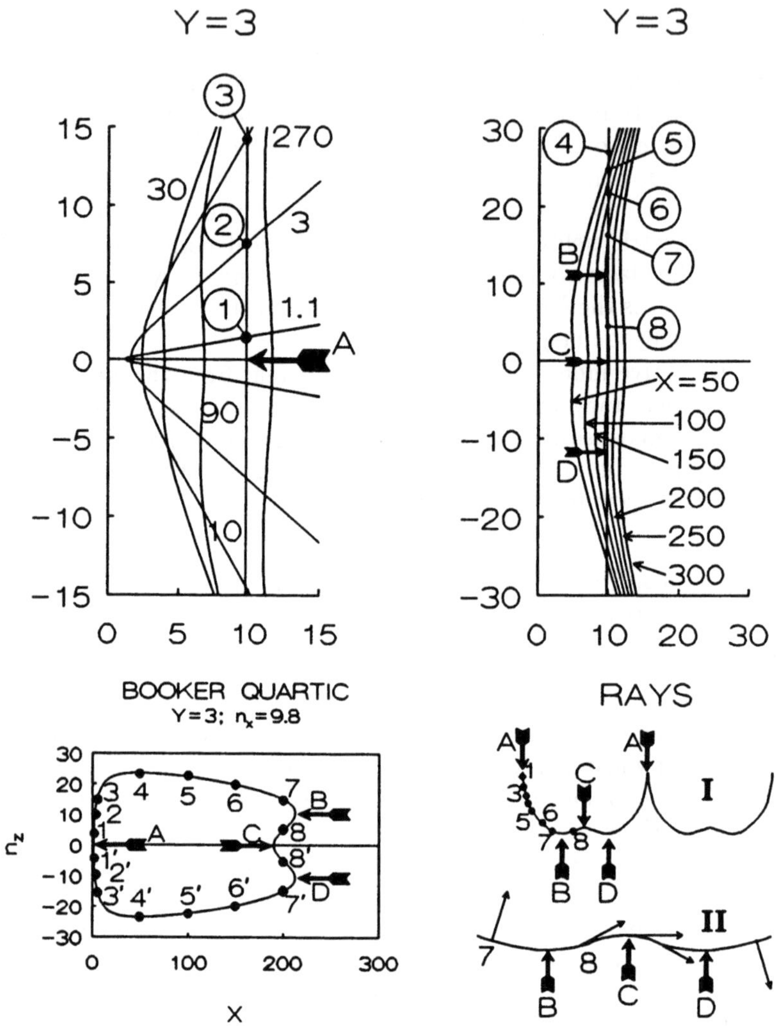

Figure 11.7: Whistler propagation in a medium with plane stratifications perpendicular to the magnetic field.

the field to making a small negative angle we can see that the ray becomes more and more nearly perpendicular to the wave normal and hence to the magnetic field. When the wave normal is exactly parallel to the field the ray is at right angles to the field and changes direction discontinuously through 180^0. The ray thus has a cusp at point A where $X = 1$. This is an example of the phenomenon called the *spitze* [17].

The Booker quartic equation (§4.2.3) can be used to find the z-component of the refractive index as a function of the x-component. In this case $l_x = 1$, $l_y = l_z = 0$. The coefficients β and δ of (4.1) are zero and the other coefficients are

$$\alpha = 1 - X - Y^2 \tag{11.7}$$
$$\gamma = -2(1 - X - Y^2)(1 - n_x^2 - X) + XY^2[1 + n_x^2] \tag{11.8}$$
$$\epsilon = (1 - X)\left\{(n_x^2 - 1)(1 - Y) + X\right\}\left\{(n_x^2 - 1)(1 + Y) + X\right\} \tag{11.9}$$
$$\gamma^2 - 4\alpha\epsilon = X^2Y^2\left[(1 - n_x^2)^2Y^2 + 4n_x^2(1 - X)\right] \tag{11.10}$$

The quartic thus becomes a quadratic in q^2.

The behavior of the root of the quartic corresponding to the whistler mode is shown in the lower left panel of Figure 11.7. The numbers labeling values of n_z correspond to the labels on the refractive index surfaces. The heavy arrows again mark the reflection levels where the ray is horizontal. These occur where two branches of the quartic become equal. This happens if $\epsilon = 0$, giving rise to three conditions

$$X = 1; \;\; X = (1 - n_x^2)(1 - Y); \;\; X = (1 - n_x^2)(1 + Y) \tag{11.11}$$

or if the discriminant, $\gamma^2 - 4\alpha\epsilon$, is zero. This condition is satisfied where

$$X = 1 + \left\{\frac{Y(1 - n_x^2)}{2n_x}\right\}^2 \tag{11.12}$$

For the case shown in Figure 11.7 the reflection level A corresponds to $X = 1$, B and D to $X = 1 + \{Y(1 - n_x^2)/2n_x\}^2$ and C to $X = (1 - n_x^2)(1 - Y)$.

The ray path can be found from (4.12) which, in this case, becomes

$$\frac{dx}{dz} = \frac{\frac{\partial\gamma}{\partial n_x}n_z^2 + \frac{\partial\epsilon}{\partial n_x}}{2n_z(2\alpha n_z^2 + \gamma)} \tag{11.13}$$

or can be estimated using the Poeverlein construction.

Some possible rays for the case where X is a monotonic function of z are shown in the lower right panel of Figure 11.7. The general form of the ray is labeled I. There is a spitze type of reflection at A. The ray then passes through the points labeled 1 – 7 on the refractive index surfaces before being reflected at B. It then passes through point 8 and is reflected again at C. Thereafter its path is repeated as a mirror image until it undergoes another reflection of the type labeled A. Its detailed shape near the level $X = 200$ is labeled II.

We thus see that, in principle, a whistler wave can be guided by a medium stratified parallel to the magnetic field with the electron density having a monotonic dependence on z. In practice conditions like this are not found in the magnetosphere. The $X = 1$ reflection level occurs, in this case, where $\omega_P < \Omega$. Such conditions only appear at high latitudes and low altitudes where the stratifications are not parallel to the field. The guiding of whistlers is generally due to conditions favoring reflection levels of the type B, C and D with the plasma having field aligned enhancements or depletions of electron density.

11.6.2 Propagation in a Broad, Plane-Stratified, Bell Shaped Duct

In order to understand the guiding of whistler energy by ducts let us consider reflections of the type occurring at B, C and D in Figure 11.7. Recall that this figure applies to conditions typical of regions 7c, 8b and 8c of the CMA diagram. On the unfolding boundary between regions 8a and 8b and between 7b and 7c the points of inflection on the whistler refractive index coalesce. In regions 7a, 7b and 8a the surface has no points of inflection. From Table 3.1, when heavy ions are neglected, the formula for this boundary is

$$X = \frac{2[2 - Y^2 \pm Y]}{(2 - Y)(2 + Y)}$$

When $X \gg Y$, as it is in all cases of interest, this becomes

$$Y \simeq 2 \quad \text{or} \quad \omega \simeq \frac{1}{2}\Omega \tag{11.14}$$

The treatment that follows therefore applies to the case

$$\omega < \frac{1}{2}\Omega \tag{11.15}$$

Near point C in Figure 11.7 the ray is reflected as it encounters a decreasing electron density (decreasing X); it therefore has curvature away from regions of lower density. In such circumstances the signal may be trapped in an enhancement of electron density. If it is started at

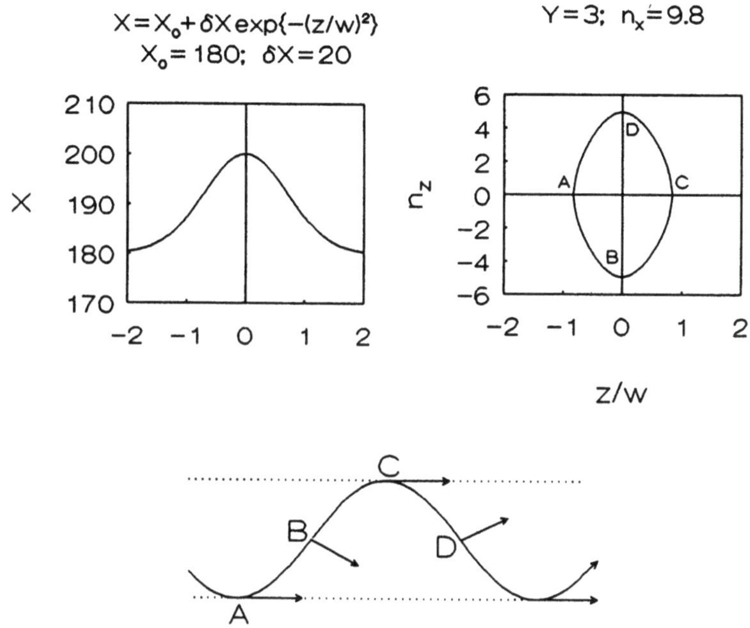

Figure 11.8: Ducting by an enhancement of ionization. Top left: Profile of duct. Top right: n_z as a function of z. Bottom: Qualitative nature of ray.

the plane of maximum electron density, making a small angle with the plane of the stratifications, it encounters a region of decreasing electron density and therefore curves back towards the maximum of the duct. On the other side of the maximum it again encounters a negative electron density gradient. Its curvature is thus reversed and it meanders back and forth across the duct. Such trapping was first discussed by Smith [97, 98].

For reflections of the type B and D the reverse is true. A ray is reflected when it encounters an increasing electron density (increasing X). In such conditions it can be trapped in a depletion of electron density.

Trapping in Electron Density Enhancements

Consider a model of a duct which consists of a plane-stratified Gaussian enhancement of ionization on a uniform background. We assume a constant magnetic field (Y = constant) in the x-direction, and electron density depending only on z such that

$$X = X_0 + \Delta X \exp\{-(z/w)^2\} \qquad (11.16)$$

If ΔX is positive this represents a Gaussian enhancement and if it is

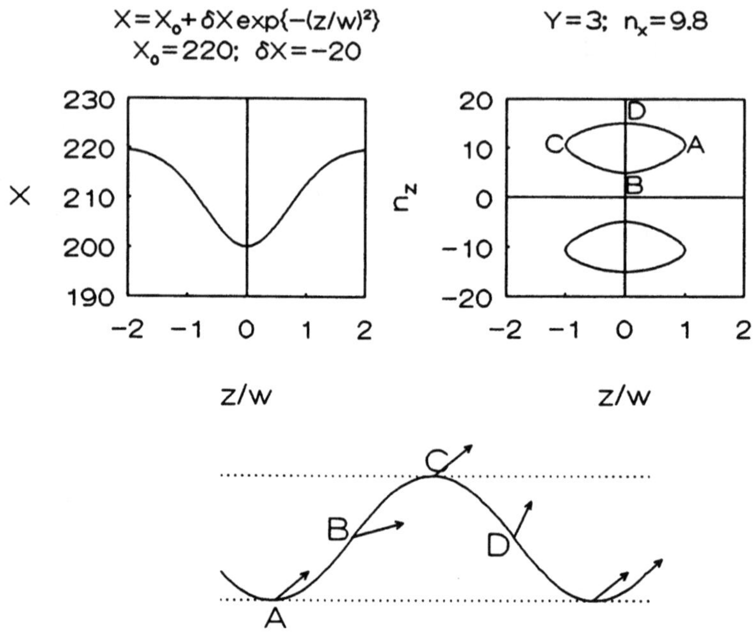

Figure 11.9: Ducting by a depletion of ionization. Top left: Profile of duct. Top right: n_z as a function of z. Bottom: Qualitative nature of ray.

negative a Gaussian depletion of electron density. We assume that n_X is fixed by Snell's law. The reflection level C occurs where $X = (1-n_x^2)(1-Y)$. Thus, in order for trapping to occur, we require $X_0 < (1-n_x^2)(1-Y)$. Furthermore the ray must not encounter a B- or D-type reflection level. These occur at values of X greater than at the C-level such that $X = 1 + \{Y(1-n_x^2)/2n_x\}^2$ and thus necessary and sufficient conditions for ducting are

$$X_0 < (1-n_x^2)(1-Y) \text{ and } 1 + \left\{\frac{Y(1-n_x^2)}{2n_x}\right\}^2 < X_0 + \Delta X$$

An example of such trapping is shown in Figure 11.8. The value of n_x is chosen to meet the conditions for ducting. The top right hand panel shows the behavior of n_z. It is real in the center of the duct between the reflection levels. It is maximum at the plane of symmetry where the wave normal makes the maximum angle with the stratifications. Reflection takes place where $n_z = 0$. The ray is shown qualitatively at the bottom. The letters A, B, C and D show corresponding points on the ray and the curve for n_z. When it is reflected the wave normal is horizontal. The arrows indicate the direction of the wave normal. When the wave normal makes a positive angle with the stratifications the ray makes a negative angle with them.

Trapping in Electron Density Depletions

The same model may be taken with a negative value of ΔX. The wave must be propagated in the middle of the duct and must encounter a reflection of type B or D satisfying $X = 1 + \{Y(1-n_x^2)/2n_x\}^2$ Thus the condition for guiding is

$$X_0 > 1 + \left\{\frac{Y(1-n_x^2)}{2n_x}\right\}^2$$

Figure 11.9 shows guiding in a depletion where this condition holds and, in addition, n_x is chosen such that

$$X_0 + \Delta X < (1-n_x^2)(1-Y)$$

In this case there are two branches of the curve for n_z. For one of them n_z is always positive and the wave normal always makes a positive angle with the magnetic field. The ray corresponding to this case is shown in the figure. There is an identical ray with the wave normal always making a negative angle with the field.

If, on the other hand,

$$X_0 + \Delta X > (1-n_x^2)(1-Y)$$

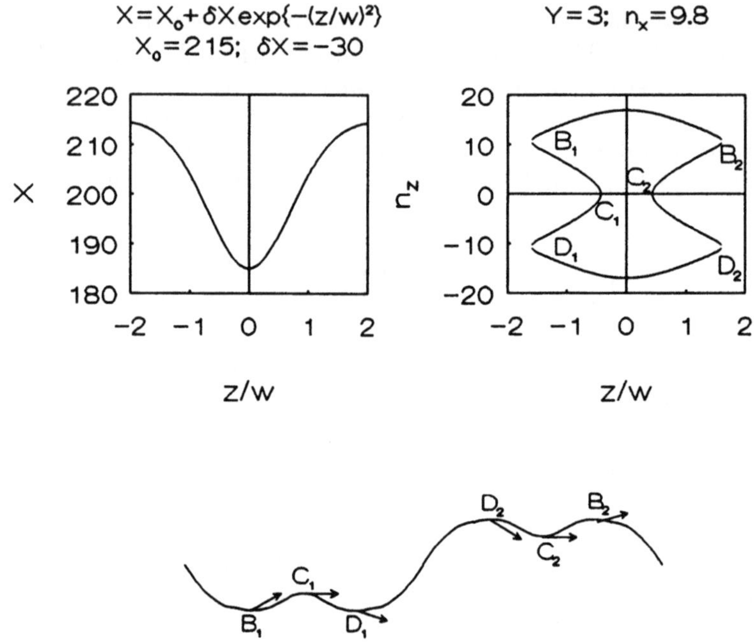

Figure 11.10: Ducting by a depletion of ionization. Top left: Profile of duct. Top right: n_z as a function of z. Bottom: Qualitative nature of ray.

the wave encounters a reflection level of type C and the two branches are connected as shown in Figure 11.10. The ray path is then of the form shown in the bottom panel of this figure. It crosses the plane of symmetry, is reflected at a B-type reflection level as the density increases, then is reflected at a C-type level as it moves to lower density, returns to the B-type level, is reflected, crosses the symmetry plane and repeats a mirror image of this path on the other side of the duct.

In a depletion of ionization the wave normal makes a relatively large angle with the wave normal. A whistler emerging from such a duct after being propagated from the opposite hemisphere would not generally be favorably oriented to penetrate the ionosphere. In addition, conditions for this type of ducting are not favorable if $Y \gg 1$ because the condition for the B- and D-type reflection levels cannot easily be satisfied. Ducting by a depletion of ionization is thus likely to play only a minor part in modifying the path of higher frequency whistlers near the equatorial plane. In the following we limit ourselves to ducting in enhancements of ionization.

Group Velocity of a Ducted Wave

A ducted whistler ray, as it is reflected back and forth, travels a longer path than one which is aligned with the magnetic field. Does this lead to inaccuracy if we assume longitudinal propagation when analysing the observations? In this section we find expressions for the longitudinal component of the group velocity and show that it does not differ significantly from the value for parallel propagation.

In the approximation $X \gg Y \gg 1$ where the refractive index is given by $n^2 = X/(Y\cos\theta - 1)$ the coefficients of the Booker quartic (11.10) can be further simplified. It is easiest to note that $\cos\theta = n_x/n$ and use the simplified expression for n^2 to show that

$$\alpha \simeq 1 \tag{11.17}$$
$$\gamma \simeq (2 - Y^2)n_x^2 + 2X \tag{11.18}$$
$$\epsilon \simeq (n_x^2 + X)^2 - n_x^4 Y^2 \tag{11.19}$$
$$\gamma^2 - 4\alpha\epsilon \simeq n_x^2 Y^2 (n_x^2 Y^2 - 4X) \tag{11.20}$$

Thus

$$n_z^2 = \frac{1}{2}\left\{(Y^2 - 2)n_x^2 - 2X \pm n_x Y\sqrt{n_x^2 Y^2 - 4X}\right\} \tag{11.21}$$

The x- and z-components of the group velocity are $\partial\omega/\partial k_x$ and $\partial\omega/\partial k_z$ where $k = \omega n/c$. They can be found by differentiating the formula for n_z. The result is

$$\frac{1}{c}V_{G,x} = \frac{2}{Yn_x\left[Yn_x \pm \frac{n_x^2Y^2 - 2X}{\sqrt{n_x^2Y^2 - 4X}}\right]} \tag{11.22}$$
$$\frac{1}{c}V_{G,z} = -\frac{2n_z}{Yn_x\left[Yn_x \pm \frac{n_x^2Y^2 - 2X}{\sqrt{n_x^2Y^2 - 4X}}\right]} \tag{11.23}$$

These are plotted in Figure 11.11 for the parameters corresponding to Figure 11.8. The upper panel shows the dependence of normalized electron density X on z. The lower panel shows the two components of the group velocity. The z-component is easy to understand. It is a closed loop. The z-component of the group velocity is alternately positive and negative as the ray is reflected back and forth between the two reflection levels.

At first sight the x-component is more puzzling. It is larger than the z-component and varies slightly in magnitude across the region within which the ray is guided. This shows that the ray is nearly aligned with the field. Outside the region of guiding this component is real, although, according to ray theory, there is no propagation outside the guiding region.

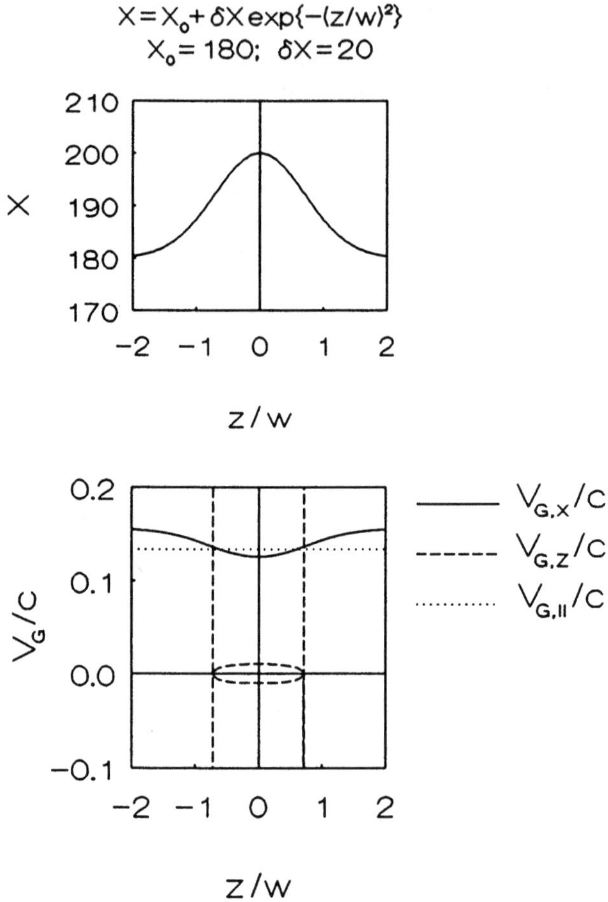

Figure 11.11: Components of the group velocity of a ducted wave compared with the group speed of a longitudinal wave.

This can be interpreted in terms of the inhomogeneous waves described in §2.5.3. If the duct were a uniform slab of enhanced electron density the wave would be totally internally reflected at the boundaries leading to an easily understood form of guiding. Outside the boundaries the wave would be inhomogeneous, with planes of constant amplitude parallel to the boundary and planes of constant phase perpendicular to it. The Poynting vector is parallel to the boundary and the amplitude of the wave drops off exponentially with distance from the boundary. In the case of a smoothly varying electron density distribution a proper solution of the reflection problem shows that the wave fields exhibit an Airy function behavior in the neighborhood of the reflection [17]. Outside the ducting region there is indeed a non-zero wave field which can be understood by analogy with the inhomogeneous wave picture. It drops off rapidly with distance and becomes negligible at a few wavelengths from the reflection. This distance is small compared with the length scale on which the ray theory holds.

Also shown in Figure 11.11 is a horizontal dotted line representing the group velocity of a whistler wave which is propagated parallel to the field in a uniform medium with a density equal to that at the maximum of the duct. The amount by which this differs from the mean velocity in the duct is small and allows us to make the assumption of parallel propagation when analysing whistlers. Smith [97] has shown that, over a wide variety of conditions, the error in making this assumption is less than 1%.

The above treatment only holds if the dimensions of the duct are large compared with the wavelength so that ray theory applies. They are almost certainly applicable to the strong traces shown, for example, in Figure 10.5. The "swishy" background of this signal suggests a fine structure of much smaller ducts. In such ducts ray theory may not apply and waveguide mode theory is appropriate. This is beyond the scope of this book but is treated for plane stratified ducts by Walker [111, 112] and for ducts of circular cross section by Laird and Nunn [72]

Half Gyrofrequency Duct Cut-Off

We have noted in (11.15) that this treatment applies to the case $\omega < \Omega/2$. Guiding in an enhancement of ionization only occurs for frequencies below half the electron gyrofrequency. Above this frequency a depletion is required. Suppose a whistler is trapped in a duct at a relatively low altitude and guided along it towards the equatorial plane. The electron density and magnetic field change with distance along the duct. If we assume this change is slow compared with the change transverse to the duct the theory of the previous section can be assumed to hold locally. The lowest gyrofrequency encountered along the path is at the equatorial plane. All frequencies above half this minimum gyrofrequency encounter a region where ducting is no longer possible and they escape from the duct. All frequencies below half this minimum are ducted across the equatorial plane and into the opposite hemisphere. Those which escape cannot reach the ground; those that remain ducted can. The consequence is that whistlers observed on the ground exhibit an upper waveguide cut-off at a frequency equal to half the minimum gyrofrequency on their path. The whistler traces in Figure 10.5 clearly show this cut-off. In the figure the lower the nose frequency the greater is the L-value of the path along which they are ducted. The greater the L-value the smaller is the equatorial gyrofrequency. Thus the diminishing cut-off frequency is correlated with a diminishing nose frequency of the multiple traces.

11.6.3 The Effect of Field Curvature – One-Sided Ducts

Real ducts are not bell shaped enhancements on a uniform background. The background plasma density and magnetic field decrease with increasing radius and irregularities are superimposed on the background. In addition the magnetic field line is curved. In particular the field curvature can introduce additional guiding effects.

Visitors to the famous whispering gallery in St Paul's Cathedral in London are entertained by the whispering gallery effect. In the hemispherical dome of the cathedral a whisper near the wall is transmitted round the inside of the dome and can clearly be heard near the wall on the other side[2]. It is inaudible a small distance away from the wall. The mode of transmission of the sound is illustrated in Figure 11.12. A ray started at A is reflected successively round the dome at near grazing incidence with little loss. It thus reaches the opposite side at B as shown. There is little attenuation because all rays near grazing incidence are focused along the same path. The reflection coefficient for rays which are not near grazing incidence is smaller and such rays have their intensity reduced by the inverse square law. There is thus only a weak signal reaching points far from the wall. A proper understanding requires waveguide mode theory rather than the simple ray picture [18]. It is often convenient to treat such a problem in a Cartesian geometry. This is done by the method of *modified refractive index* where the wall is regarded as flat and a gradient of refractive index gives a compensating curvature to the ray as shown on the right of the figure.

[2]The similar effect beneath the dome of the Capitol in Washington D. C. has more to do with focusing than the whispering gallery effect.

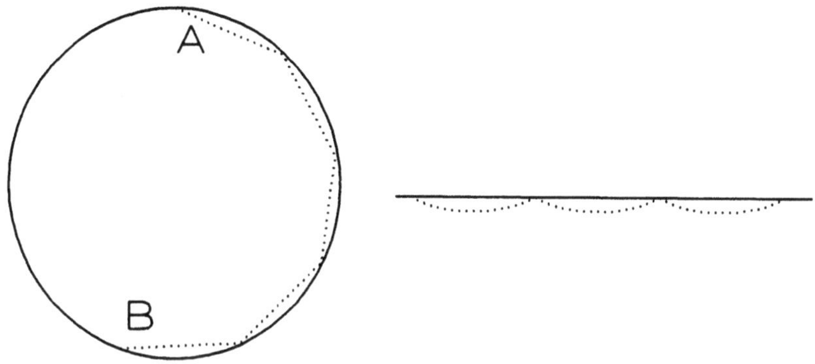

Figure 11.12: Propagation in a whispering gallery

Effect of Duct Curvature

When the magnetic field lines are curved it is convenient to use a generalization of the method of modified refractive index [110]. The generalization is necessary because of the anisotropy of the medium. We assume that locally the field lines can be approximated by circles of radius r. We also assume that the stratifications of the medium are cylindrical. In a local cylindrical coordinate system we can use a genralization of Snell's law. If the refractive index vector has components n_r, n_θ then, at a cylindrical boundary between two regions of different density, the phase of incident and transmitted waves must vary with θ in the same way. The phase depends on θ as $(\omega/c)\int n_\theta \, r \, d\theta$ = constant. The factor r is the appropriate scale factor for the coordinate system. This implies that $rn_\theta = constant$. We may write

$$\frac{rn_\theta}{r_0} = n_x \tag{11.24}$$

where r_0 is the radius of curvature on some reference field line in the duct. If $|r - r_0| \ll r_0$ we may write

$$n_\theta = \left(1 - \frac{r - r_0}{r_0}\right) = \left(1 - \frac{z}{r_0}\right) \tag{11.25}$$

where $z = r - r_0$. If the medium is slowly varying the phase variation of the waves is of the form

$$\frac{\omega}{c}\int \mathbf{n}.d\mathbf{r} = \frac{\omega}{c}\left\{\int n_r \, dr + n_\theta r\theta\right\} \tag{11.26}$$

The relationship between n_r and n_θ is the Booker quartic equation. Thus we can treat problems of curvature by the same techniques as in a plane geometry simply by replacing n_x by its modified value (11.25).

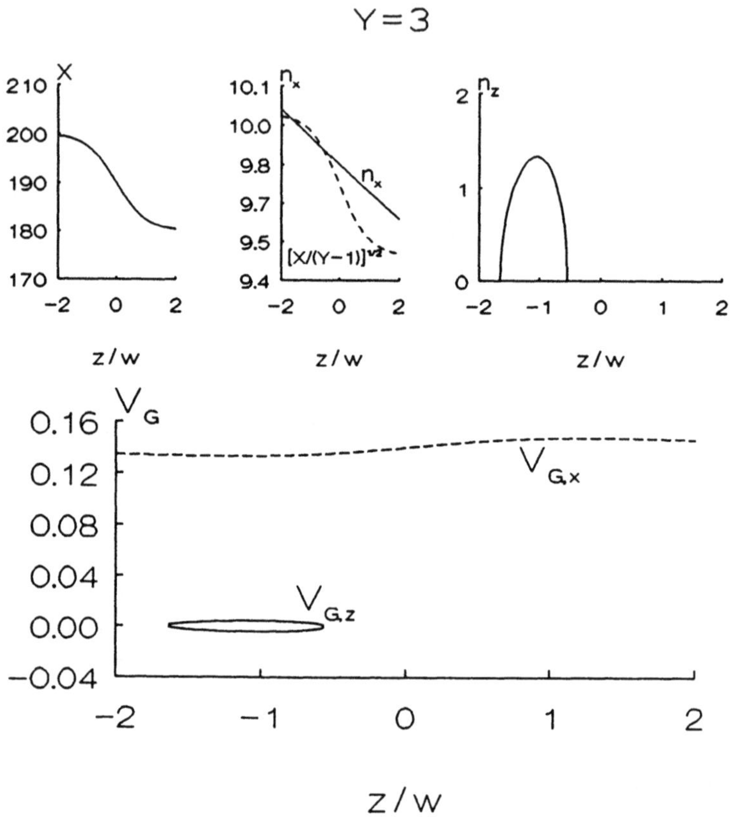

Figure 11.13: Propagation in a one-sided duct. Top left: Duct profile. Top center: Intersection of curve for curvature modified n_x and condition for reflection. Top right: Modified n_z. Bottom: x- and z-components of the group velocity.

Whispering Gallery Propagation in One-Sided Ducts

Consider a field aligned duct in which the electron density drops smoothly from a high value to a low value across the duct. A suitable model is

$$X = X_{max} - \frac{e^{2z}}{1+e^{2z}}\delta X \tag{11.27}$$

Such a duct profile is shown on the top left of Figure 11.13

The condition for reflection of a whistler wave at point C is

$$n_x = \sqrt{1 - \frac{X}{1-Y}} \simeq \sqrt{\frac{X}{Y-1}} \tag{11.28}$$

The top center panel of Figure 11.13 shows this condition plotted as a dashed curve against z. The modified value of n_x (11.27) is also plotted as a full line. The curves intersect at two points representing reflection. When $n_x > \sqrt{X/Y - 1}$ propagation in the whistler mode is possible. It can thus be seen that there is a region of propagation between two reflection points which implies the possibility of ducting of the same type as in a whispering gallery.

In the plasmasphere both the electron density and the magnetic field magnitude decrease with radius. Decreasing Y implies increasing refractive index; decreasing X implies decreasing refractive index. Whispering gallery propagation can take place in regions where the decrease in electron density with radius is faster than the decrease in magnetic field strength. It is of interest to note that the plasmapause is such a region so that whistlers tend to be guided in the plasmapause even in the absence of a duct fine structure.

11.6.4 Ray Tracing in Models with Ducts Included

A number of authors have carried out detailed ray tracing studies of whistler waves in various models of the plasmasphere. In some of these the effect of ducts have been included [104]. Such ray tracing can show how the waves get trapped in ducts. A particularly detailed model has been used by Ramlall [K. J. Ramlall, private communication]. This consists of a number of Gaussian enhancements of random amplitude, width and position, superimposed on the Rycroft-Alexander [103] model of the plasmasphere. Such ray tracing shows that the simple picture of a wave propagated in an isolated duct is an oversimplification. A ray may be trapped in a duct, leak into a different duct, escape and be recaptured. The plasmapause plays a role in the guiding. The result is that there is a general tendency for the signal to be guided leading to a "swishy" background. On this are superimposed the intense traces arising from signals which are completely trapped in strong ducts.

11.7 Summary

- A lightning stroke is an efficient source of electromagnetic radiation in the v. l. f. range.

- Much of the radiation from a lightning stroke travels substantial distances in the earth–ionosphere waveguide, being reflected back and forth between lower ionosphere and ground. The waveguide cut-off is between 1 and 2 kHz. Frequencies near the cut-off suffer a greater delay, producing the characteristic shape of a spheric.

- Some of the energy penetrates the ionosphere. Snell's law shows that, because of the large ionospheric refractive index, the wave normal lies within an acceptance cone making small angles with the vertical. Rays which reach the opposite hemisphere can only emerge from the ionosphere if their downward pointing wave normals lie within the acceptance cone.

- Ray tracing studies in a smooth plasmasphere model show that, although the Storey mechanism maintains the ray close to the direction of the magnetic field, the wave normal makes large angles with it. Rays reaching the opposite hemisphere make large angles with the wave normal which does not fall within the acceptance cone. Ultimately the resonance cone is encountered and the wave may be coupled to a plasma wave which is Landau damped. Under the right conditions positive ion effects may instead cause it to be magnetospherically reflected. Such whistlers may bounce back and forth, moving to higher latitudes with each reflection until they encounter the plasmapause.

- To explain the ground observations it is necessary to invoke ducts of enhanced or depleted electron density. Theoretical study of the nature of propagation in such ducts shows that the signals which reach the ground must have been propagated in enhancements of ionization.

- When curvature is taken into account it is possible to produce guiding which is analogous to whispering gallery propagation. If the density decreases sufficiently fast with radius the ray curves towards the earth with a curvature greater than the field line curvature; as it reaches regions with smaller gradients the curvature is less than the field line curvature and it moves back to the center of the duct.

Chapter 12

Waves in the Plasmasphere – 3. Ion Cyclotron Whistlers

12.1 Introduction

A wave propagation theorist, on encountering the ion cyclotron whistler, cannot fail to be fascinated. Precise observational evidence and a satisfying theoretical treatment involving the understanding of several esoteric features of plasma wave propagation theory justify the devotion of a separate chapter to this phenomenon. Although the ion cyclotron whistler is not a major denizen of the plasma wave jungle it is one of the most interesting.

12.2 Observations

Ion cyclotron whistlers [43] are observed in the geomagnetic latitude range $10^0 < |\theta| < 70^0$ by low altitude satellites. They occur as a modification of the trace of an ordinary fractional hop whistler which has been generated by lightning on the ground. They are more common during local night than local day. They are also more common in the northern hemisphere.

A typical example with a number of proton whistlers is shown in Figure 10.6. Another example is shown in Figure 12.1. An idealization of the phenomenon, labeled with the nomenclature used in discussing it, is shown in Figure 12.2.

The frequency range of the signal is a band of a few hundred hertz. It consists of a short descending tone followed by a longer ascending tone. The descending tone can be identified as a short fractional hop whistler which has traveled from the ground to the satellite. This is labeled in the figure as the electron whistler. Generally a small band of frequencies in the neighbourhood of the ion gyrofrequency is missing or weak. The lower part of the electron whistler trace merges with a trace whose delay increases rapidly with frequency, becoming infinite at the ion gyrofrequency. This portion of the trace is labeled in the figure as the ion cyclotron whistler.

Although electron whistlers which have traveled from the opposite hemisphere are observed, these are never associated with ion cyclotron whistlers. Ion cyclotron whistlers have always traveled upwards from

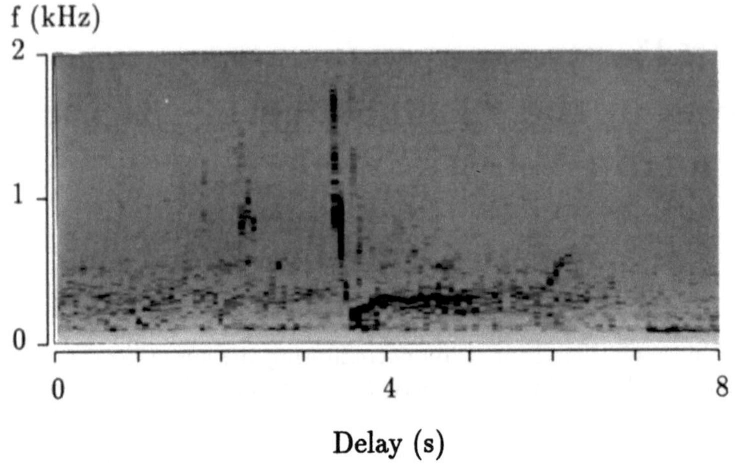

Figure 12.1: Example of a proton whistler from ISIS2.

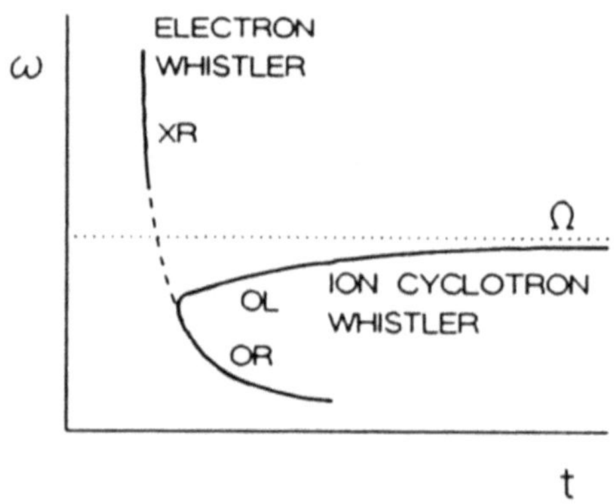

Figure 12.2: Idealization of ion cyclotron whistler.

ground to satellite. Polarization measurements show that the polarization label of the trace identified as the electron whistler is R and that of the trace labeled ion cyclotron whistler is L. Bear in mind that, in the northern hemisphere, an upward traveling wave is propagated approximately in the opposite direction to the magnetic field. The waves with polarization labels R and L are respectively left- and right-hand elliptically polarized in such circumstances. If the angle between wave normal and magnetic field direction is not too large then the electron whistler wave is left-hand circularly polarized in the northern hemisphere and right-hand circularly polarized in the southern hemisphere. The opposite is the case for the ion cyclotron whistler.

There have also been observations of helium whistlers [9]. These show an additional trace similar to the proton whistler which occurs just below the local helium gyrofrequency.

The observation that ion cyclotron whistlers appear to be more common in the northern hemisphere is thought to arise from the fact that, because of the much greater land mass, thunderstorms are much more frequent in the northern hemisphere than in the south.

12.3 Ion Cyclotron Whistler Propagation in a Uniform Medium

Ion cyclotron whistlers, in the same way as whistlers, are generated by lightning. The complicated nature of the frequency–time dynamic spectrum is a consequence of propagation effects in the ionosphere between the source and the satellite. Some of these properties are a direct consequence of the dependence of the electron density and the magnetic field on the height. Nevertheless, in order to understand the phenomenon, it is useful to consider the propagation of electromagnetic waves in a uniform medium. The occurrence of ion cyclotron whistlers requires at least two ion species to occur.

12.3.1 Dispersion Relation

Consider a cold plasma consisting of two heavy ion species, H^+ and He^+, each with a fractional abundance of 50%. The dispersion relation, as described in §3.4, is shown in Figure 12.3. This figure should be studied in conjunction with Figure 10.2 which shows the refractive index surfaces in different regions of the CMA diagram. First consider the dashed lines which represent longitudinal propagation. These curves have been labeled with the polarization labels L and R. As we shall see, X and O-labels are ambiguous and they have not been included. The lowest frequencies correspond to region 13 of the CMA diagram. For the L-wave ck/Ω

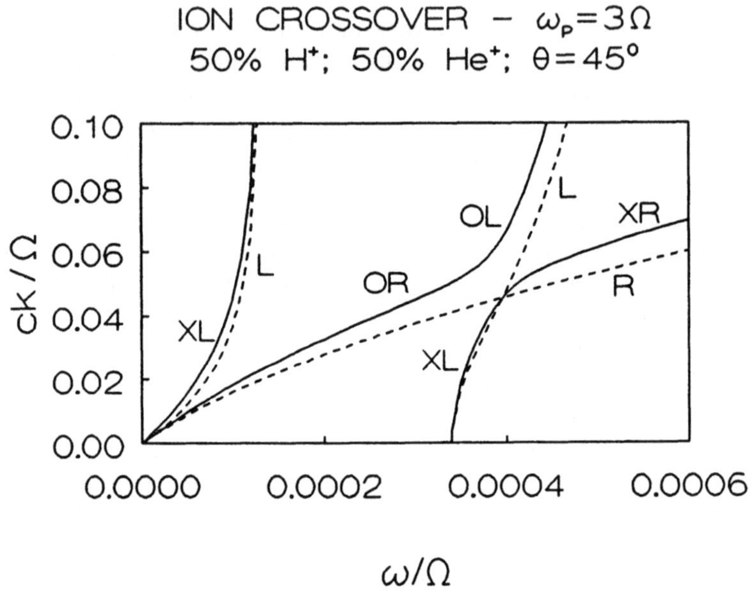

Figure 12.3: Dispersion relation for waves near ion cyclotron resonance. Dashed lines: Longitudinal propagation. Solid lines: Oblique propagation at 45^0 to the magnetic field.

increases rapidly and it experiences a resonance at the He^+ gyrofrequency $1.36\times10^{-4}\Omega$, representing the transition between regions 13 and 15b. The R-wave has a smaller value of ck/Ω and rises steadily without resonance or cut-off. The unfolding between regions 15b and 15a produces no effects in this case. At the cut-off between regions 15a and 14b the L-wave again encounters a propagation region. The value of ck/Ω for this wave rises rapidly and is equal to the R-wave value at the cross-over between regions 14b and 14a. The value of ck/Ω then rises rapidly until it encounters the resonance at the proton gyrofrequency, representing the boundary between regions 14a and 11.

The unusual feature of this dispersion relation is that the two curves cross at the cross-over frequency. At this frequency both waves have the same phase speed. This point corresponds to the point where the two refractive index surfaces touch in the third panel of Figure 10.2. For strictly longitudinal propagation the two waves are propagated independently even at the cross-over.

Now consider the full curves in Figure 12.3. These represent the dispersion relation when the wave normal makes an angle of 45^0 with the magnetic field. In Figure 10.2 a line making an angle of 45^0 with the x-axis intersects the refractive index surface at the appropriate point. It can be seen that, at points remote from the cross-over the dispersion relation is very similar to that for longitudinal propagation. Near the cross-over the picture is very different. The connection between the curves changes. The R-curve merges with the L-curve and *vice versa.*

The nomenclature can lead to confusion. We have chosen, in accordance with standard practice, that the whistler wave shall be defined as the wave with R-polarization label and the ion cyclotron wave as the wave with L-polarization. For longitudinal propagation the labeling is maintained at all points on the curves. It is, however, the longitudinal case which is anomalous. The two curves cross because the refractive index surfaces touch for longitudinal propagation. At all other angles of propagation the waves are associated either with the open S-surfaces or the closed F-surface.

It is often said that the whistler wave is coupled to the ion cyclotron wave for oblique propagation. This is a matter of semantics rather than physics. Near the cross-over the polarization label of the continuous curve changes and, indeed, its sense of polarization changes. Because of our definition of a whistler wave in terms of the sense of its polarization we change the labels on a continuous curve from whistler to ion cyclotron wave at the cross-over. It would be more accurate to say that the strictly longitudinal whistler below the cross-over is coupled to the strictly longitudinal whistler above it. At other angles of propagation the whistler branch below the cross-over is not connected to that above it. Indeed coupling is important in understanding ion cyclotron whistlers; when collisions are taken into account there is coupling between the two whistler branches for a cone of angles near the magnetic field direction. This is considered later.

12.3.2 Group Refractive Index

As described in §3.4 the slope of the curves in Figure 12.3 is $n' = c\,dk/d\omega = c/V_L$, where V_L is the component of the group velocity in the wave normal direction. The quantity n' is the *group refractive index.* A short pulse of infinite plane waves moves parallel to the wave normal with speed V_L. Its delay after moving a distance s is $(c/V_L)s = n's$. The delay is proportional to n' and thus a plot of ω against n' has the shape of a frequency–time dynamic spectrum in a uniform medium.

Clearly

$$n' = n + \omega \frac{\partial n}{\partial \omega} \tag{12.1}$$

Figure 12.4 is such a plot for the conditions of Figure 12.3.

It can be seen that this diagram shows a number of similar features to the dynamic spectrum of an ion cyclotron whistler. There are also a number of differences. The diagram represents what would be seen if a plane wave were to be launched at a particular angle to the field in a uniform medium.

For propagation parallel to the magnetic field there are three separate branches represented by the dashed lines. The lowest branch represents a helium cyclotron wave with L-polarization label; the sense of the polarization is the same as the sense of rotation of a positive ion – left-handed for propagation parallel to the field and right-handed for propagation antiparallel to the field. The branch to the left of the figure is the whistler with R-polarization label having circular polarization with the sense of rotation of a negative electron. The isolated branch to the right is the proton cyclotron wave with L- polarization – again with the sense of rotation of a positive ion.

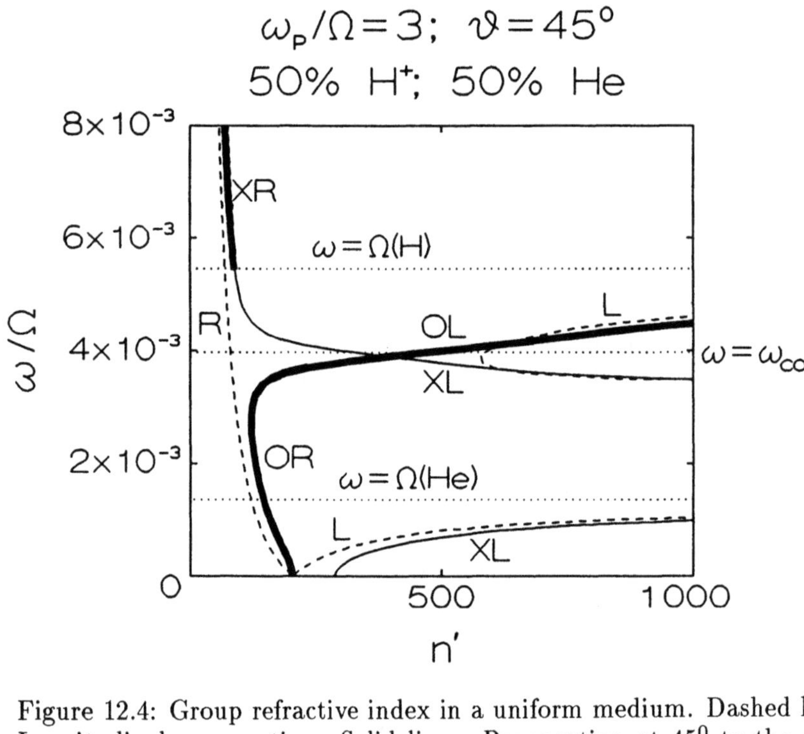

Figure 12.4: Group refractive index in a uniform medium. Dashed lines: Longitudinal propagation. Solid lines: Propagation at 45^0 to the magnetic field. Dotted lines: Resonance frequencies and cross-over.

When the wave normal makes an angle with the field the three branches are connected differently as shown by the dotted lines. If the angle with the field were very small they would lie very close to the dashed curves and the horizontal line representing the cross-over frequency. The part of the curve which was associated with the lower part of the whistler curve in the case of longitudinal propagation is connected to the portion of curve which was associated with the upper part of the proton cyclotron wave. Its labeling changes from OR to OL at the cross-over. The part of the curve which was associated with the upper part of the whistler curve is connected to the portion of the curve which was associated with the lower part of the proton cyclotron wave. Its labeling changes from XR to XL at the cross-over.

The shape of parts of the solid curves strongly suggests the shape of an idealized ion cyclotron whistler. These parts are shown as heavier lines in the diagram. The lower XL branch and the lower part of the XR–XL branch never appear in the data. In some examples the gap between the upper part of the XR branch and the OR–OL branch is partly filled by a signal which appears more like the longitudinal R branch.

The phenomenon cannot be fully understood without considering the whole propagation path through the ionosphere.

12.4 Nature of Ion Cyclotron Whistler Generation

12.4.1 Properties of the Ionosphere

The nature of an ion cyclotron whistler depends on the properties of the ionosphere below the satellite on which it is observed. Such satellites are in orbits between a few hundred and one or two thousand kilometers. Below this level the ionosphere may, without serious inaccuracy, be regarded as horizontally stratified. We shall be dealing with frequencies near the He^+ and H^+ gyrofrequencies. Aside from these two species the lightest ion species is O^+. All the features we wish to illustrate are exhibited by a plasma which consists of a mixture of light H^+ and He^+ ions with fractional abundances η_1 and η_2 and infinitely massive ions with fractional abundance η. In the lower parts of the ionosphere the positive ions are heavy. Immediately above the $F2$ maximum O^+ dominates. At a few hundred kilometers the dominant ion is He^+ and above this, near an altitude of 1000 km H^+ becomes the most important ion.

The electron gyrofrequency at the base of the ionosphere near the surface of the earth at mid-latitudes is of the order of 1 MHz. Thus the proton gyrofrequency is roughly 600 Hz and the He^+ gyrofrequency roughly 150 Hz. The plasma frequency is always much greater than the electron gyrofrequency in any region of interest. The magnetic field, and hence the gyrofrequencies, drop off with altitude as the inverse cube of the radius (Appendix C). In a dipole field the direction of the magnetic field does not change with altitude.

12.4.2 Cut-off, Resonance and Cross-over

The CMA diagram is that for a three ion plasma and is thus more complicated than any discussed in Chapter 3. The heavy ions dominate at low altitudes, there is then a layer of helium and at high altitudes protons are dominant. The model of the ion distribution is shown in Figure 12.5a.

The discussion of the CMA diagram in Chapter 3 is not adequate because the relative proportions of the ions are changing with height. The regions could be discussed in terms of a three-dimensional CMA diagram with the third axis representing the proportions of the three types of ion. Instead we shall plot the CMA boundaries in a graph of frequency as a function of height in the model.

The resonances occur at the ion gyrofrequencies and are shown by solid lines in Figure 12.5. The only condition for a cut-off which occurs in this frequency range is $\kappa_L = 0$ which is given by

$$1 - \frac{X}{1-Y} - \frac{X_1}{1+Y_1} - \frac{X_2}{1+Y_2} = 0 \tag{12.2}$$

The values of X and Y for the heavy ions are zero because we have assumed that they are infinitely massive. Since $X \gg Y \gg 1$, $X_i = \eta_i X Y_i / Y$, we can write this in the form

$$\frac{X}{Y}\left[1 - \frac{\eta_1 Y_1}{1-Y_1} - \frac{\eta_2 Y_2}{1-Y_2}\right] = 0$$

The common factor X/Y cancels and thus, at these frequencies the cut-off does not depend on the electron density. In terms of frequency this can be written as a quadratic equation

$$\omega^2 + [(\eta_1 - 1)\Omega_1 + (\eta_2 - 1)\Omega_2]\omega + (1 - \eta_1 - \eta_2) = 0 \tag{12.3}$$

The cut-offs are shown by dashed lines in Figure 12.5.

The cross-overs for the transition from R- to L-polarization are given by $\kappa_2 = 0$. The same approximations apply and this condition may be written

$$\frac{X}{Y}\left[1 + \frac{\eta_1 Y_1^2}{1-Y_1^2} + \frac{\eta_2 Y_2^2}{1-Y_2^2}\right] = 0 \tag{12.4}$$

This yields a quadratic in the square of the frequency

$$\omega^4 + \left[(\eta_1 - 1)\Omega_1^2 + (\eta_2 - 1)\Omega_2^2\right]\omega^2 + \Omega_1^2\Omega_2^2(1 - \eta_1 - \eta_2) = 0 \tag{12.5}$$

The cross-overs are shown by dotted lines in Figure 12.5.

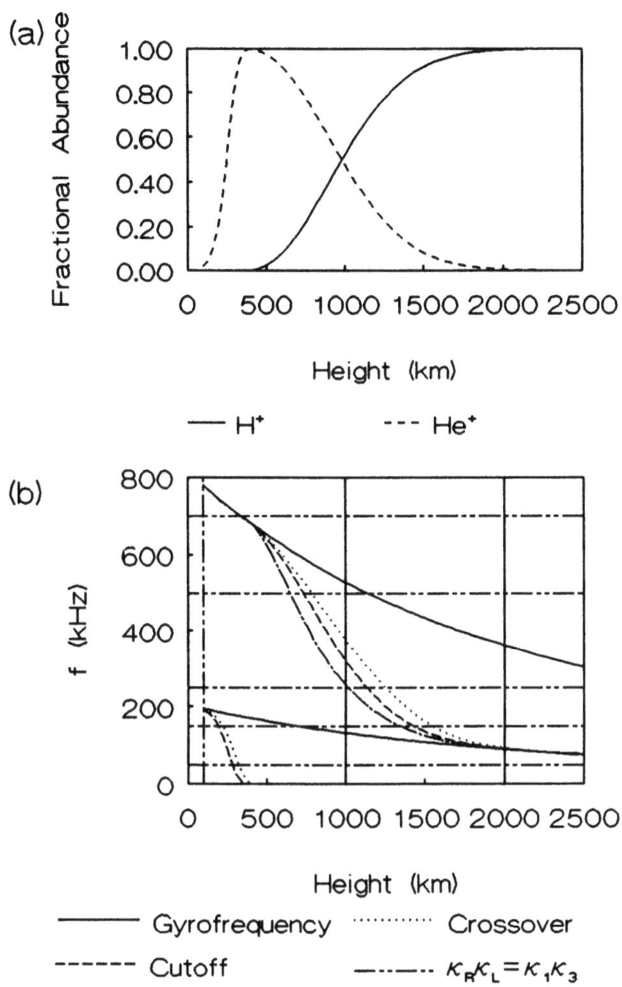

Figure 12.5: CMA boundaries in the model. (a) Fractional abundance of He^+ and H^+ ions. (b) Solid lines: Resonances. Dashed lines: Cut-offs. Dotted lines: Cross-overs.

It is also of interest to plot the cross-over $\kappa_R\kappa_L - \kappa_1\kappa_3 = 0$ which represents the transition from O- to X-polarization. The same approximation gives

$$\frac{\eta_1 Y_1}{1-Y_1^2} + \frac{\eta_2 Y_2}{1-Y_2^2} = 0 \tag{12.6}$$

or

$$\omega^2 = \frac{\Omega_1\Omega_2(\eta_1\Omega_2 + \eta_2\Omega_1)}{\eta_1\Omega_1 + \eta_2\Omega_2} \tag{12.7}$$

This is shown as a chain line in Figure 12.5.

12.4.3 Properties of Waves in the Model When There Is No Coupling

Because the plasma frequency is everywhere much greater than the gyrofrequency, the refractive index is very large over the whole region of interest. The acceptance cone (§11.4) is very narrow and the wave normal is almost vertical. No serious error is made by considering exactly normal incidence so that the wave normal makes the same angle with the magnetic field at all heights. The properties of the wave at various heights may be understood by considering the nature of the refractive index as a function of height. Of course the ray may make a large angle with the vertical and the wave packet may not travel vertically. We assume that the signal does not travel a horizontal distance which is so great that the change in the magnetic field direction or the curvature of the stratifications needs to be considered.

Consider the propagation of waves at various frequencies which are propagated at normal incidence from the ground and reach a satellite. Two vertical solid lines in Figure 12.5b represent the positions of two satellites at different altitudes of 1 000 km and 2 000 km. We consider waves at 700 Hz, 500 Hz, 250 Hz, 150 Hz and 50 Hz as represented by the horizontal broken lines in the figure. The refractive index is shown as a function of height for each of these frequencies in Figure 12.6. Let us first consider the signal which would reach the satellite on the assumption that the medium is slowly varying and strict ray theory applies. The waves at each frequency pass through a different set of regions of the CMA diagram.

The 700 Hz wave is an example of a wave whose frequency is above the highest gyrofrequency at all altitudes. At lower altitudes there are no protons and the frequency is above the helium gyrofrequency; by the time there is an appreciable proton density it is above the proton gyrofrequency. It is a whistler wave with XR polarization label and reaches both satellites without a change in polarization. Clearly the signal received by either satellite for all such frequencies is a normal fractional hop whistler.

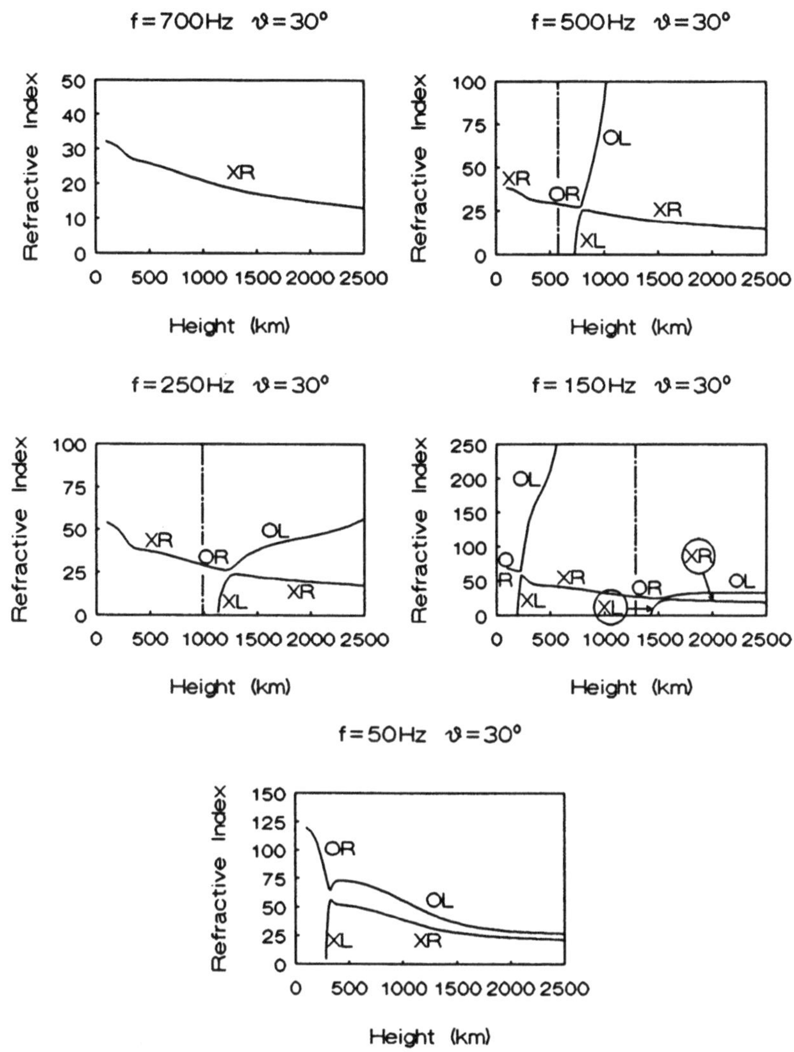

Figure 12.6: Refractive index as a function of height for the ion cyclotron whistler frequencies shown in Figure 12.5.

The 500 kHz wave has a much more interesting history. It starts as a whistler wave with XR-label, having a frequency above the maximum gyrofrequency. It then encounters the cross-over $\kappa_L\kappa_R = \kappa_1\kappa_3$ where its polarization label changes to OR. There is no physical effect but the label change is necessary for continuity as described in §3.3.2. It then encounters the cross-over $\kappa_2 = 0$ where its polarization label changes from OR to OL. Here there is a physical change; although the wave normal is not parallel to the field it is in general elliptically polarized. The sense depends on the magnetic field direction and thus there are opposite senses in northern and southern hemispheres. In the southern hemisphere the wave is right-hand polarized below the cross-over and left-hand polarized above it. The reverse is true in the northern hemisphere. It passes through a level where it is linearly polarized at the cross-over. The wave then encounters the proton gyroresonance above which it cannot be propagated. The satellite at 1000 km can receive this wave as an ion cyclotron whistler. The satellite at 2000 km altitude does not receive it. Just below the $\kappa_2 = 0$ cross-over is the ion cyclotron cut-off. Above this an ion cyclotron wave can be propagated with XL-label. At the cross-over it changes the sense of polarization and above this is a whistler with XR label. On the basis of ray theory this wave cannot reach either satellite from a source on the ground.

The reader should follow the propagation of the waves for the remainder of the curves in Figure 12.6. Each shows a different behavior. Note that, for the waves with frequencies below the largest helium gyrofrequency, in the model used there is a significant fraction of helium ions where the wave enters the ionosphere. The wave thus has a frequency between the helium gyrofrequency and the gyrofrequency of the heavy ions. It is thus in region 15 of the CMA diagram and has an OR polarization label.

It is clear that, according to strict ray theory, waves with frequencies between the helium and proton gyrofrequencies do not reach the satellite. Such a theory cannot, therefore, explain the appearance of ion cyclotron whistlers between these frequencies. To do this it is necessary to invoke coupling between the branches of the refractive index at the cross-over frequencies. This is described in §12.5

12.5 The Effect of Collisions on Coupling

As we have seen, as the angle θ between wave normal and magnetic field approaches zero, the two branches of the dispersion relation approach closer and closer at the cross-over frequency. At $\theta = 0$ they touch and the two curves connect in a different way (§12.3.1). Whereas, when $\theta \neq 0$, the polarization labels L and R are interchanged as the cross-over is passed, at $\theta = 0$ the two curves cross and the L- and R-labels are the same on either side of the boundary. This occurs at a *coupling point* where the discriminant of the quadratic equation for the refractive index is zero so that the refractive index and polarization of each wave are the same. The coupling point is a branch point of the multi-valued function n.

Physically it is possible to understand that, when two independent waves with different polarizations encounter a region where their polarization and phase speed are the same, they remain in phase over a number of wavelengths and can be coupled together. The full mathematical treatment is difficult and beyond the scope of this book, involving a good understanding of the theory of multivalued functions of a complex variable. We shall present a heuristic treatment.

In a collisionless plasma coupling only occurs for propagation exactly along the field. Coupling takes place for oblique propagation in a lossy medium when the effect of collisions with neutral particles is taken into account. To understand how this occurs we study the nature of the discriminant D near a cross-over when collisions are taken into account. It is given by (2.62) with the elements of the dielectric tensor being given by (2.45), (2.46), (2.47), (2.48), (2.49). In this case, where the refractive index is complex, two roots of the equation are equal at a cross-over frequency which is in general complex. This occurs where $D = 0$ or where

$$\frac{\sin^2\theta}{\cos\theta} = \pm i\frac{2\kappa_2\kappa_3}{\kappa_L\kappa_R - \kappa_1\kappa_3} = \pm iG^{-1} \tag{12.8}$$

where G is given by (2.93). From the quadratic for the polarization (2.92) the condition requires the polarization to be ± 1 which implies that it is linearly polarized at the cross-over. When there are no collisions G^{-1} is real and is zero at the cross-over. The condition is thus only satisfied for $\theta = 0$. When G^{-1} is complex with a small imaginary part its real part passes through zero near the cross-over frequency. It is then possible to find a real value of θ, θ_{CO}, say, which satisfies the condition. There is thus a pair of coupling points for $\theta = \pm\theta_{CO}$. These points coalesce when $\theta = 0$. The full theory of coupling in such circumstances is difficult and given, for example, by Budden [19, Chap. 16]. We can, nevertheless, understand it in principle by studying the behavior of the refractive index in the neighborhood of a coupling point.

The quantities κ_1, κ_2, κ_3, κ_R, κ_L are complex functions of height. For a wave with vertical wave normal making angle θ with the magnetic field we can study the behavior of the real and imaginary parts of the refractive index near the height where the coupling condition is satisfied for an angle θ_{CO}. It turns out that the behavior is different depending on whether θ is less than or greater than θ_{CO}. Figure 12.7 is a qualitative plot showing the nature of the refractive index in these two cases. The condition $D = 0$ is never satisfied exactly for real values of the height z.

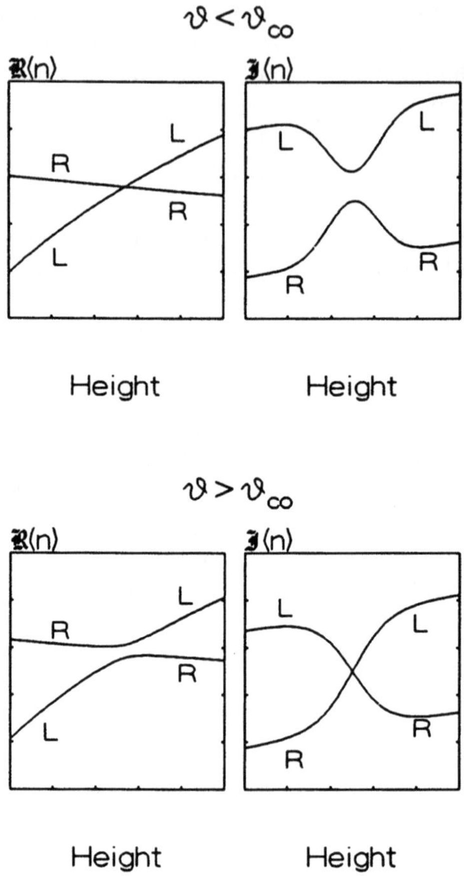

Figure 12.7: Qualitative behavior of the real and imaginary parts of the refractive index as a function of height near a coupling region.

The diagrams illustrate the general result that, for $\theta < \theta_{CO}$ the real parts are equal at a critical value of z in the coupling region while the small imaginary parts of the two roots do not become equal. The behavior of the real part is similar to that for exactly longitudinal propagation in the case of no collisions. For $\theta > \theta_{CO}$ the reverse is true. The imaginary parts are equal at the critical value of z and the real parts are not. In this case the behavior of the real part is similar to that for oblique propagation in the case of no collisions.

In the case of small collision frequency the propagation of the wave is governed by the real part of the refractive index. The imaginary part gives a small correction which represents damping of the wave due to collisions. We now see that, instead of coupling being limited to the case of a vertical magnetic field, we can define a *coupling cone* with vertical axis and semi-angle θ_{CO}. If the magnetic field direction lies within this cone of angles the two waves are coupled.

The angle θ_{CO} is a function of frequency and the height at which the cross-over occurs. Its value increases with increasing frequency. We are now in a position to explain all the features of the data.

Consider Figure 12.6. The 50 Hz and 150 Hz waves encounter a cross-over at relatively low altitude. This is associated with the change from heavy ions to He^+ ions. The coupling cone has a wide angle for these frequencies and altitudes and thus for most latitudes the OR-branch is coupled to the XR-branch and the XL- to the OL-branch. Thus at all frequencies the wave propagated upwards from the lower altitudes towards the satellite is a whistler wave with R polarization label.

Let A be the satellite at 1 000 km and B the satellite at 2 000 km. Consider each of the frequencies in turn. The 700 Hz wave reaches both satellites directly as a whistler wave with polarization label R. Frequencies in this range give rise to the uppermost portion of the idealized trace in Figure 12.2. The 500 Hz wave encounters a cross-over region near 750 km. The coupling cone is wide for the higher frequencies in this range and the OR-branch is coupled to the XR-branch so that these frequencies also contribute to the uppermost portion of the idealized trace. For lower frequencies in this range the coupling cone is narrower so that $\theta > \theta_{CO}$. The two branches are not coupled and these frequencies do not reach satellite B. The result is the missing portion of the electron whistler trace shown dotted in Figure 12.2. The OL wave does however, reach A. Frequencies in this range reach the lower satellite and form the ion cyclotron whistler on the idealized trace. In this case it is a proton whistler. The 250 Hz wave reaches A directly as a whistler with OR-label. Frequencies in this band contribute to the lowest part of the idealized trace. For satellite B, since the coupling cone is narrow at these frequencies, this frequency range forms the proton whistler. The same is true for the 150 Hz wave. The 50 Hz wave reaches both satellites as a whistler, forming part of the lowest portion of the trace.

Other scenarios are possible. Whether coupling occurs or not depends on whether θ_{CO} is less than or greater than θ. This in turn depends on the satellite altitude, the relative abundances of ions, the collision frequencies and the magnetic dip. Jones [62] discusses a number of different cases, including He^+ whistlers, which have been reported.

12.6 Summary

- Ion cyclotron whistlers are observed at medium latitudes by satellites at relatively low altitudes.
- The signal has the appearance of a fractional hop whistler, modified below the ion gyrofrequency by an ion cyclotron trace with a much longer delay.
- The theory of wave propagation in a uniform medium in a multi-ion plasma yields waves which have some of the properties of the observations.
- A ray theory treatment in a model of the ionosphere shows that, in a collisionless medium, the signal in the frequency range of interest cannot reach the satellite.
- The observations are explained by allowing for the effect of collisions. These permit coupling between different wave modes at levels where the frequency is equal to the cross-over frequency.

Chapter 13

Waves in the Plasmasphere – 4. Doppler Shifted Cyclotron Resonance of Electrons with Whistlers

13.1 Introduction

In the outer regions of the plasmasphere and in the plasmapause there is, as well as the cold plasma, a substantial population of energetic plasma of ring current origin. The highest energy particles of the van Allen belts are found on the innermost L-shells. In such regions v. l. f. waves are propagated in the whistler mode with the conditions being almost entirely determined by the cold plasma. Resonant hot electrons from the ring current population can cause amplification of existing waves or can be the source mechanism which generates waves in the whistler mode. The predominant interaction is Doppler shifted cyclotron resonance.

In this chapter we outline a number of different types of observations of whistler mode waves which require a wave-particle interaction to explain them. These include the amplification of lightning induced whistlers, hiss which is associated with a weakly turbulent spectrum of whistler mode waves, triggered emissions, chorus and discrete emissions. All these phenomena involve a Doppler shifted cyclotron resonance interaction between whistler mode waves and electrons with energies of a few tens of keV. Although the discrete emissions actually occur outside the plasmapause they involve the interaction of ring current electrons and cold plasma modes through Doppler shifted cyclotron resonance and are conveniently considered together with plasmasphere phenomena.

We then discuss the growth rate of whistler mode waves in a uniform medium and the consequent resonant diffusion of electrons with an arbitrary velocity space distribution. The dominant effect of the waves on the particles is pitch angle diffusion. It is shown that the particle distribution is driven towards marginal stability no matter what the initial distribution.

Next a non-uniform magnetosphere is discussed. An equilibrium is set up. Wave growth due to wave–particle interaction balances wave loss at the boundaries of the medium; pitch angle diffusion of particles into the loss cone balances the particle loss by precipitation. This leads to an understanding of the generation of broad band plasmaspheric hiss as well as an appreciation of the importance of this mechanism for understanding the dynamics of ring current electrons.

The amplification of lightning induced whistlers is discussed and mechanisms for generating a number of other whistler mode phenomena are outlined.

13.2 Some Relevant Observations

As a whistler is propagated from source to receiver it undergoes a number of loss processes. A relatively small fraction of the energy is transmitted through the boundary from the earth–ionosphere cavity to the ionosphere, there is absorption in the ionosphere and a small fraction of the signal reaches the receiver. It has been estimated that these losses may amount to 20 dB along the path [53, §3.11]. The observed intensity of whistlers suggests an amplification process which offsets these losses. Occasionally this amplification is so efficient that the whistler bounces many times back and forth between hemispheres. An example of such a multihop whistler is shown on a compressed time scale in Figure 13.1. Note the

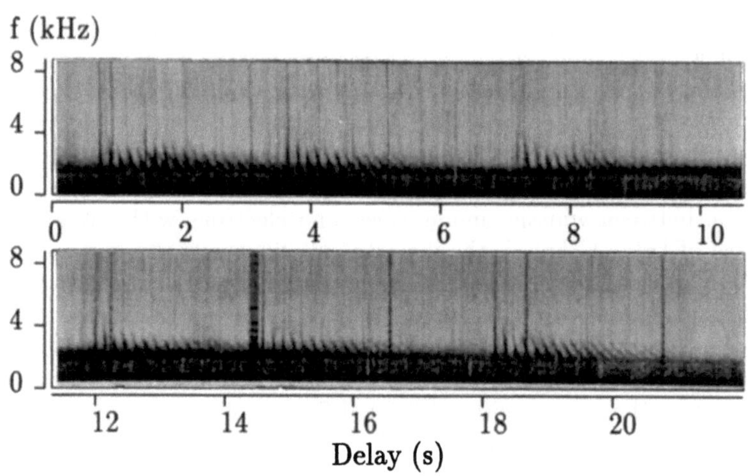

Figure 13.1: Multihop whistler from Sanae, Antarctica.

increasing dispersion with successive bounces. Such whistlers can either be observed in the same hemisphere as the lightning stroke, in which case successive traces have undergone an even number of bounces, or in the opposite hemisphere, in which case they have undergone an odd number of bounces.

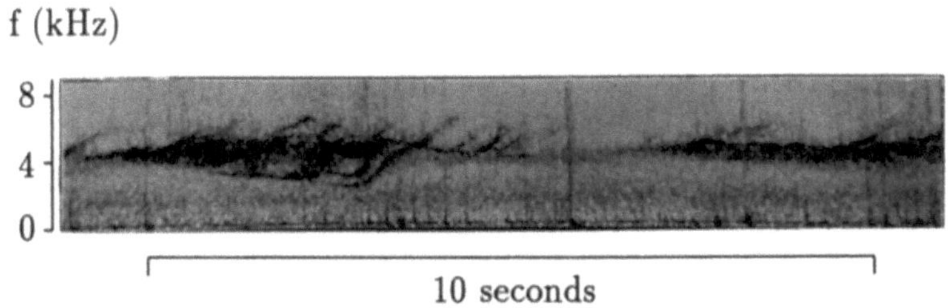

Figure 13.2: Dynamic spectrum of multiple emissions from Sanae Antarctica.

A second phenomenon which is regularly observed on the ground at times of ring current activity is *hiss*. This a broad of incoherent v. l. f. noise at frequencies of a few kilohertz. This featureless signal is associated with weak turbulence in the ring current plasma. The wave particle interaction which is responsible for it is a key mechanism in determining the dynamics and lifetime of the ring current particles. A band of such hiss can be seen in Figure 13.1 with an upper cut-off at about 2.2 kHz. Satellite observations show that whistler mode activity is dominated by a band of *plasmaspheric hiss* which fills the whole plasmasphere at frequencies of a few hundred Hz below the equatorial lower hybrid frequency.

Other discrete v. l. f. emissions occur in the region outside the plasmasphere. The shapes of their dynamic spectra give rise to names such as *hooks, risers* and *fallers*. In Figure 13.2 an example of complicated multiple emissions including hooks and risers is shown.

Discrete events may occur in clusters. Regular sequences of discrete events or clusters of discrete events may occur with periods of a few seconds. These are called periodic emissions. Closely spaced discrete events [60] produce an audio signal which reminded early workers of the sound of birds as the sun rose. As a result the phenomenon was named dawn chorus. This led to misunderstanding as there was no suggestion that the phenomenon occurred at dawn. It is now simply known as *chorus*.

Other v. l. f. signals appear to be triggered by external sources. In Figure 13.3 a multiple nose whistler is shown. Immediately following it

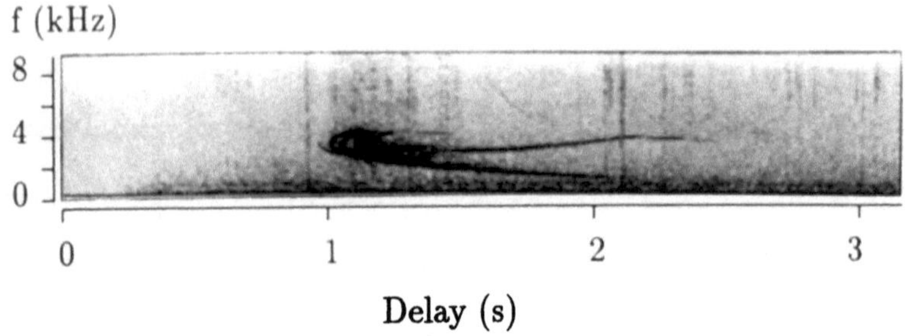

Figure 13.3: Triggered emission observed at Sanae Antarctica.

is a narrow band tone which rises slowly in frequency and appears to be triggered near the nose frequency of the whistler. It is an example of a *triggered emission*. A wide variety of such phenomena occur.

Triggered emissions can also be induced by man-made signals. Helliwell [53] has observed that Morse code signals are a good source. The dashes cause emissions and the dots do not. This suggests that the signal needs to endure for a sufficiently long time to organize the particles in velocity space so that the distribution becomes locally unstable. Such emissions have also been triggered by the Siple transmitter [55]

All these phenomena are associated with interactions between whistler waves and electrons through the Doppler shifted cyclotron resonance.

13.3 Whistler Wave–Particle Interaction in a Uniform Medium

13.3.1 Doppler Shifted Cyclotron Resonance

The interaction between whistlers and electrons through the Doppler shifted cyclotron resonance is described in §8.6.2. Resonance occurs when electrons have a parallel component of the particle velocity given by

$$v_{\|} = \frac{\omega - \Omega}{k_{\|}} \tag{13.1}$$

For the whistler mode $\omega < \Omega$ and the waves and particles move in opposite directions. In the Storey approximation

$$v_{\|} \simeq -\Omega/k_{\|} \tag{13.2}$$

The seminal paper on this topic is by Kennel and Petschek [70]. We shall base substantial parts of our treatment on theirs. Much of the importance of this work relates to its importance in understanding the dynamics of the ring current particles and is thus somewhat outside the scope of this work. This topic is covered in detail by Lyons and Williams [76, Chap. 5] who treat many of the extensions of the work. Our objective is to provide an illustrative example of the importance of wave–particle effects without necessarily making the models very realistic.

13.3.2 Growth Rate

In §7.4.4 we briefly studied the growth of a whistler wave in an anisotropic distribution. We saw that a loss-cone distribution was a suitable distribution for growth by Doppler shifted cyclotron resonance. In this section we shall make a further study of the growth rate, emphasizing the manner in which growth depends on the pitch angle distribution. No specific model is adopted for the distribution function. Instead parameters characteristic of the distribution function are defined in terms of integrals over velocity space and the growth rate and other important processes are discussed in terms of these parameters.

The plasma in the ring current region consists of cold ions and electrons originating from the ionosphere and energetic ions and electrons which have been injected from the tail. The injection process generally leads to a hot plasma distribution with $T_\perp > T_\parallel$. Particles with small pitch angles penetrate to low altitudes and are lost to the neutral atmosphere. There is thus also a loss cone. We shall be considering frequencies that are large compared with the lower hybrid frequency. The ions can then be assumed to form an infinitely massive background. The zero order distribution function f_0 describes all the electrons. It is expected to be an anisotropic loss-cone distribution including contributions from the cold and hot electron species.

A more detailed treatment than ours shows that the growth is maximum for parallel propagation. As the angle between the wave normal and the magnetic field increases the cyclotron resonance becomes weaker and, in addition, the longitudinal electric field leads to Landau damping which can dominate for large angles. For this reason we shall only consider parallel propagation with $k_\parallel = k$ and $\mathbf{k}_\perp = 0$.

We consider a plasma which is not far from equilibrium so that the growth is small. The growth rate is then given by (6.34)

$$\omega_i = -\frac{D_i(\omega_r, k)}{[\partial D_r/\partial\omega]}$$

The dispersion relation is found from (7.43). If $\omega_p^2 \gg \Omega\omega$, which is always the case in the regions of interest, this may be written

$$D \equiv \frac{c^2k^2}{\omega^2} + \frac{\omega_p^2}{\omega(\omega-\Omega)} - \frac{i}{\epsilon_0\omega}\Re(\sigma_R) = 0 \tag{13.3}$$

The plasma frequency ω_p is found using the total electron density of both hot and cold electrons. The derivative of the real part of the dispersion relation is easily shown to be

$$\frac{\partial D_r}{\partial \omega} = -\frac{\omega_p^2\Omega}{\omega^2(\omega-\Omega)^2} \tag{13.4}$$

The imaginary part of the dispersion relation arises from the contribution of electrons to the conductivity for the whistler wave (with polarization label R) which is found from (6.25)

$$\sigma_R = \frac{\pi q^2}{im}\int_{-\infty}^{\infty}\int_0^{\infty}\frac{v_\perp^2}{\omega}\frac{(\omega-kv_\parallel)\frac{\partial f_0}{\partial v_\perp}+kv_\perp\frac{\partial f_0}{\partial v_\parallel}}{\omega-kv_\parallel-\Omega}\,dv_\perp\,dv_\parallel \tag{13.5}$$

If α is the pitch angle of a particle then

$$v_\perp = v\sin\alpha \text{ and } v_\parallel = v\cos\alpha \tag{13.6}$$

and the operator appearing in the numerator of the integrand of (13.5) is

$$v_\parallel\frac{\partial}{\partial v_\perp} - v_\perp\frac{\partial}{\partial v_\parallel} = \frac{\partial}{\partial\alpha} \tag{13.7}$$

From an integration by parts we see

$$\int_0^{\infty} v_\perp^2\frac{\partial f_0}{\partial v_\perp}\,dv_\perp = -2\int_0^{\infty} v_\perp f_0\,dv_\perp \tag{13.8}$$

Thus we may write

$$\sigma_R = \frac{\pi q^2}{imk}\int_{-\infty}^{\infty}\frac{2\int_0^\infty v_\perp f_0\,dv_\perp}{v_\parallel-\frac{\omega-\Omega}{k}}dv_\parallel + \frac{\pi q^2}{im\omega}\int_{-\infty}^{\infty}\frac{v_\parallel\int_0^\infty v_\perp\tan\alpha\frac{\partial f_0}{\partial\alpha}\,dv_\perp}{v_\parallel-\frac{\omega-\Omega}{k}}dv_\parallel \tag{13.9}$$

The real part of σ_R is πi times the residue at the pole. Define the anisotropy,

$$A(v_{\text{res}}) = \left[\frac{\int_0^\infty v_\perp\tan\alpha\frac{\partial f_0}{\partial\alpha}\,dv_\perp}{2\int_0^\infty v_\perp f_0\,dv_\perp}\right]_{\text{res}} \tag{13.10}$$

where the subscript *res* implies that the quantity has been evaluated at $v_\parallel = v_{\text{res}} \equiv (\omega-\Omega)/k$. This is a measure of the degree of pitch angle anisotropy at the resonance. Also define the *fractional density of resonant particles*

$$\eta = \frac{2\pi(\Omega - \omega)}{Nk} \left[\int_0^\infty v_\perp f_0 \, dv_\perp \right]_{\text{res}} \tag{13.11}$$

These may be substituted in the expression for the conductivity and the result used in (13.3) and (6.34) to obtain the growth rate

$$\omega_i = \pi\Omega \left(\frac{\omega}{\Omega} - 1 \right)^2 \eta \left\{ A - \frac{1}{(\Omega/\omega) - 1} \right\} \tag{13.12}$$

The quantity A is a measure of the anisotropy of the plasma. For the type of distribution leading to whistler instability $\partial f_0/\partial\alpha$ is positive for $0 < \alpha < \pi/2$ and negative for $\pi/2 < \alpha < \pi$. Since $\tan\alpha$ changes sign at $\alpha = \pi/2$, A is positive for distributions for which instability is possible.

The total electron density is

$$N = 2\pi \int_{-\infty}^{\infty} \int_0^\infty v_\perp f_0 \, dv_\perp \, dv_\parallel$$

The density of electrons with parallel velocity components within a range $\Delta v_\parallel$ centered on the resonance velocity is $\Delta N_{\text{res}} = 2\pi\Delta v_\parallel \int_0^\infty v_\perp f_0 \, dv_\perp$. The resonant velocity is $V_{\text{res}} = (\Omega - \omega)/k$. Thus η is a measure of the density of electrons with parallel velocity in a range $\Delta v_\parallel = v_{\text{res}}$ centered on v_{res} and expressed as a fraction of the total electron density. For whistlers $\Omega > \omega$ and we have carried out our derivation on the assumption that k is positive. We have thus ensured that η is positive.

We are led to a condition for instability. If the growth rate is to be positive for waves of a particular frequency then, for this frequency,

$$A > \frac{1}{(\Omega/\omega) - 1} \tag{13.13}$$

The integrals in the expressions for A and η are evaluated along the resonance line in a region in velocity space such as that shown in Figure 8.3. In this case there are no resonances at the harmonics of the gyrofrequency. The resonance condition $v_\parallel = (\omega - \Omega)/k$ is represented for one frequency by a single vertical line. On different parts of the line $\partial f_0/\partial\alpha$ may be positive or negative. If when the expression for A has been integrated over all the resonant particles it is positive then there is net growth at that frequency. The position of the resonance line depends on frequency and, for an appropriate distribution function, electrons in a large region of velocity space may interact with the waves.

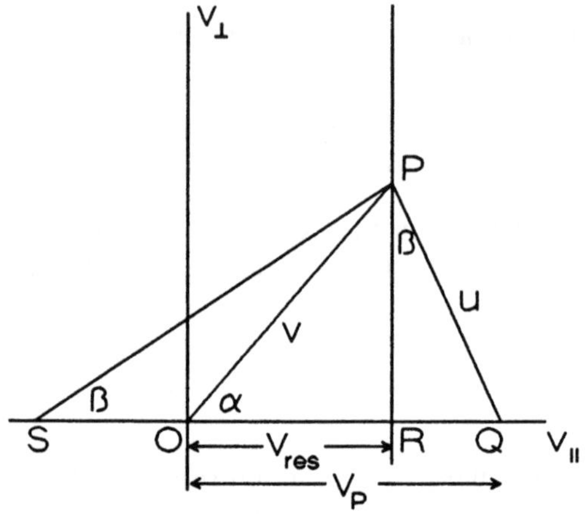

Figure 13.4: Relationship between orientation of single wave characteristic and pitch angle.

13.3.3 Diffusion of Resonant Particles in Velocity Space

We note that, in this type of situation, the single wave characteristics make small angles with the contours of constant energy. As energy increases the single wave characteristics become nearly coincident with the contours of constant energy. The major change which a resonant particle experiences as it diffuses along a single wave characteristic is a change in pitch angle. The change in energy can then be regarded as a second order effect.

To see how this comes about we consider diffusion along a single wave characteristic using the ideas of §8.4. A general diffusion equation in the $v_\parallel$–$v_\perp$ space would be a tensor equation of the form

$$\frac{\partial f_0}{\partial t} = \frac{\partial}{\partial v_j}\left[D_{jk}\frac{\partial f_0}{\partial v_k}\right]$$

The diffusion equation (8.28) is expressed in a different coordinate system in which the radius of the single wave characteristics and β are the coordinates. Rather than performing a formal transformation to velocity space coordinates to get the diffusion tensor we choose to find an equation for diffusion along single wave characteristics expressed in terms of pitch angle α and particle speed (or equivalently energy).

In Figure 8.7 β is the angle between the single wave characteristic and the $v_\parallel$-axis at resonance. To see how it is related to the pitch angle α consider Figure 13.4. In this diagram Q, a distance $V_P = \omega/k$ from the origin, is the center of a single wave characteristic which is a circle of radius u. PR is a portion of the line along which particles are resonant. It is parallel to the $v_\perp$-axis and a distance $v_\parallel = V_{\rm res} \equiv (\omega - \Omega)/k$ from it. PS is a tangent to the single wave characteristic and makes an angle β with it. The resonant particles at P have speed v and pitch angle α. Since $S\hat{P}Q$ is a right angle, $R\hat{P}Q = \beta$. From the diagram

$$
\begin{aligned}
u\cos\beta &= v\sin\alpha && (13.14)\\
u\sin\beta &= V_P - V_{\rm res} = \Omega/k && (13.15)\\
v\cos\alpha &= V_{\rm res} = (\omega-\Omega)/k && (13.16)
\end{aligned}
$$

Hence

$$\tan\alpha = \frac{ku\cos\beta}{\omega-\Omega} = \frac{\Omega\cot\beta}{\omega-\Omega} \qquad (13.17)$$

The rate of change of α with respect to β, evaluated along the single wave characteristic, is then easily shown to be

$$\alpha' \equiv \frac{d\alpha}{d\beta} = \cos^2\alpha + \left(1-\frac{\omega}{\Omega}\right)\sin^2\alpha \qquad (13.18)$$

In (8.26) B_L, $E_\parallel$ and $k_\perp$ are zero for a whistler wave propagated along the magnetic field. For the Doppler shifted cyclotron resonance $l = -1$ and thus the rate at which a resonant particle diffuses along the single wave characteristic is

$$\Theta_{-1} = \frac{qv\sin\alpha}{m\cos\beta}B_R = \frac{\Omega v\sin\alpha}{\cos\beta}\frac{B_R}{B_0} = \frac{qu}{m}B_R = \frac{\Omega v\sin\alpha}{\cos\beta}\frac{B_R}{B_0} \qquad (13.19)$$

The diffusion time, from (8.15), is

$$\tau = \frac{k\Omega}{\Delta k(\Omega-\omega)(\Omega+2\omega)} \qquad (13.20)$$

Thus the diffusion coefficient is

$$D = \frac{1}{2}\tau\Theta^2 = \frac{\Omega u^2}{(1-\omega/\Omega)(1+2\omega/\Omega)}\frac{k}{\Delta k}\left(\frac{B_R}{B_0}\right)^2 \qquad (13.21)$$

In the diffusion equation (8.28)

$$\frac{\partial}{\partial s} = -\frac{1}{u}\frac{\partial}{\partial\beta} = -\frac{\alpha'}{u}\frac{\partial}{\partial\alpha}$$

It therefore becomes

$$\frac{\partial f_0}{\partial t} = \frac{1}{(1-\omega/\Omega)(1+2\omega/\Omega)}\frac{(\alpha')^{3/2}}{\sin\alpha}\frac{\partial}{\partial\alpha}\left[\frac{k\Omega}{\Delta k}\left(\frac{B_R}{B_0}\right)^2(\alpha')^{1/2}\sin\alpha\frac{\partial f_0}{\partial\alpha}\right] \qquad (13.22)$$

This is a form which represents diffusion which is constrained to be along the single wave characteristic but expressed in terms of the rate of change with respect to the pitch angle. The energy change results because of the constraint that the particles must follow the characteristic. We have written the equation in this form because it clearly suggests a simple approximation. In the Storey approximation $\omega \ll \Omega$. Then $\alpha' \simeq 1$ and this may be written

$$\frac{\partial f_0}{\partial t} = \frac{1}{\sin\alpha}\frac{\partial}{\partial\alpha}\left\{\Omega\frac{k}{\Delta k}\left(\frac{B_R}{B_0}\right)^2 \sin\alpha\frac{\partial f_0}{\partial\alpha}\right\} \tag{13.23}$$

In this approximation the single wave characteristics are centered on the origin and the diffusion is pure pitch angle diffusion. Changes in energy only occur when the next order of approximation is taken into account. This is the form of the diffusion equation used by Kennel and Petschek [70]. Note that, even at the whistler nose frequency, the error in making the approximation is less than 10%; at the half gyrofrequency cut-off it is 25%. It should be borne in mind that errors in neglecting oblique propagation may also be as large as this. Nevertheless the approximation is valuable for illustrating the techniques to be used in studying such problems.

For small Δk the quantity $\tilde{B}(k)^2 \equiv B_R^2/\Delta k$ is $2\mu_0$ times the energy per unit wave number at resonance. The diffusion coefficient may then be written

$$D = \Omega\frac{k\tilde{B}(k)^2}{B_0^2} = -\frac{\Omega^2}{v\cos\alpha}\frac{\tilde{B}(k)^2}{B_0^2} \equiv \frac{D^*}{|\cos\alpha|} \tag{13.24}$$

Note that, for positive k, $\alpha > \pi/2$ and D is positive. The diffusion coefficient is a measure of the ratio of the total wave energy in a bandwidth k to the background magnetic field energy[1].

13.3.4 Effect of Resonant Diffusion on the Particle Distribution

The effect of resonant diffusion on an unstable distribution is to drive it towards stability. In Chapter 8 it was seen that, for instability the gradient of the distribution function was in such a direction that, on average, particles lost energy and the wave gained energy during diffusion. In this process they moved in a direction such as to reduce the unstable gradient in the distribution function.

In order to study the way in which the distribution function is modified consider Figure 13.4. This corresponds to a single wave frequency

[1] Kennel and Petschek [70] obtained the same formula. In their formula $k^* \equiv \Omega/v = -k\cos\alpha$.

which resonates with particles located along the line PR in velocity space. For this frequency, from (13.17) the single wave characteristics cross the resonance line at angles making angles β with the $v_{\|}$-axis such that

$$\frac{dv_{\perp}}{dV_{\text{res}}} \equiv \tan\beta = -\frac{\Omega V_{\text{res}}}{(\Omega - \omega)v_{\perp}} \tag{13.25}$$

In the case of Doppler shifted cyclotron resonance with whistler waves it can be anticipated that there is a broad band spectrum of waves with frequencies below the gyrofrequency. Each component wave in the spectrum resonates with a different V_{res}, given by

$$V_{\text{res}} = -\frac{(\Omega - \omega_{\text{res}})}{k} \tag{13.26}$$

Equation (13.25) can be integrated to define a path in velocity space along which the resonant particles diffuse. At each point along the path particles resonate with waves of different frequency. For reasons which will become apparent this curve is called a *contour of marginal stability.* It is not a single wave characteristic which corresponds to a locus on which particles can resonate with a wave of a particular frequency. In a given situation the only points on a single wave characteristic which actually correspond to resonance are the intersections of the single wave characteristic with the resonance condition.

The contour of marginal stability is so named because, if it is a contour of constant f_0, then the gradient of the distribution function along this contour is zero. The net flux of resonant particles is proportional to df_0/ds, where s is measured along a single wave characteristic. At all points along the contour of marginal stability the contour is tangent to the single wave characteristic for the corrresponding resonance frequency. Thus, along this contour, the flux of resonant particles is zero.

If the contours of constant f_0 correspond everywhere with the contours of marginal stability then the distribution function is everywhere marginally stable. A small disturbance in the distribution function will produce a local gradient along the contour. Depending on the direction of the gradient the distribution function becomes unstable or overstable in this region. If it becomes unstable the resulting local flux of particles in velocity space produces wave growth and the distribution function will relax to marginal stability. If it becomes overstable there is no further change unless a wave from outside is propagated through the plasma with the right frequency to resonate with the overstable particles. Such a wave will be damped; energy will be transferred to the particles and they will diffuse in such a direction as to force the distribution towards marginal stability. In space there are many sources of such waves and plasmas are often close to marginal stability.

It is easy to show that the marginal stability condition is satisfied everywhere if f_0 is constant along contours of marginal stability. We can use (13.14), (13.15) and (13.16) to show that, along a single wave characteristic,

$$\frac{\partial f_0}{\partial s} = \sin\beta\frac{\partial f_0}{\partial v_\perp} + \cos\beta\frac{\partial f_0}{\partial v_\parallel} = \frac{1}{ku}\left\{(\omega - kv_\parallel)\frac{\partial f_0}{\partial v_\perp} + kv_\perp\frac{\partial f_0}{\partial v_\parallel}\right\} \quad (13.27)$$

The numerator of the integrand of (13.5) is proportional to this. Thus, if $\partial f_0/\partial s = 0$ along all contours of marginal stability, then the condition (13.13) is satisfied automatically.

For whistlers we can explicitly find the contours of marginal stability from (13.25). To do this we express V_{res} and $v_\perp$ as functions of the parameter ω_{res}. We use the real part of the approximate dispersion relation (13.3) to replace k in (13.26) with the result

$$\left(\frac{V_{\text{res}}}{c}\right)^2 = \frac{(\Omega-\omega)^3}{\omega\omega_p^2} \quad (13.28)$$

Define normalized velocity space coordinates

$$x = \frac{\omega_p V_{\text{res}}}{\Omega c}; \quad y = \frac{\omega_p v_\perp}{\Omega c} \quad (13.29)$$

The expression for x can be differentiated and substituted in (13.25) to get an expression for y as a function of ω_{res}. The result is

$$y\frac{dy}{d\omega_{\text{res}}} = -\frac{1}{2\Omega}\left\{2 - \frac{\Omega}{\omega_{\text{res}}} - \frac{\Omega^2}{\omega_{\text{res}}^2}\right\} \quad (13.30)$$

This can be easily integrated. We thus have two equations for x and y with ω_{res} as a parameter

$$x^2 = \frac{\Omega}{\omega}\left(1-\frac{\omega}{\Omega}\right)^3 \quad (13.31)$$

$$y^2 = -\frac{\Omega}{\omega}\left(1 - \frac{\omega}{\Omega}\ln\frac{\omega}{\Omega} + \frac{2\omega^2}{\Omega^2}\right) + a^2 \quad (13.32)$$

where a is an arbitrary constant which defines the contour. A set of such contours for various values of a is shown in Figure 13.5. It can immediately be seen that the marginally stable distribution is anisotropic with $T_\perp > T_\parallel$. It can also be seen that for larger values of x and y they approximate closer and closer to circles. In the Storey approximation the contours become circles $x^2 + y^2 = a^2$. It is this case that was discussed by Kennel and Petschek [70].

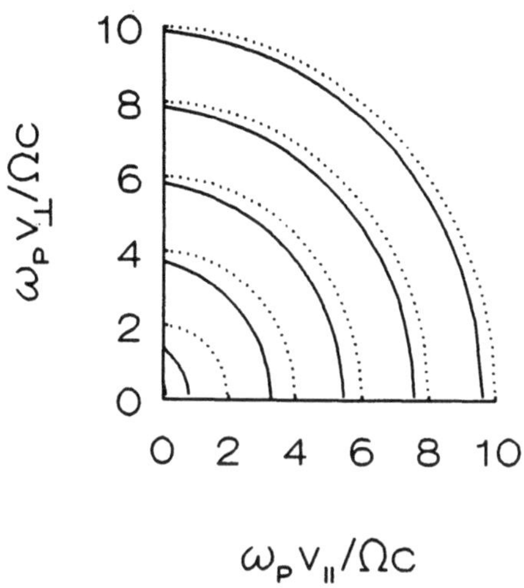

Figure 13.5: Contours of marginal stability in normalized velocity space. The dotted curves are circles centered on the origin.

13.4 Whistler-Mode Noise in a Non-Uniform Medium

The actual situation in the magnetosphere is much more complicated than in a uniform medium. The energetic particles bounce back and forth along the field lines. Those particles with small equatorial pitch angles penetrate down to the neutral atmosphere and are lost by precipitation; in the absence of replenishment the loss cone would be empty. Such a distribution is unstable to whistler waves which therefore grow in amplitude, causing resonant particles to diffuse towards the loss cone, partially filling it. These particles are, in due course, precipitated and lost. The whistler waves are propagated along the field lines and are partially reflected at the ionosphere; some of the wave energy is transmitted and lost. It is assumed that the particles are continually replenished by an unspecified injection mechanism from the higher latitude ring current. The objective of this section is to study the steady state situation in which the energy lost, both through wave transmission out of the region and the loss of precipitated particles, is balanced by the energy of injected particles. In this way we find the expected steady state distribution function and estimate a limit on the trapped particles.

The treatment involves several simplifications. The first is that the Storey approximation is adequate for describing the waves so that the diffusion is purely in pitch angle. The second is that the injected particles are all injected with a 90^0 pitch angle. This is a clear oversimplification but adequate to illustrate the technique.

There are three characteristic times associated with the process. The particle lifetime τ_L is a pitch angle diffusion time, the mean time from injection until it has reached the loss cone. The particle escape time τ_E is the time taken for the particle to be precipitated once it has reached the loss cone. The wave propagation time τ_W is the time for a wave packet to be propagated from equator to the ionosphere, where it is partially reflected and transmitted, and back to the equator. An order of magnitude estimate of the field line half-length is LR_E and thus $\tau_E \simeq LR_E/V_{\text{res}}$ and $\tau_W \simeq 2LR_E/V_G$. The lifetime τ_L is determined by the equilibrium.

13.4.1 Wave Growth and Loss

The waves are assumed to grow with a growth rate ω_i given by (13.12). This depends on the position along the field line. As a wave is propagated from equator to ionosphere its amplitude grows by a factor $\exp\langle\omega_i\rangle\tau_W$ and its intensity by the square of this. Since most of the growth occurs and the group speed is smallest in the equatorial region, the mean value of the growth rate is close to its equatorial value. Let the reflection coefficient for the wave energy at the ionosphere be R. Then, if the total wave energy is to remain constant, $R\exp 2\langle\omega_i\rangle\tau_W = 1$ or

$$\langle\omega_i\rangle = \frac{1}{2\tau_W}\ln\frac{1}{R} \tag{13.33}$$

13.4.2 Diffusion into the Loss Cone

The loss cone is the cone of equatorial pitch angles $\alpha < \alpha_0$ within which particles are lost to the atmosphere. Because α_0 is small within the loss cone $\sin\alpha \simeq \alpha$ and $\cos\alpha \simeq 1$. The diffusion coefficient (13.24) is thus $D \simeq D^*$. If the wave spectrum $\tilde{B}(k)$, which depends on α through $k = (\omega - \Omega)/v\cos\alpha$, is fairly flat then we may assume that D^* is independent of α. The pitch angle diffusion equation must be modified to allow for the particle loss. We can approximate $(\partial f_0/\partial t)_{\text{loss}}$ by $-f_0/\tau_E$. The diffusion equation can be written

$$\frac{\partial f_0}{\partial t} = \frac{D^*}{\alpha}\frac{\partial}{\partial\alpha}\left(\alpha\frac{\partial f_0}{\partial\alpha}\right) - \frac{f_0}{\tau_E} \tag{13.34}$$

For equilibrium $\partial f_0/\partial t = 0$ and thus

$$\frac{1}{\alpha}\frac{\partial}{\partial\alpha}\left(\alpha\frac{\partial f_0}{\partial\alpha}\right) = \frac{f_0}{D^*\tau_E} \tag{13.35}$$

This equation has the solution (B.7)

$$f_0(v,\alpha) = F(v) I_0\left(\frac{\alpha}{\sqrt{D^*\tau_E}}\right) \tag{13.36}$$

where I_0 is the modified Bessel function of the first kind and $F(v)$ is an arbitrary function of v. In this region, since $I_0' = I_1$ (B.9),

$$\frac{\partial f_0}{\partial \alpha} = \frac{F(v)}{D^*\tau_E} I_1\left(\frac{\alpha}{\sqrt{D^*\tau_E}}\right) \tag{13.37}$$

Outside the loss cone particle injection is assumed to occur at $\alpha = \pi/2$. There is thus a source term

$$\left(\frac{\partial f_0}{\partial t}\right)_S = S(v)\delta\left(\alpha - \frac{\pi}{2}\right)$$

in the diffusion equation which, in a steady state, may be written

$$\frac{\partial}{\partial \alpha}\left(D^* \tan\alpha \frac{\partial f_0}{\partial \alpha}\right) = S(v)\sin\alpha\,\delta\left(\alpha - \frac{\pi}{2}\right) \tag{13.38}$$

This may be integrated twice to yield

$$\frac{\partial f_0}{\partial \alpha} = \frac{S(v)}{\sqrt{D^* \tan\alpha}} \tag{13.39}$$

and

$$f_0(v,\alpha) = \frac{S(v)\ln\sin\alpha}{D^*} + G(v) \tag{13.40}$$

where $G(v)$ is an arbitrary function of α.

At the boundary of the loss cone, $\alpha = \alpha_0$, both f_0 and $\partial f_0/\partial\alpha$ must be continuous. If these conditions are applied we can solve for the arbitrary functions $F(v)$, $G(v)$. The distribution function then becomes

$$f_0(v,\alpha) = \begin{cases} \dfrac{S(v)}{D^*\tan\alpha_0} \dfrac{\sqrt{D^*\tau_E}}{I_1(\xi_0)} I_0(\xi) & \alpha < \alpha_0 \\ \dfrac{S(v)}{D^*}\left[\ln\dfrac{\sin\alpha}{\sin\alpha_0} + \dfrac{\sqrt{D^*\tau_E}}{\tan\alpha_0}\dfrac{I_0(\xi_0)}{I_1(\xi_0)}\right] & \alpha > \alpha_0 \end{cases} \tag{13.41}$$

where

$$\xi = \frac{\alpha}{\sqrt{D^*\tau_E}}; \quad \xi_0 = \frac{\alpha_0}{\sqrt{D^*\tau_E}}$$

To this should be added the distribution function for cold electrons,

$$f_c(v) = N_c\delta(v)$$

It can thus be seen that, in this approximation, the dependence of f_0 on v is entirely determined by the distribution of the injected particles. The pitch angle distribution, however, is entirely determined by the diffusion coefficient, the pitch angle and the particle lifetime τ_E.

There are two extreme cases, depending on the value of ξ_0. These correspond to $\alpha_0 \gg \sqrt{D^*\tau_E}$ and $\alpha_0 \ll \sqrt{D^*\tau_E}$. We can use the asymptotic expansion for the Bessel functions in the first case and the approximation for small argument in the second. It is also realistic to assume a small angle for the loss cone so that $\tan\alpha_0 \simeq \alpha_0$. For $\alpha_0 \gg \sqrt{D^*\tau_E}$ we get

$$f_0(v,\alpha) = \begin{cases} \dfrac{S(v)}{D^*}\dfrac{e^{(\xi-\xi_0)}}{\sqrt{\xi_0\xi}} & \alpha < \alpha_0 \\ \dfrac{S(v)}{D^*}\left[\ln\dfrac{\sin\alpha}{\sin\alpha_0} + \dfrac{1}{\xi_0}\right] & \alpha > \alpha_0 \end{cases} \tag{13.42}$$

and for $\alpha_0 \ll \sqrt{D^*\tau_E}$

$$f_0(v,\alpha) = \begin{cases} \dfrac{S(v)}{D^*}\dfrac{2}{\xi_0^2}\left(1 - \dfrac{\xi_0^2}{8} + \dfrac{\xi^2}{4}\right) & \alpha < \alpha_0 \\ \dfrac{S(v)}{D^*}\left[\ln\dfrac{\sin\alpha}{\sin\alpha_0} + \dfrac{2}{\xi_0^2}\left(1 + \dfrac{\xi_0^2}{8}\right)\right] & \alpha > \alpha_0 \end{cases} \tag{13.43}$$

The pitch angle distribution is shown for several values of α in Figure 13.6.

13.4.3 Energy Balance

We have so far balanced the particle injection and loss budget on the assumption that the wave energy density is given. It remains to be seen how this wave energy density could be predicted from theory.

The growth rate depends both on the anisotropy A (13.10), and the fractional density of resonant particles (13.11). The predicted pitch angle distribution depends on the magnitude of D^* which is directly proportional to the wave intensity. The pitch angle distribution determines A but not η. From the pitch angle distribution the growth rate ω_i could be computed as a function of wave intensity apart from a multiplying factor which is the amplitude of η and depends on the ratio of the number of electrons being injected and the number of cold electrons which do not participate in the process. If the reflection coefficient of the ionosphere is known then, for self consistency, the growth rate may also be found from (13.33). Thus the amplitude of η may be computed. The magnitude of the injected particle distribution $S(v)$ is thus determined as a function of wave intensity. This function can then be inverted to give the wave intensity as a function of the rate at which particles are being injected by an unspecified mechanism. This will not be done in detail because a number of generalizations of the pitch angle diffusion theory would be necessary to make such a calculation realistic.

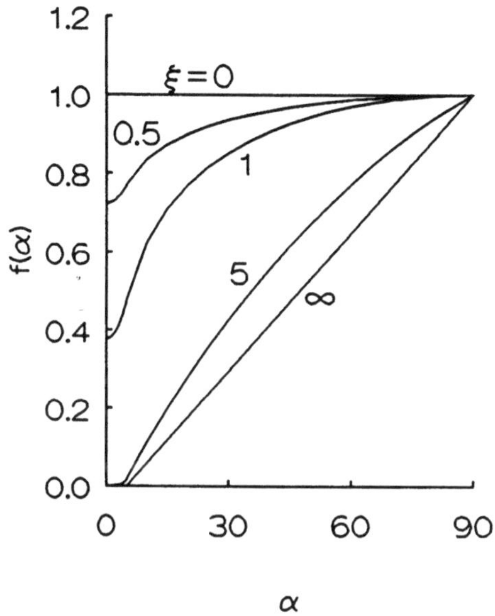

Figure 13.6: Pitch angle distribution as a function of α. The parameters shown are $\xi = \alpha/\sqrt{D^*\tau_E}$.

13.4.4 Generalizations of the Theory

It must be emphasized that this treatment is a considerable oversimplification of the actual situation in the plasmasphere. It does, however, indicate in principle how an understanding of the behavior of energetic particles interacting with whistler waves can be reached.

Considerable success has been achieved in generalizing the theory so that it can be compared with observations of particle distributions in the plasmasphere. Lyons and Williams [76] give a very complete survey. Some of the important generalizations are

- In the theory we have used mean values of growth rate and diffusion coefficient which are essentially equal to the equatorial values. A full theory would evaluate the balance at all points along the field line.

- The restriction that $\omega \ll \Omega$ can be relaxed. In this case diffusion in both pitch angle and energy must be taken into account.

- If the restriction on propagation parallel to the field is removed then, for wave normals making a relatively large angle with the field, it can be seen (8.20) that interactions leading to growth take place with both R- and L-waves as well as with the parallel electric field through Čerenkov resonance [71]. These resonances occur at all the harmonics of the Doppler shifted gyrofrequency. A more general diffusion equation takes all these resonances into account [68]. A particle at a single point in velocity space may be simultaneously resonant with different resonances at different frequencies. These give rise to competing growth and loss processes which must all be taken into account in evaluating the budgets of particle and energy growth and loss.

- The assumption that all the injection occurs at a pitch angle of 90^0 is clearly not accurate. A realistic form for the distribution of the injected particles requires an understanding of the injection process. This process arises from interactions between the particles and ultra-low frequency hydromagnetic waves on a time scale related to the bounce and drift periods of the particles. This leads to resonance on a different time scale. Instead of pitch angle diffusion there is diffusion across L-shells. The balance between this radial diffusion and the loss processes due to pitch angle diffusion has been studied by a number of authors. The results are summarized by Lyons and Williams [76].

- The effect of positive ions on propagation has not been taken into account. At frequencies below the lower hybrid frequency the wave is in region 11 of the CMA diagram. As we have seen, in this region, the refractive index surface is closed and there is no resonance cone. The formulae which we have used are accurate for propagation along the field but do not allow propagation perpendicular to the field. This has a two-fold effect. On the one hand resonances for waves propagated perpendicularly to the field must be taken into account and on the other hand the waves are not confined to the region in which they are generated. This is described by Lyons *et al.* [78]. Magnetospheric reflection similar to that described for the MR whistler in §11.5.2 takes place. The wave can thus be propagated far from the field line on which it originates. Ray tracing shows that it bounces back and forth until it reaches the plasmapause. It then encounters a steep decreasing density gradient which reflects it as described in §11.6.3. The energy thus spreads throughout the plasmasphere [107]. This is the explanation of the plasmaspheric hiss which fills the plasmasphere below the equatorial lower hybrid frequency.

When all such effects are taken into account in order to model the pitch angle distributions of energetic radiation belt electrons throughout the plasmasphere impressive agreement between theory and observations has been obtained [78].

13.4.5 Ion Cyclotron Waves and Proton Pitch Angle Diffusion

Just as electrons interact through electron cyclotron resonance with whistler mode waves so positive ions interact with ion cyclotron waves in the region below the ion gyrofrequency. The sense of polarization is reversed but otherwise the theory is essentially the same. The ray direction of the ion cyclotron mode is more nearly parallel to the magnetic field and thus the theory above is more closely applicable as the waves are confined to the region where the pitch angle diffusion occurs and do not fill the plasmasphere. The theory is consistent with observations of proton pitch angle distributions [66].

13.4.6 Amplification of Whistlers

No further theory is needed to understand the amplification of lightning induced whistlers. The pitch angle distribution of ring current electrons is determined through the balance set up as electrons are injected, undergo pitch angle diffusion by interacting with whistler mode noise and are precipitated. The resulting equilibrium demands a pitch angle distribution which allows positive growth of waves. If a strong, externally generated wave such as a whistler passes through such a region in time τ its intensity increases by an amount $2\omega_i\tau$. Even when the turbulence is weak this may be sufficient to compensate for the losses to which it is subject. On rare occasions conditions may be favorable enough for multihop propagation as shown in Figure 13.1.

13.5 Whistler-Mode Signals Generated by Energetic Particles

The wide variety of discrete emissions, triggered emissions, and chorus are all thought to be initiated by Doppler shifted cyclotron resonance. A general anisotropic distribution cannot produce discrete signals such as these. Any generation mechanism must require that the particles have well-defined energy and are somehow localized in space and time. The variety of spectral shapes arises both because of the motion of the particle source along the field line and because of whistler wave propagation effects. The manner in which the source particles are organized into appropriate bunches requires a suitable non-linear mechanism. We shall discuss such mechanisms only briefly and in outline.

13.5.1 Propagation Effects

Dowden [32] was able to reproduce the spectral shapes of a variety of emissions by considering the combined effect of a source moving along the field line and the delay due to propagation effects. Such a theory requires a *deus ex machina* to inject a bunch of electrons with the same energy and pitch angle somewhere near the equatorial plane. Suppose such a bunch initially moves along the field line towards the observer on the ground, is mirrored and moves back to the opposite mirror point before losing coherence. The Doppler shifted cyclotron mechanism causes radiation in the opposite direction to the particle motion as shown in §13.3.1. The observer thus sees nothing until the particle is mirrored.

Just after the particle bunch is mirrored it begins to radiate towards the observer. The radiation frequency is

$$\omega = \Omega + kv_{\parallel} = \Omega + \frac{\omega n}{c}v_{\parallel} \tag{13.44}$$

where $v_{\parallel}$ is negative. Initially the gyrofrequency is high and the parallel velocity small so that a high frequency is radiated. As the particle recedes the frequency gets lower and the delay larger, reaching a minimum at the equator. On the other side of the equator the frequency rises and the delay increases. Here, however, there is a region of smaller gyrofrequency between source and observer; as the wave frequency approaches the equatorial gyrofrequency its delay increases without limit. Frequencies above the equatorial frequency cannot reach the observer. The result is a hook-like spectral form.

The refractive index is given by

$$n^2 = \frac{\omega_p^2}{\omega(\Omega - \omega)} \tag{13.45}$$

The parallel velocity is determined by the conditions that the energy should remain constant, $v_{\parallel}^2 + v_{\perp}^2 = v_{\rm eq}^2$, and that the magnetic moment should be constant, $v_{\perp}^2/\Omega = v_{\rm eq}^2 \sin^2 \alpha_{\rm eq}/\Omega_{\rm eq}$, where the subscript *eq* refers to the equatorial value of the variable. When these conditions are substituted in (13.44) we get

$$(\Omega - \omega)^3 = \omega_p^2 \Omega \frac{v_{\rm eq}^2}{c^2}\left(1 - \sin^2 \alpha_{\rm eq}\frac{\Omega}{\Omega_{\rm eq}}\right) \tag{13.46}$$

The plasma and gyrofrequencies are functions of latitude. Dowden [32] computed the frequency as a function of latitude in a dipole model with ω_p^2 proportional to Ω. The time delay is the sum of the time taken for the particle to reach the point at which a particular frequency is radiated and

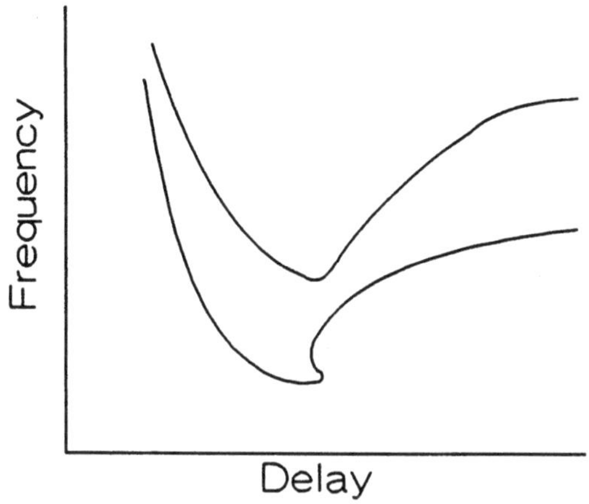

Figure 13.7: Combined effects of particle and propagation delay for a particle bunch.

the wave propagation time from that point to the observer. Clearly such computations will depend somewhat on the model. The chief features are shown qualitatively in Figure 13.7. The frequency range is of the order of 10 kHz and the time scale two or three seconds. The upper curve shows the typical shape for particles with energies of some 5 keV. The lower curve shows the typical shape for particles with energies of a few tens of keV. By invoking different initial conditions a variety of spectral shapes can be produced. For example a particle which initially is propagated away from the observer and is precipitated at the mirror point will yield a rising tone. One which moves towards the observer, is mirrored and then lost at the equatorial plane produces a falling tone. Multiple bounces produce multiple hooks and so on.

It is clear that there is a certain amount of *ad hoc* argument; any full theory must predict how such particle bunches are formed and lost. The number of free parameters allows a great deal of liberty in producing spectral shapes such as those observed. Nevertheless it is highly probable that such propagation effects are important in the production of discrete and periodic emissions. In the following we shall see how non-linear effects which lead to particle bunching can be used to explain the initiation of such phenomena. On the whole the observations do not provide enough information to constrain the speculations of the theorists.

13.5.2 Non-Linear Effects

In §8.5.2 we have seen that, if the wave and particles interact for a sufficiently long time, non-linear effects eventually may lead to the bunching of particles in phase. Such bunching forms the basis of theories of the generation of v. l. f. emissions. The non-linear processes do not lend themselves to easy analysis. The major attacks have been either phenomenological [54, 56] or by computer simulation [80, 81]. There is no universal agreement on the details of the process [82]. In this section we indicate the physical processes which must be taken into account in such a theory but do not perform any detailed calculations.

A crucial observation is that Morse code signals can induce triggered emissions. When this occurs the emissions originate from the dashes but not from the dots. The implication is that the coherent signal must last a sufficiently long time in order to bunch the particles in phase so that they can begin to radiate. The key features of any viable theory are

- The interaction takes place in a region of space, usually assumed to be at the equatorial plane, within in which waves and resonant particles following their unperturbed paths would remain in phase.

- A triggering signal in the whistler mode, such as a Morse code dash or a strong whistler or a previously generated emission which has been reflected from the ionosphere, is propagated through the interaction region. It undergoes a Doppler shifted cyclotron resonance with electrons spiralling in the opposite direction. It grows as it is propagated through the interaction region.

- As it grows non-linear effects become important and phase bunching of the electrons takes place. The phase bunched electrons move in the opposite direction to the wave. They constitute an oscillating circularly polarized transverse current which is expected to radiate further whistler mode waves.

- The phase bunched electrons leaving the interaction region at the point where the initiating wave left it constitute a feedback mechanism. If the non-linear effect is sufficiently strong the process may become self-sustaining. The interaction region acts like an oscillator, producing an approximately monochromatic signal such as the triggered emission shown in Figure 13.3.

- Even if feedback is not important the bunched electrons continue down the field line and may radiate a varying frequency as described in §13.5.1 thus generating discrete emissions.

Understanding of the non-linear processes is not complete and further work remains to be done.

13.6 Summary

- A variety of v. l. f. wave phenomena which are observed appear to involve wave amplification or generation through Doppler shifted cyclotron resonance between whistler waves and electrons spiralling in the opposite direction along the field. These include whistler amplification, discrete emissions, hiss and triggered emissions.

- Computations of the growth rate for whistlers show that growth takes place when the anisotropy of the distribution function exceeds a critical value. This can occur for anisotropic distributions with $T_{\perp} > T_{||}$ or loss cone distributions.

- The main effect of the wave on the particles is pitch angle diffusion. The change in energy is a smaller effect. In a uniform plasma, wave generation drives the distribution towards marginal stability. Waves from outside the region of interest would drive an overstable distribution towards marginal stability.

- In a more realistic magnetospheric model an equilibrium can be set up for interaction between ring current electrons and whistler noise. Particles are injected from higher L-values by cross-L diffusion and removed from the loss cone by precipitation in the neutral atmosphere. Waves cause pitch angle diffusion of the injected particles into the loss cone. The diffusion coefficient is proportional to the wave intensity. The waves are partly lost and partly reflected at the ionosphere. In a steady state the equilibrium requires conservation of the particle number density and of wave and particle energy. The steady state distribution function and the wave intensity can be predicted and are determined by the rate of particle injection. This is a crucial mechanism in understanding ring current dynamics.

- The distribution function thus found is unstable to waves propagated from outside such as lightning-generated whistlers. They are thus amplified. This amplification may be sufficient to counterbalance propagation losses so that the whistler undergoes many bounces.

- Bunched particles may generate whistler mode waves of varying frequency as they spiral along a field line. The combined wave and particle delay times give rise to signals showing the features of discrete emissions.

- Non-linear effects due to a triggering wave give rise to phase bunching of the electrons. The radiation from these may be fed back to the system to provide the self-sustaining oscillation which is needed to cause a triggered emission. Alternatively the bunched particles may be the source of discrete emissions.

Chapter 14

Waves in the Auroral Region

14.1 Introduction

In this chapter we consider a number of phenomena which occur in the region which maps down to the auroral oval. Roughly speaking, this can be defined by the Region III field lines between the plasmapause and the magnetopause. Statistically the auroral oval is the region of maximum auroral occurrence. As seen in Figure 9.8, this is the only magnetospheric region in which the plasma frequency may be less than the gyrofrequency over significant volumes. Thus wave modes which are not seen elsewhere may be generated.

The region is filled with a wide spectrum of waves with frequencies ranging from a few hundred hertz to to one or two megahertz. Some are localized near their source while others fill the whole region. Those with frequencies above the solar wind plasma frequency escape from the magnetosphere into the solar wind; the earth is an astronomical radio source. The chief sources of the waves are located at low altitudes above auroral arcs and at high altitudes in the equatorial plane just outside the plasmapause.

All these waves can be characterized as cold plasma modes which have been excited by high energy particles. In this chapter we describe the nature of the wave modes which exist and investigate some propagation phenomena which are important in understanding them. In particular coupling between Z-mode and ordinary or whistler modes through the so-called Ellis or Jones windows may be important features of their propagation. The generation mechanisms of the waves are outlined.

14.2 Observations

The earliest phenomenon associated with auroral zone waves to be observed was received on the ground as *auroral hiss* [53]. The same phenomenon was seen by low altitude satellites as illustrated in Figure 14.1. This is an example of a *v. l. f. saucer*, a band of hiss which occurs as the satellite passes over a region of auroral activity. The frequency of the band drops to a minimum when the satellite is directly over the source region.

Since these observations a wide variety of wave types have been iden-

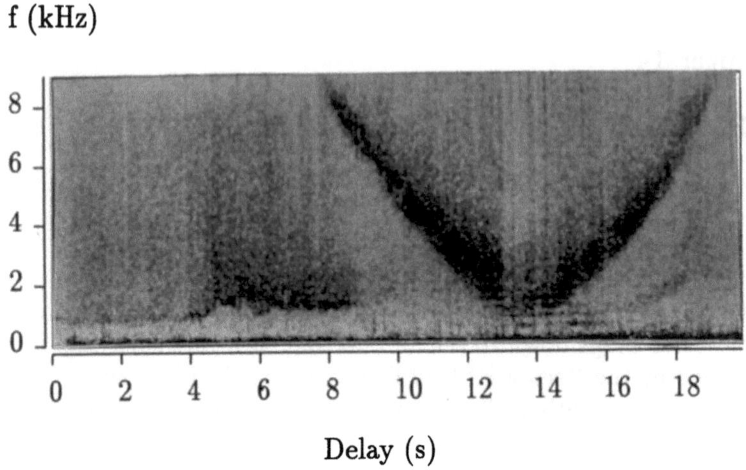

Figure 14.1: V. l. f. saucer observed on ISIS 2.

tified by many satellites. In the following we summarize the results of numerous workers.

14.2.1 Waves Originating from Regions of Auroral Activity

A fairly comprehensive picture is now available of the waves originating from regions of auroral activity. Above such regions the dependence of the CMA boundary frequencies on altitude is shown in Figure 9.8. It is also of interest to note how the cut-off and resonance frequencies behave as a function of latitude. This is shown in a schematic way in Figure 14.2. At low L-values the electron density, or equivalently the plasma frequency, is large in the plasmasphere. As L increases the plasma density decreases dramatically in the plasma trough. It rises somewhat at high L-values over the polar cap where it is very irregular although it is shown as smooth in this schematic diagram. There is thus a region of low electron density above the auroral zone where the plasma frequency is less than the gyrofrequency.

One of the most significant advances has been the availability of a comprehensive set of wave experiments on the DE 1 satellite [45] which allow the measurement not only of spectral information but also of wave polarization and direction of arrival of the signal. As a result of such measurements Calvert and Hashimoto [20] have been able to present a good picture of the source and propagation paths of such waves. The waves can be classified as follows:

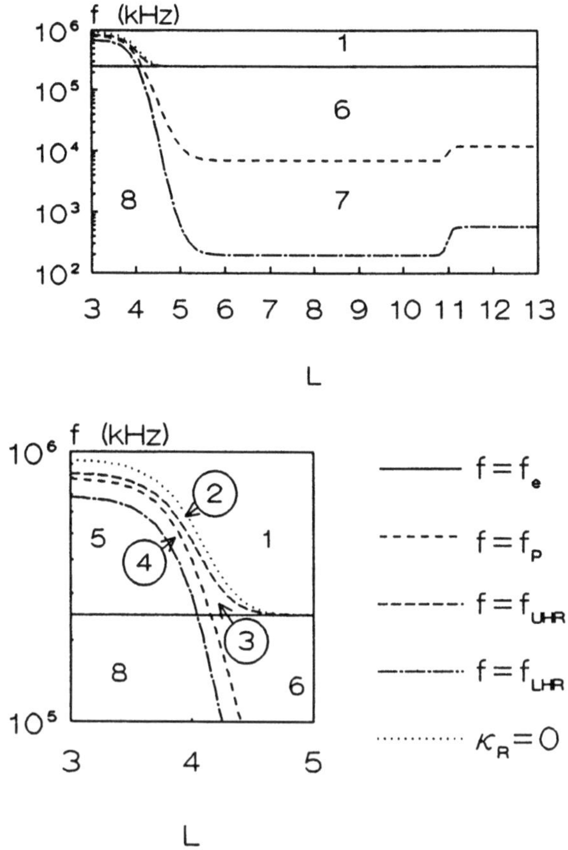

Figure 14.2: CMA boundary frequencies as a function of L in the auroral region.

Auroral Kilometric Radiation

This intense source of radiation is generated in association with discrete auroral arcs [41]. Its source is on the field line above the arcs at about two earth radii. The total power emitted by the source may be as large as 10^9 W [94]. It occurs in a frequency band in the range 100 - 400 kHz, with wavelengths of the order of kilometers, hence the name *auroral kilometric radiation* or AKR. The wave mode in which it is propagated is predominantly the extraordinary mode with polarization label RX [45]. A small component of the LO mode may also occur.

The radiation fills the region above the auroral arcs. Its frequency is above the plasma frequency of the solar wind so that it can escape through the magnetopause into interplanetary space. As a result the earth is an intense astronomical radio source.

Auroral Whistler Mode Hiss

This is generated at relatively low altitudes above the location of the auroral arcs. The mechanism is thought to be Čerenkov resonance between an oblique whistler wave and the electron beams associated with the aurora. Auroral hiss is observed on a low altitude satellite as upward or downward propagated v. l. f. waves in the whistler mode. The example of a v. l. f. saucer in Figure 14.1 from ISIS 2 is an upward propagated wave. The frequency band has a minimum above the location of the source. The saucer shape is the result of propagation effects which are discussed later. On a high altitude satellite such as DE 1 the radiation is always propagated upwards. The lower boundary of the frequency band shows a similar saucer or funnel shape. The upper frequency cutoff is the local plasma frequency at the satellite, above which the whistler mode cannot be propagated.

Z-Mode Radiation

The Z-mode radiation appears to originate in the same region as the whistler mode hiss [52, 20]. When observed at high altitude it occurs in region 6 of Figure 14.2 above the plasma frequency and below the gyrofrequency. It is likely to be observed poleward of the field line on which the auroral activity occurs. It seems very probable that it is generated as higher frequency whistler mode radiation at low altitudes and is converted to Z-mode radiation at the level at which it reaches the plasma frequency by a mechanism known as the Z-mode window which is studied in more detail below.

14.2.2 Waves Originating near the Equatorial Plane

At frequencies above the local gyrofrequency and plasma frequency there is a band of radiation observed in the auroral cavity which appears to originate just outside the plasmapause near the equatorial plane. Its wavelength is of the order of tens of kilometers and thus Jones [64] has called it *terrestrial myriametric*[1] *radiation* or TMR. The current picture of its generation and propagation is built up from a variety of observations on many different spacecraft. We summarize it here, giving reference to some of the observations which have contributed to formulating this picture.

- Near the equatorial plane just outside the plasmapause there is a region of intense wave generation. These are electrostatic waves generated at the upper hybrid frequency and at half gyrofrequency harmonics [91] by the Harris instability mechanism [7] described in §7.4.3. Recall that this mechanism requires a cold plasma background and a hot plasma component with a loss cone distribution. Waves are produced near the Z-mode resonance cone.

- These Z-mode waves are confined to a region within a few degrees of the equatorial plane by propagation effects [64].

- There exists a small range of propagation directions for which the Z-mode is coupled to the ordinary wave and the radiation escapes through a "window" (§14.5.2). Theory predicts [64] that it is emitted in a beam making an angle $\arctan\sqrt{\Omega/\omega}$ with the equatorial plane. Experimental tests [65] give support to this prediction.

- If the frequency is below the plasma frequency in the magnetosheath the ordinary mode waves are multiply reflected back and forth between magnetopause and auroral cavity. Frequencies emitted from different parts of the source are not the same and overlap; this radiation forms a continuum in the auroral cavity which was originally called the nonthermal continuum [42].

- If the frequency is above the plasma frequency in the magnetosheath it escapes from the magnetosphere and may be observed in the solar wind [15].

14.3 Propagation of Whistler Mode Hiss

Whistler mode hiss is produced by a Čerenkov resonance with beams of auroral electrons traveling along the magnetic field. Near the whistler

[1] Greek $\mu\upsilon\rho\iota\alpha\varsigma$ or *murias* = ten thousand

resonance cone a whistler wave can be described by the electrostatic approximation; the electric field component is polarized approximately in the direction of the wave normal and the ray is almost perpendicular to the wave normal. The parallel phase velocity component is sufficiently small for resonance with an electron beam having energies of the order of several keV.

Suppose such a mechanism is operative in a localized region at some level below a satellite. In Figure 14.3 we show a set of refractive index surfaces for different frequencies at a level where the plasma frequency is greater than the gyrofrequency. For simplicity the magnetic field is taken as vertical which is a reasonable approximation in the auroral region. Depending on the frequency the wave may be in region 7, 8a or 8b of the CMA diagram and the surface may or may not have points of inflection. The direction of the ray near the resonance cone is shown as an arrow. This is the initial ray direction of the energy generated at the source. The rays are propagated upwards, being refracted by the decreasing density gradient so that they flare outwards. This is shown schematically in the lower part of the diagram. A satellite passing over the source follows a path indicated by the dashed line. Clearly it first encounters the higher frequencies. The frequency is minimum as it passes over the source and then rises again as it continues on its path. If the satellite altitude is low then the band of frequencies observed is determined by the characteristics of the source. The result is the characteristic saucer shape of Figure 14.1. If the satellite is at higher altitudes the higher frequencies encounter the plasma frequency before reaching the satellite and are reflected. In this case the upper cut-off is the local plasma frequency at the satellite and the shape is that of a funnel with a flat top. This allows the use of the observations to study the electron density [85, 86].

14.4 Z-Mode Radiation

14.4.1 Refractive Index Surfaces for $Y > 1$

Consider a plane stratified medium with varying electron density and a uniform magnetic field such that $Y > 1$. Sets of refractive index surfaces for such a medium are shown for the Z-mode in Figure 14.4a and for the whistler and O-mode surfaces in Figure 14.4b. Consider the Z-mode in (a). Unlike the case of $Y < 1$, when $Y > 1$ the Z-mode has access to free space. Above the $F2$ maximum it is, however, screened from the ground by the higher electron density in the ionosphere. At $X = 0$ in free space the surface is a unit sphere. For $X = 0.5$ it is a prolate spheroid with $n > 1$. For $X = 0.9$ it has developed points of inflection so that it has a "bump" for propagation nearly parallel to the field. If X is increased slightly to $X = 1.1$ the bump is reversed and becomes

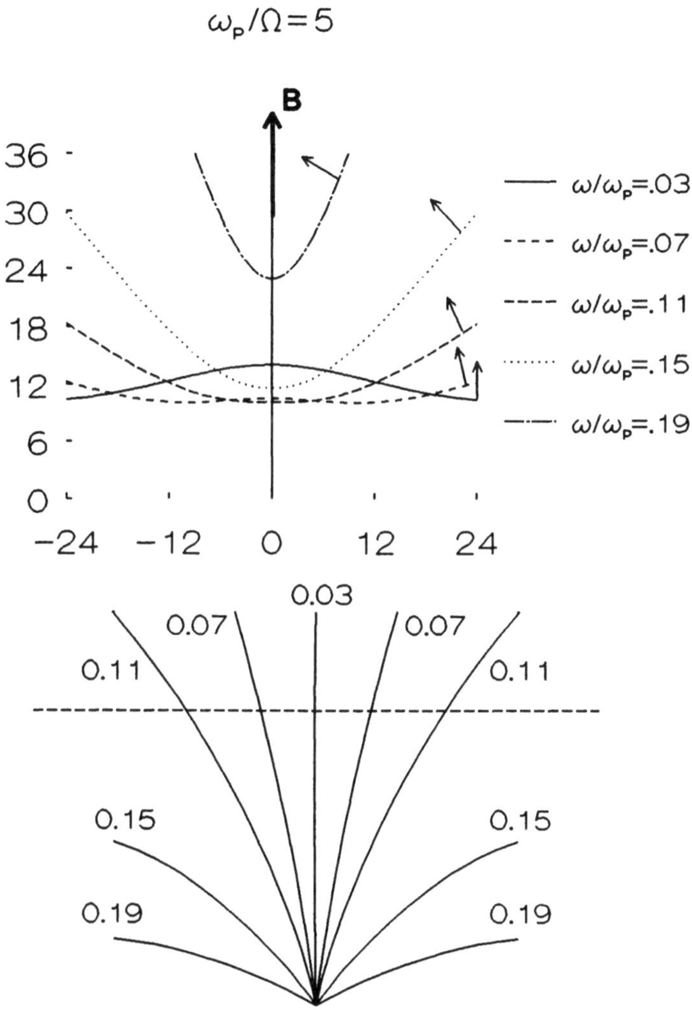

Figure 14.3: Refractive index surfaces for a set of frequencies normalized in terms of the plasma frequency. The magnetic field is vertical. The arrows show the ray direction near the resonance cone. Ray paths for the same frequencies are shown schematically below. The dashed line represents the path of a satellite which receives the radiation.

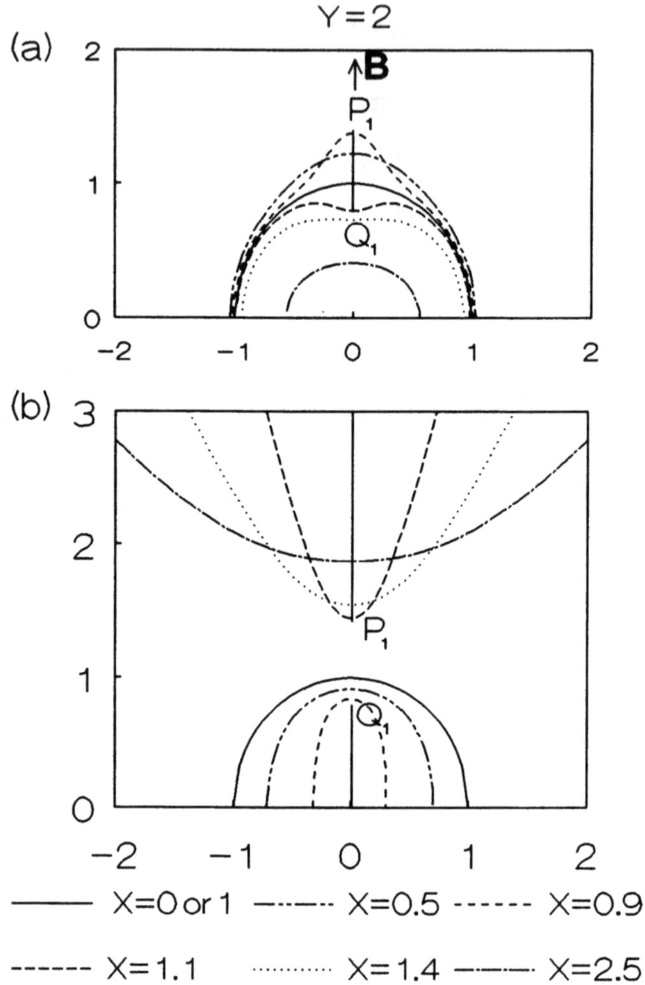

Figure 14.4: Refractive index surfaces for $Y = 2$. The magnetic field is vertical. (a) Z-mode. (b) Whistler and ordinary modes.

a "dimple". Between these two values, at $X = 1$, where the frequency is equal to the plasma frequency, the surface is degenerate. It collapses to the unit sphere and two segments of a straight line. One segment is shown in the diagram as P_1Q_1. The other is its mirror image below the n_x-axis. As X increases to 1.4 the dimple is smoothed out. For $X = 2.5$ the surface is an oblate spheroid. It becomes a point and the wave is cut off at $X = 3$ where the frequency is equal to the lower hybrid frequency.

The ordinary mode surface in (b) is also, of course, a unit sphere at $X = 0$. As X increases it becomes a prolate spheroid. As X approaches unity the spheroid becomes more and more elongated, shrinking to a line segment Q_1Q_2, half of which is shown in the figure, at $X = 1$. Above $X = 1$ the O-mode is cut off but the whistler mode is propagated. At large values of X this mode has a wide resonance cone and the shape characteristic of regions 7 and 8a of the CMA diagram. For the value of Y shown it nowhere has points of inflection but for $Y > 2$ it may be in region 8b of the CMA diagram for larger values of X. As X is decreased it shrinks to a pair of line segments, one extending from P_1 to infinity as shown in the diagram and the other its mirror image below the n_x-axis.

14.4.2 The Jones and Ellis Windows

In Figure 14.5 the whistler mode surfaces and those Z-mode surfaces for which $X \leq 1$ have been superimposed. When $X = 1$ the degenerate line segments which represent the whistler and Z-mode surfaces touch at the points P_1 and P_2. In the same way, if the ordinary mode surfaces for $X \leq 1$ and the Z-mode surfaces for $X \geq 1$ are plotted on the same diagram the line segments P_1Q_1 and P_2Q_2 connect with the line segment Q_1Q_2 which represents the degenerate ordinary mode surface at $X = 1$. The points P_1, P_2, Q_1, Q_2 are examples of *window points.* Similar window points for the case $Y < 1$ were noted by Ellis [38]. The window points Q_1, Q_2 are called *Ellis windows.* The importance of the windows at P_1 and P_2 for auroral mode Z-propagation was first noted by Jones [63]; we shall refer to these as the *Jones windows.* It is easy to show that the Jones windows correspond to values of the refractive index vector

$$\mathbf{n} = \pm\hat{\mathbf{e}}\sqrt{\frac{Y}{Y-1}} \tag{14.1}$$

while the Ellis windows occur for

$$\mathbf{n} = \pm\hat{\mathbf{e}}\sqrt{\frac{Y}{Y+1}} \tag{14.2}$$

where $\hat{\mathbf{e}}$ is a unit vector in the direction of the magnetic field.

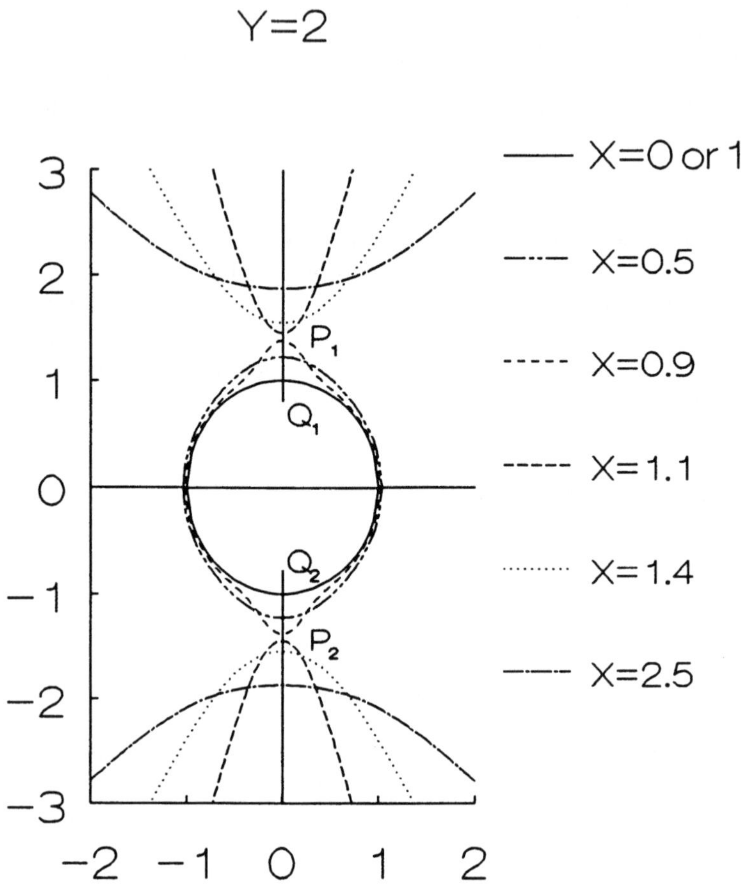

Figure 14.5: Superposition of the whistler mode surfaces for $X \geq 1$ and the Z-mode surfaces for $X \leq 1$. The magnetic field is vertical.

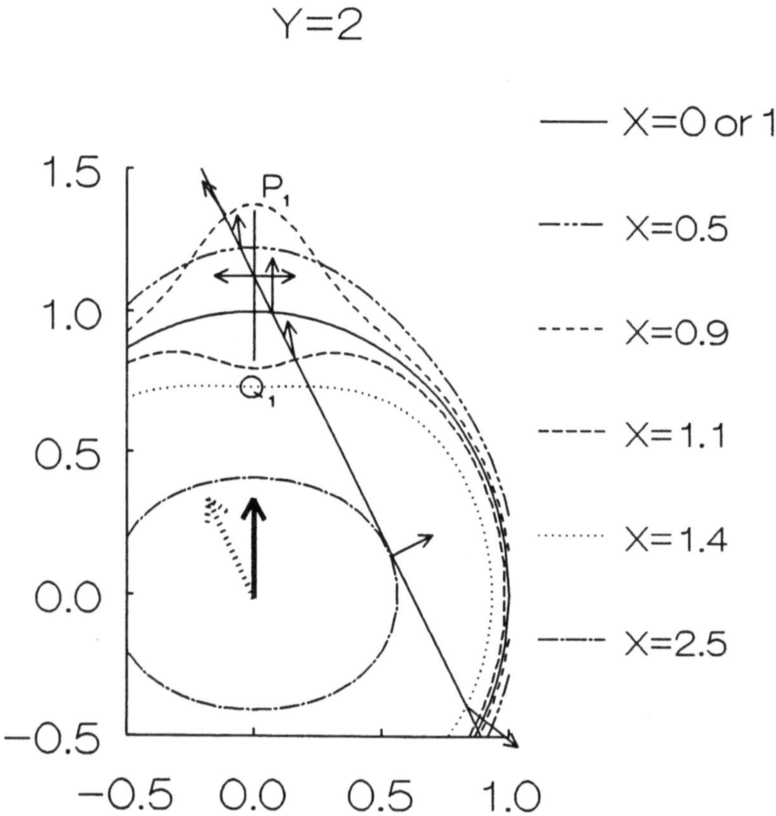

Figure 14.6: Poeverlein construction for a ray in the Z-mode. The heavy arrow shows the magnetic field direction. The broken arrow shows the normal to the stratifications. The small arrows show the direction of the ray at each value of X.

Consider now the propagation of a ray in the magnetic meridian which, when it reaches the level $X = 1$ has a refractive index near, but not exactly equal to, the Jones window value. A Poeverlein construction (§4.2.4) for such a ray is shown in Figure 14.6. Snell's law requires that in a coordinate system (x', z') in which the z'-axis is normal to the stratifications $n_{x'}$ is constant. The small arrows show the direction of the ray at each value of X; this condition defines a Poeverlein construction line parallel to the broken arrow. The wave normal direction is determined by the points where this line intersects the refractive index surface. The corresponding ray is shown on the left in Figure 14.7. Note that the coordinates in Figure 14.7 are the (x', z') coordinates with z' parallel to the broken arrow in Figure 14.6. It is assumed that the electron density

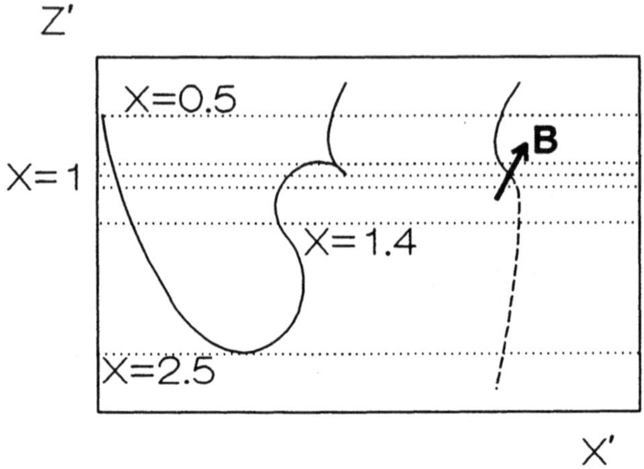

Figure 14.7: Rays near the Jones window.

increases with decreasing altitude and that the magnetic field is constant. This is not realistic above the auroral zone where magnetic field and electron density decrease at comparable rates. Since, however, the important features of propagation near the Jones window are controlled by the change from $\omega_p < \omega$ to $\omega_p > \omega$, the discussion is qualitatively correct. At high altitudes the ray direction is downward. At $X = 2.5$ the Poeverlein construction line is tangent to the refractive index surface, the ray is horizontal and reflection takes place. The ray reaches the level at which points of inflection of the refractive index surface occur, crosses $X = 1$ on the portion of the degenerate refractive index surface which is a unit circle and continues upwards. Slightly above this level it again becomes horizontal and is reflected downwards until it again encounters the level $X = 1$ where the degenerate refractive index surface is the straight line segment P_1Q_1. At this level its limiting direction is normal to the magnetic field. It then flips through 180^0 to form a *spitze* [19, §10.9] and is reflected upwards again. It then is propagated upwards without further reflection.

If the angle of incidence of the ray is changed only slightly the situation changes dramatically. Suppose the angle of incidence is such that

$$n'_x = \cos\theta\sqrt{\frac{Y}{Y-1}} = l_{x'}\sqrt{\frac{Y}{Y-1}} \qquad (14.3)$$

Then the Poeverlein construction line passes exactly through the Jones window and the situation is as shown in Figure 14.8. Consider now the

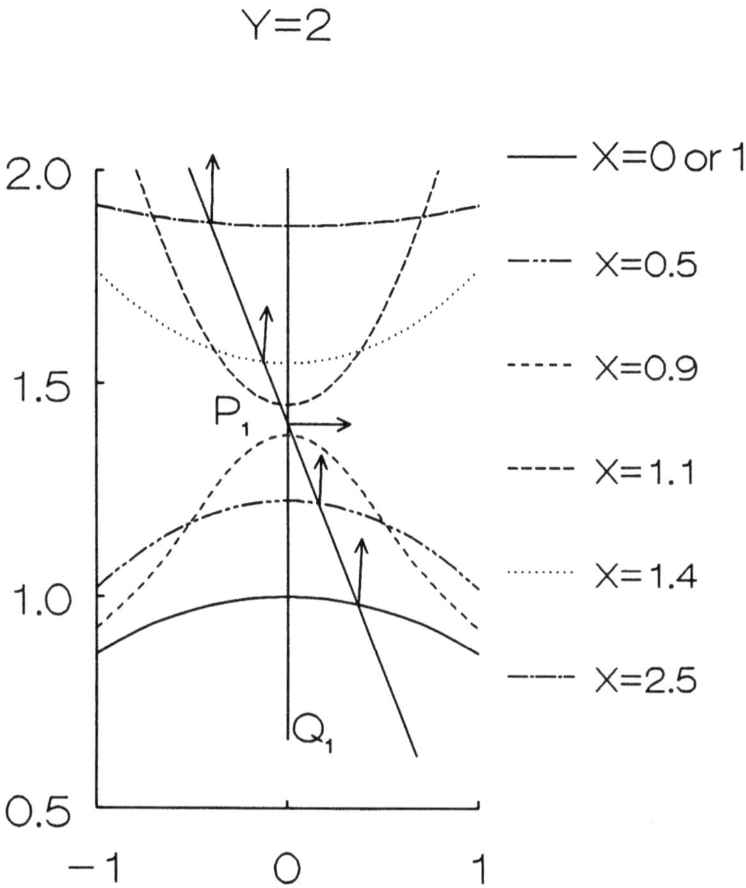

Figure 14.8: Poeverlein construction for a ray passing through the Jones window.

whistler wave starting at low altitude. The corresponding ray is shown on the right-hand side of Figure 14.7. As z' increases it passes through the levels $X = 2.5$, $X = 1.4$, $X = 1.1$ until it encounters the level $X = 1$ at the Jones window point. Instead of being reflected here it is converted into the Z-mode and continues upwards.

Further understanding comes from considering the nature of the solutions of the Booker quartic equation (4.1) shown in Figure 14.9. The angle between the magnetic field and the normal to the stratifications is $\theta = 30^0$ and the value of n_x at which the ray passes through the Jones window is then

$$n_{x,\text{crit}} = \sqrt{\frac{Y}{Y-1}} \cos\theta = \frac{1}{\sqrt{2}} \tag{14.4}$$

The diagram shows the z' component of the refractive index as a function of normalized electron density X. In the WKB approximation the phase of a wave component varies as

$$\exp\left\{\frac{\omega}{c}\left[n_{x'}x' \int^{z'} n_{z'}\, dz'\right]\right\}$$

As described in Chapter 4 reflection takes place where two roots of the quartic become equal. Near this level the WKB approximation breaks down and solutions of the equations must be appropriately matched across the reflection level. In the diagrams the ordinary wave corresponds to the branch which lies between $X = 0$ and a reflection level for which $X < 1$. It will not be further considered here. The Z-mode is represented by the branch which starts at $X = 0$ and is ultimately reflected at a value of X well above $X = 1$. The whistler mode is represented by the branch in the upper right-hand corner of each diagram and by another branch below the x'-axis, for which $|n_{z'}|$ is always greater than 1 so that it does not appear on this diagram.

In the upper panel $n_{x'} < n_{x',\text{crit}}$. The shape of the Z-mode branch for the upgoing wave can be related to the behavior which can be deduced from Figure 14.4. The z'-component of $\mathbf{n}$ increases initially. The ray then encounters a reflection level a little above $X = 1$, returns to the reflection level at $X = 1$, where it has a spitze, and is reflected again. The curves then show that propagation occurs for increasing values of X until reflection takes place and the wave returns to $X = 0$. The whistler mode has a reflection level where two roots of the quartic become equal near one of the Z-mode reflection levels. One of the roots becomes infinite where the coefficient $\alpha = 0$; the other is real for all values of X above the reflection level.

The behavior in the bottom panel, where $n_{x'} > n_{x',\text{crit}}$, is very similar. The Z-mode, however, is not reflected at $X = 1$. Instead the whistler

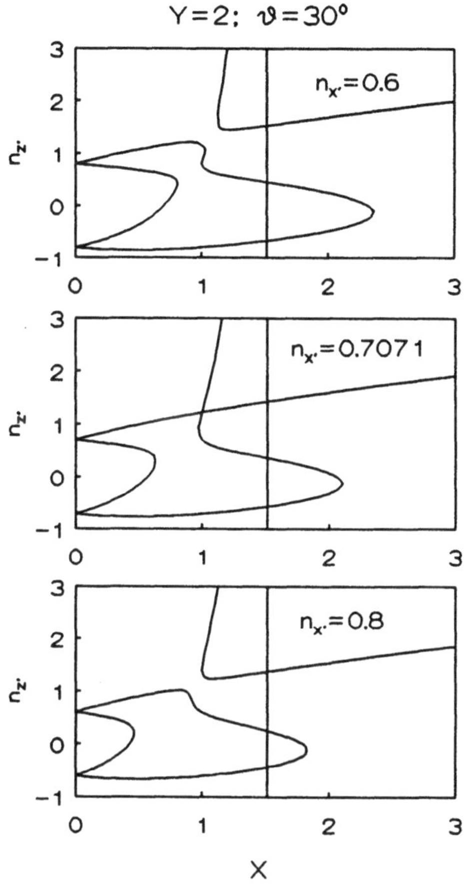

Figure 14.9: Roots of the Booker quartic near the ray passing through the Jones window, for which $n_x = 0.7071$. Another branch of the curves occurs in each case for values of $n_x < -1$.

mode reflection occurs at $X = 1$ and consideration of a Poeverlein construction in this case would show a spitze-like behavior for the whistler ray.

In the middle panel, where $n_{x'} = n_{x',\mathrm{crit}}$, there is a degeneracy at $X = 1$. The whistler mode and Z-modes coalesce. The Z-mode for $X < 1$ is continuous with the whistler mode for $X > 1$ and *vice versa.* This is the case of transmission through the Jones window.

There is a range of values of $n_{x'}$ on either side of $n_{x',\mathrm{crit}}$ where the Z-mode and whistler mode reflection levels are very close. Beween these levels the corresponding roots of the quartic are complex and the wave is inhomogeneous. If the barrier is narrow enough then evanescent barrier penetration can take place. The amount of penetration depends on the ratio of the width of the barrier and the e-folding distance of the inhomogeneous wave. Computation of the amount of penetration requires a more detailed study of the behavior of waves in the neighborhood of a coupling point than can be done here. A straightforward account is given by Budden [19, §§17.6, 17.7] who shows that the half angular width of the window is of the order of one or two degrees. Within this half angular width more than half the intensity is converted from whistler to Z-mode.

It is probable that it is by this process that whistler mode noise, generated at low altitudes by a Čerenkov interaction, is able to reach a high altitude satellite such as DE 1 in the Z-mode.

14.5 Terrestrial Myriametric Radiation

14.5.1 Confinement of Z-Mode Rays near the Equatorial Plane

Region 3 of the CMA diagram occurs between the upper hybrid frequency and the plasma frequency and in this region the Z-mode refractive index surface is open with a resonance cone as described in Chapter 3; propagation is possible in this mode perpendicular to the magnetic field but not parallel to it. Waves with frequencies between the plasma frequency, given by $X = 1$, and the upper hybrid frequency given by $X = 1 - Y^2$ or

$$\omega_{UHR} = \sqrt{\omega_p^2 + \Omega^2} \tag{14.5}$$

have refractive index surfaces in this region. TMR is generated near the resonance cone of the Z-mode near the equatorial plane just outside the plasmapause. In this region the plasma frequency is substantially larger than the gyrofrequency and so the band of frequencies satisfying the condition for waves to be in this portion of the CMA diagram is narrow. The normal to the stratifications is radial and the magnetic field is horizontal.

Sets of refractive index surfaces for both the ordinary mode and the Z-mode are shown in Figure 14.10. Consider the Z-mode surfaces on the right. The upper hybrid condition in this case is $X = 0.75$ and there are no Z-mode waves for values of $X < 0.75$. At $X = 0.8$, in region 3a of the CMA diagram, the surface is a simple open surface with a resonance cone. At $X = 0.9$, in region 3b, it has developed points of inflection. These are more pronounced at $X = 0.99$. The mode is continuous between regions 3 and 4 but, at $X = 0.101$ in region 4, the surface is closed. At the boundary between regions 3 and 4 the surface is degenerate consisting of a unit sphere and the straight line segments shown as heavy lines in the diagram extending from the Ellis window points at $n_{x'} = \pm\sqrt{Y/(Y+1)}$ to infinity. It touches the ordinary mode refractive index surface at these points. The ordinary mode is continuous throughout regions 1, 2 and 3 but is cut off in region 4. It has an ovoid shape, decreasing in radius with increasing X and becoming more and more elongated until, at the cutoff $X = 1$ it degenerates to the heavy line segment between the Ellis window points.

Now consider the ray path of a Z-mode wave which is generated near the upper hybrid frequency. We can use the Poeverlein construction (§4.2.4) to sketch such rays. The vertical broken construction lines in the figure are such lines through the Ellis windows. Other construction lines are parallel to these. Three typical rays are sketched in Figures 14.11a, 14.11b and 14.11c. In Figure 14.11a n_x is large; the ray starts near the level $X = 1 - Y^2$, is propagated downwards to greater electron densities, reverses direction and is reflected, with a spitze-like behavior at $X = 1$. It returns to the level where it was generated and is likely to be absorbed. In Figure 14.11b $|n_x|$ is close to $\sqrt{Y/(Y+1)}$, the Ellis window value. The behavior near $X = 1$ is modified; the ray penetrates below $X = 1$ and the reflection at the spitze is inverted. Otherwise the path is similar. In Figure 14.11c where n_x is large the ray is reflected above the $X = 1$ level but otherwise is similar.

Two possible ordinary mode rays are shown in Figures 14.11d and 14.11e. Such a wave , started at a high altitude, is propagated downwards and reflected. If $|n_x| < \sqrt{Y/(Y+1)}$ the reflection is at $X = 1$ and it shows a spitze; otherwise it is reflected at a higher level.

The important feature to note about the Z-mode ray paths is that they are confined in a horizontal direction. The result is that most of the radiation generated by such a mechanism is confined to within a few degrees of the equatorial plane. This is the explanation of the intense upper hybrid emissions observed in this region.

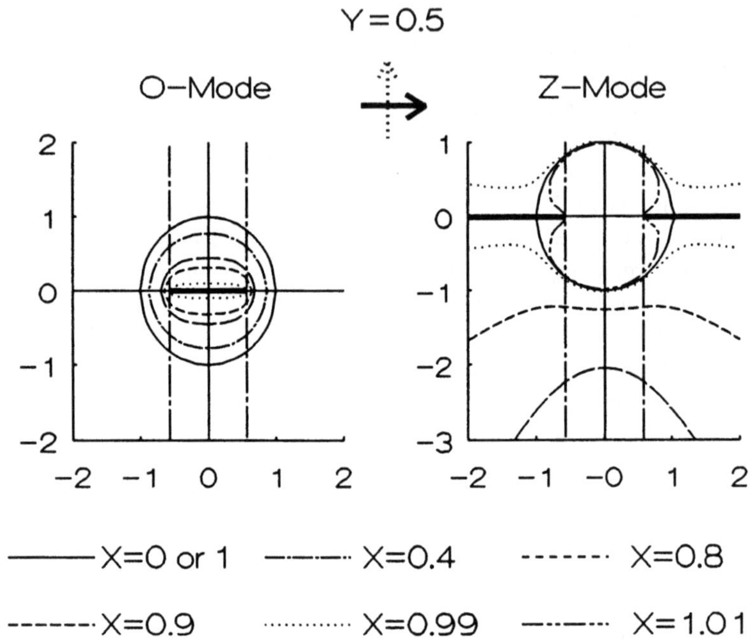

Figure 14.10: Sets of refractive index surfaces for the ordinary and Z-modes at the equatorial plane. The heavy arrow shows the magnetic field direction and the broken arrow the normal to the stratifications.

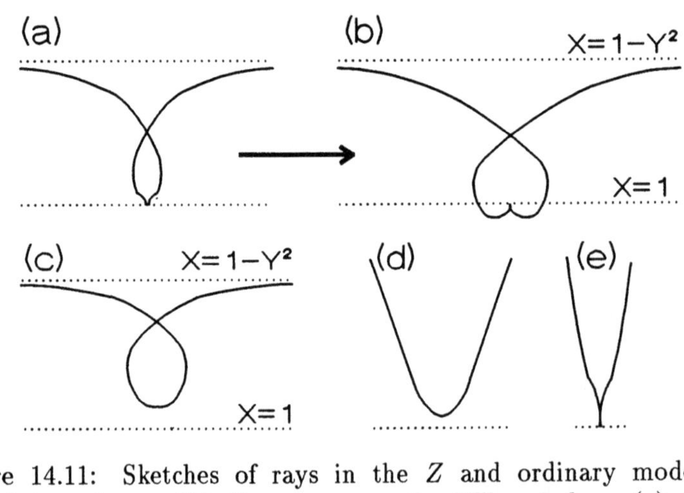

Figure 14.11: Sketches of rays in the Z and ordinary modes. (a) Z-mode, n_x large, (b) Z-mode, near the Ellis window, (c) Z-mode, $|n_x| < \sqrt{Y/(Y+1)}$, (d) ordinary mode $1 > |n_x| > \sqrt{Y/(Y+1)}$, (e) ordinary mode, $|n_x| < \sqrt{Y/(Y+1)}$.

14.5.2 The Ellis Window

Z-mode rays with $n_x = \sqrt{Y/(Y+1)}$ escape through the Ellis window. When they reach the level $X = 1$, instead of being reflected with a spitze-like behavior they are converted into ordinary mode waves which are propagated out into the cavity between the plasmapause and magnetopause with ray paths like one half of the path shown in Figure 14.11e. The mechanism is exactly the same as that of auroral Z-mode noise passing through the Jones window. Again the angular half width of the window is typically a few degrees.

The direction of the ray makes an angle $\psi = \arctan n_x/n_z$ with the radial direction. The electron density decreases rapidly and the refractive index approaches unity so that $n_z = \pm\sqrt{1 - n_x^2}$. If we set n_x equal to the Ellis window value this means that

$$\psi = \pm \arctan \sqrt{Y} = \pm \arctan \sqrt{\Omega/\omega} \tag{14.6}$$

The radiation thus emerges from the Ellis window in two beams, symmetrically disposed about the equatorial plane. This prediction is supported by the experimental evidence [65].

The higher frequencies, greater than the magnetosheath plasma frequency, generated in this way may escape through the magnetopause; the lower frequencies are trapped between the magnetopause and plasmapause and give rise to a lower intensity TMR continuum.

14.6 Auroral Kilometric Radiation

The currently accepted mechanism for explaining the generation of AKR was first proposed by Wu and Lee [119]. The chief features of their explanation are the following:

- The electron distribution high above a region of auroral arc activity has a loss cone which contains few upward moving particles. The loss cone for particles which have come from the opposite hemisphere is filled. In the neighborhood of this loss cone there is a region of velocity space within which $\partial f_0/\partial v_\perp > 0$ and instability is possible.

- AKR is generated in a region where $\omega_p \ll \Omega$. It is generated in a band just above the electron gyrofrequency. Most of it is in the free space extraordinary mode with polarization label RX. There is a small proportion of the free space ordinary mode with LO label.

- In such circumstances, because of the large phase velocity of any propagated wave mode Čerenkov radiation is not likely. A cyclotron resonance can occur. It is important to take relativistic effects into account. Such effects are only important in modifying the mass and hence the gyrofrequency of the resonant particles. The waves which are resonant have wave normals almost perpendicular to the magnetic field.

This theory was improved by Dusenbery and Lyons [36] who used a more realistic expression for the cold plasma dispersion relation and discussed the particle resonance using a measured particle distribution. We shall give a brief treatment of the major features of the theory only.

14.6.1 Nature of the Distribution Function

Auroral arcs are caused by localized beams of particles which have been accelerated by electric fields parallel to the magnetic field. The electric field points upwards along the field line. The distribution function at the altitude where AKR is generated is strongly affected by this field. This population consists of those electrons which have mirror points below the region and which are not prevented by the electric field from gaining access to the region.

In order to understand the nature of the distribution function we study the adiabatic motion of single particles [26]. Such particles move along the field lines conserving their magnetic moment $mv_\perp^2/2B$ and energy $\frac{1}{2}mv_\parallel^2 - e\Phi$ where Φ is the electric potential at some reference level, taken in this case to be the equatorial plane. Electrons on the boundary of the loss cone are precipitated at some level at the top of the E-region of the ionosphere. Denote quantities measured at this level by a subscript L. Then, applying the conservation laws, the locus of these particles in velocity space at some other level is the hyperbola

$$v_\parallel^2 + v_\perp^2\left(1 - \frac{B_L}{B}\right) = \frac{2e}{m}(\Phi - \Phi_L) \tag{14.7}$$

which is illustrated in Figure 14.12a. It crosses the $v_\perp$-axis at $v_\perp = \sqrt{2\Omega(\Phi_L - \Phi)/(B_L - B)}$. Lower energy particles cannot penetrate the electric potential barrier and reach the opposite hemisphere. Those which just fail to do so have $v_\parallel = 0$ at the equator. Again we find, by applying the conservation laws, that the locus of these particles in velocity space is an ellipse

$$v_\parallel^2 + v_\perp^2\left(1 - \frac{B_{\text{eq}}}{B}\right) = \frac{2e}{m}\Phi \tag{14.8}$$

For relatively low altitudes $B_{\text{eq}} \ll B$ and this approximates the circle shown in Figure 14.12.

These boundaries divide velocity space into four regions. In the figure the primed labels correspond to regions where the electrons are moving

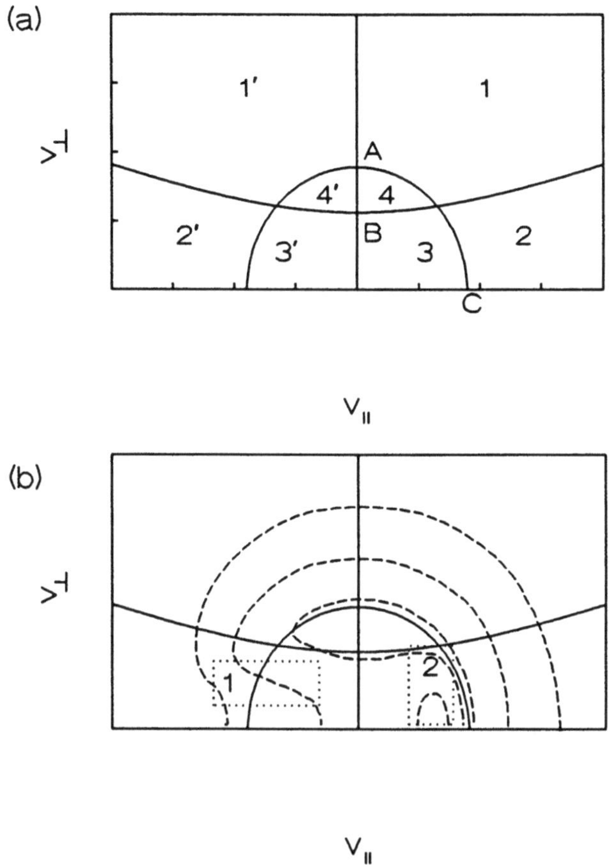

Figure 14.12: General features of electron distributions above an auroral arc. Positive values of $v_{||}$ correspond to downward moving electrons. (a) The loss cone (hyperbola) and locus of particles which do not penetrate the potential barrier (circle). (b) Sketch showing typical contours of the distribution function.

upwards and the unprimed labels to regions in which they are moving downwards. Region 1 corresponds to particles which have mirror points above the lower ionosphere in opposite hemispheres. Region 2 corresponds to particles which have mirror points within the lower ionosphere in opposite hemispheres. Region 3 corresponds to particles which have one mirror point above the lower ionosphere and another in the same hemisphere because of the electric potential barrier. Region 4 corresponds to particles with one mirror point within the ionosphere and the other in the same hemisphere because of the potential barrier.

The motion of the particles is not, however, perfectly adiabatic. In such circumstances the most energetic electrons in regions 1 and 2 bounce back and forth between mirror points in opposite hemispheres and their orbits are only slightly modified by the potential gradient. Those in region 2, falling within the loss cone, are precipitated to form the aurora and thus, in a region above the aurora there are few upward moving particles in that part of the loss cone represented by region 2'. Pitch angle diffusion processes are quite strong and particles from the opposite hemisphere in region 2 almost fill the loss cone.

The other group of less energetic particles in regions 3 and 4 is trapped between two mirror points in the same hemisphere which may be quite close together. Those lying within the atmospheric loss cone are precipitated and lost. Because of the relatively short mirror period the loss cone is not filled between two encounters with the neutral atmospheric particles so that both regions 3 and $3'$ are empty. The result is a particle distribution such as that sketched in Figure 14.12b. Such particle distributions have been observed by the S3–3 satellite [30]. Such a distribution, observed at the same time as auroral activity, was used by Dusenbery and Lyons [36] to discuss the generation of AKR.

There are two regions, denoted by dotted rectangles, in this distribution where $\partial f_0/\partial v_\perp$ is positive. One of these regions occurs for positive $v_\parallel$ at the edge of the loss cone. The other occurs for negative $v_\parallel$ in that part of the loss cone which overlaps the region occupied by the particles trapped in the same hemisphere by the potential gradient. These are regions of free energy which, in appropriate circumstances may be expected to excite waves. The electrons in these regions are weakly relativistic, having typical speeds about one tenth of the speed of light.

14.6.2 Properties of the Refractive Index

When $\omega_p \ll \Omega$ and the frequency is close to the electron gyrofrequency, $X \ll 1$ and, to a good approximation we may write

$$\kappa_R = 1 - \frac{X}{1-Y}\,; \quad \kappa_L = 1\,; \quad \kappa_3 = 1 \tag{14.9}$$

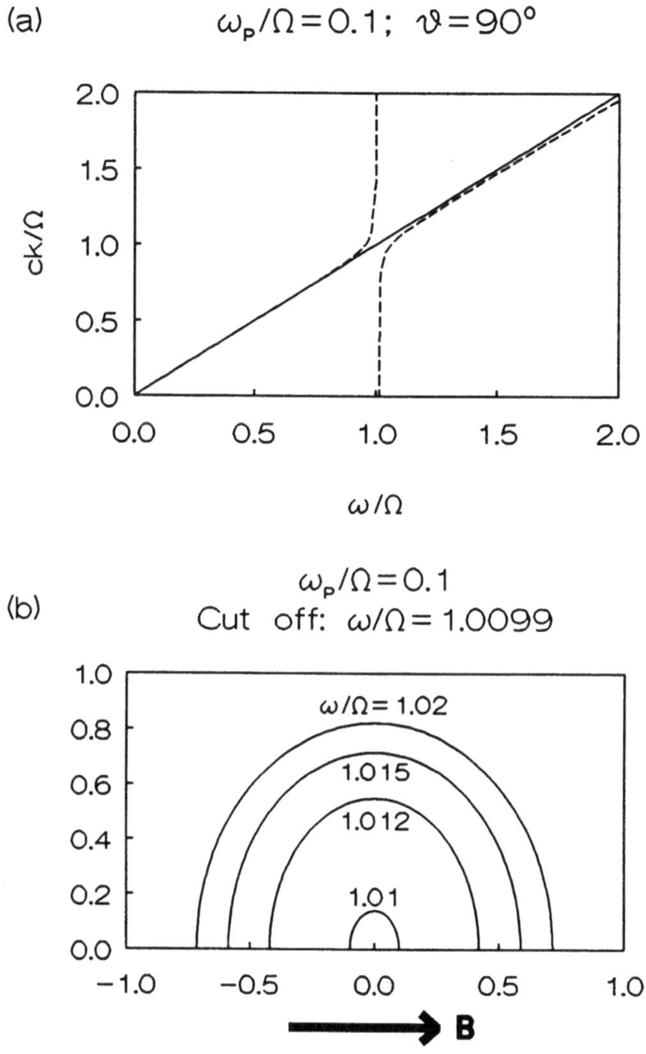

Figure 14.13: Dispersion relation and refractive index when $X \ll 1$. (a) Dispersion relation for transverse propagation. (b) Refractive index surfaces.

for the elements of the dielectric tensor (2.29, 2.19). The expression for the refractive index (2.58) then becomes

$$n^2 = \frac{1 - Y - X}{(1 - Y - \frac{1}{2}X)\sin^2\theta + (1 - Y)\cos^2\theta} \quad \text{or} \quad 1 \tag{14.10}$$

The free space ordinary mode with $n = 1$ and polarization label LO behaves as if it were in free space. The free space extraordinary mode, however, near $Y = 1$, shows the cut-off $X = 1 - Y$ and principal resonance $X = 1 - Y^2 \simeq 2(1 - Y)$, dividing regions 1, 2 and 3 of the CMA diagram. This is shown, for propagation perpendicular to the magnetic field in Figure 14.13a which illustrates the dispersion relation. There is a very narrow stop band separating the Z-mode resonance and the free space extraordinary cut-off near $\omega = \Omega$. The corresponding extraordinary mode surfaces near this value are shown in Figure 14.13b. At frequencies above this the refractive index is close to unity for all directions of propagation.

14.6.3 Single Wave Characteristics and Resonance Contours

Although the electrons which are expected to be involved with the resonance are only weakly relativistic, their relativistic mass change is crucial in understanding the nature of AKR.

The single wave characteristics in this case are elliptical as shown in Figure 8.4. In this case $\omega/k_{\parallel}$ is generally greater than the speed of light. The regions of velocity space of interest occur for relatively small electron speeds. The segment of the single wave characteristic which intersects these regions can be well approximated by a straight line parallel to the $v_{\perp}$-axis. Resonant particle diffusion along these contours will be away from the $v_{\parallel}$–axis to higher energy if $\partial f_0/\partial v_{\perp}$ is negative and towards it to lower energy if it is positive. Thus the wave gains energy from those resonant particles in regions of velocity space where $\partial f_0/\partial v_{\perp} > 0$. These are the regions labeled 1 and 2 in Figure 14.12b.

The resonance contours, on the other hand, cannot, in general, be represented by a non-relativistic approximation. The relativistic resonance contour is an ellipse given by (8.2) and centered on

$$v_{\parallel} = \frac{cn_{\parallel}}{n_{\parallel}^2 + l^2Y^2}$$

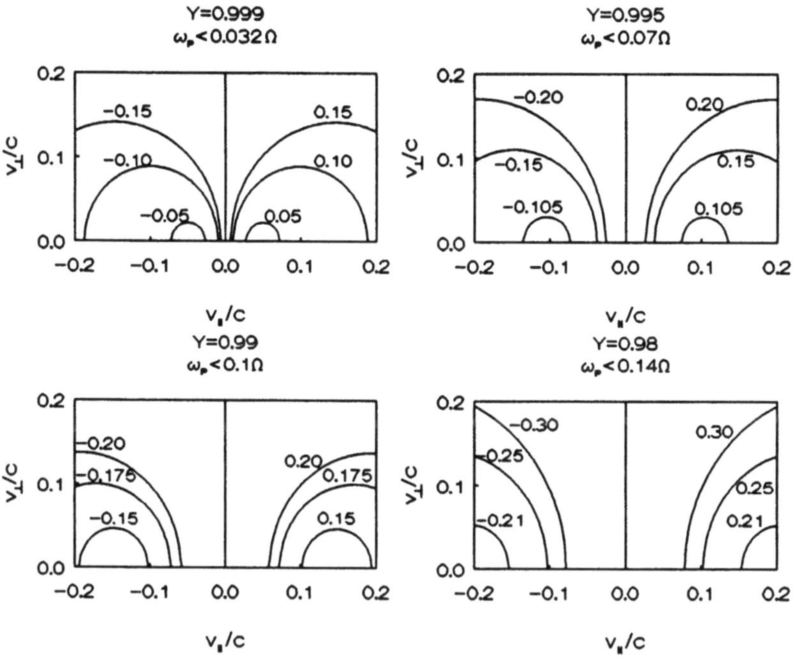

Figure 14.14: Resonance contours. Each set of curves corresponds to a different normalized frequency ω/Ω. The parameters are values of $ck_{||}/\Omega$ at resonance.

In Figure 14.14 some examples are shown for $l = -1$. Each panel corresponds to a normalized frequency close to the gyrofrequency. The parameters define the normalized parallel wave number $ck_{\|}/\Omega$ for waves which are resonant at that frequency. In each case resonance with the free space extraordinary mode will only take place if the frequency is above the cut-off frequency. The condition $X > 1 - Y$ may be written

$$\omega_p < \sqrt{\omega(\omega - \Omega)} \tag{14.11}$$

The condition is indicated for each panel of the figure.

The important features to be noted are the following:

- When the Y is very close to unity and the wave frequency is just above the gyrofrequency the relativistic resonance contours pass through regions of velocity space where the energy is quite small. As Y decreases they move to higher energies and for values such as $Y = 0.95$ resonance is only possible with highly relativistic particles which are rare.

- The contours for $v_{\|} < 0$ (electrons moving upwards) arise for $n_{\|} < 0$ and those for $v_{\|} > 0$ for $n_{\|} > 0$. The resulting waves are propagated almost at right angles to the field with a small upward component in the first case and a small downward component in the second.

- If we take $Y = 0.995$ as an example there is a frequency $\omega = 0.105\Omega$ for which all parts of the resonance contour are close to values of $v_{\|} = 0.1c$. All particles on such a contour contribute to wave growth if the contour lies within a region such as region 2 of Figure 14.12b. On the other hand a contour such as that for which $\omega = 0.15\Omega$ passes through some regions where $\partial f_0/\partial v_{\perp} > 0$ and some where $\partial f_0/\partial v_{\perp} < 0$. On such a contour the contributions to growth and loss must be integrated to find the net effect on the wave.

- The condition for cut-off limits the value of the electron density for which such an instability is possible. For example the case where $Y = 0.995$ in Figure 14.14 can only give rise to growth of a propagated wave when $\omega_p < 0.07\Omega$. It is for this reason that AKR is generated in the auroral cavity which is the only part of the magnetosphere where such conditions occur.

It can thus be expected that AKR in the free space extraordinary mode will be generated just above the local gyrofrequency in the RX extraordinary mode. Depending on which portion of velocity space is responsible for the resonance this mode may initially be propagated with an upward or downward component. If is propagated downwards it encounters a reflection level near the cut-off. The ray path which it follows depends on the model and the intitial conditions. It is, however, generally propagated outwards whatever the model. Ultimately it encounters the magnetopause and, since its frequency is certainly greater than the magnetosheath and solar wind plasma frequencies it escapes from the magnetosphere. Detailed ray tracings have been carried out by Hashimoto [51].

A certain amount of AKR appears to be generated in the free space ordinary mode with LO label. It tends to be at lower frequencies than the RX mode but has similar spectral characteristics and is never observed below the electron gyrofrequency. It has been suggested [45] that it is not generated in this mode but is converted from the RX mode by a weak mechanism such as scattering from density irregularities.

14.7 Summary

- Between the plasmapause and the magnetopause there is a cavity of low electron density. At lower altitudes it is bounded by the polar cap and known as the auroral cavity. There is a region at lower altitudes where the plasma frequency is smaller than the electron gyrofrequency. A variety of different wave phenomena are observed in this region.

- Auroral hiss is whistler mode radiation generated at low altitude by Čerenkov interaction with electron beams associated with auroral arcs. It is propagated upwards. The lowest frequencies are propagated almost vertically. The angle between the ray and the vertical increases with frequency. The result is that a satellite passing over such a source observes a band of frequencies with a minimum when the satellite is above the source.

- The auroral hiss is cut off at the plasma frequency. The Jones "window" occurs at this frequency; a narrow range of wave normal angles allows coupling between the whistler mode and the Z-mode. The result is auroral Z-mode noise which occurs in a band above the local plasma frequency.

- Intense auroral kilometric radiation (AKR) is generated at a higher altitude above the auroral activity. It has a much higher frequency and can escape into the solar wind. The earth is thus a radio source.

- Terrestrial myriametric radiation (TMR) is generated at the equatorial plane just outside the plasmapause by a Harris instability which excites the Z-mode. Because the plasma frequency is greater than the electron gyrofrequency here a narrow band of frequencies between the upper hybrid and plasma frequencies is produced.

The nature of the ray paths in this mode confines the radiation to within a few degrees of the equatorial plane. Some of this radiation is coupled into the ordinary mode through the Ellis window and is emitted in two beams symmetrically disposed about the equatorial plane. The higher frequencies escape to the solar wind. The lower frequencies are trapped between plasmapause and magnetopause and form a continuum.

- Auroral kilometric radiation is generated in the *RX* free space extraordinary mode by a relativistic electron cyclotron resonance mechanism. The distribution function above auroral arcs is complicated but has a loss cone for upward moving electrons and a local minimum for downward moving electrons. Some relativistic resonance contours lie partly or entirely in these regions of velocity space and thus waves which have appropriate values of $k_{\|}$ may experience growth. Such waves are only propagated if the plasma frequency is sufficiently low. Conditions are favorable in the auroral zone. The AKR is propagated outwards and escapes into the solar wind plasma, making the earth a strong radio source. A small amount of the *LO* mode also occurs, probably as a result of weak mode conversion from the *RX* mode.

Chapter 15

Some Final Words

15.1 Introduction

The objective of this book has been to try to develop the basic theories of plasma wave propagation and generation applicable to magnetospheric waves in as unified and concise a way as possible and to apply the theory to a representative set of observations of such waves. At the end of the exercise the author (and probably the reader) is more conscious of what has been left out than what has been included. Many phenomena which deserve consideration and the work of many authors who have made major contributions to the field have been omitted for reasons of space. We have restricted our examples to the earth's magnetosphere, making no mention of the many observations of plasma waves in other planets. In this final chapter we mention a number of topics which would have been included in a more substantial work and briefly discuss the direction in which the field is likely to move.

15.2 Man-Made Plasma Waves

Most of the experimental work on plasma waves has been passive; naturally generated waves are observed. In a few cases active experiments have been designed to elucidate the nature of such waves. The most important is the Stanford v. l. f. wave injection experiment carried out over many years at Siple, Antarctica. Helliwell [55] has reviewed this experiment in some detail. A 100 kW v. l. f. transmitter with a very large antenna which was capable of producing a variety of tailored v. l. f. signals was used systematically to excite a variety of largely non-linear phenomena. As a result much improved understanding of the non-linear problems which have been omitted from this book has been obtained.

Other man-made waves include the power line harmonics above highly industrialized and densely populated continents. The power line harmonics can be amplified by electrons of a few keV near the equator. They may cause precipitation of such electrons.

15.3 Ultra Low Frequency Pulsations

Geomagnetic pulsations are caused by a wide variety of hydromagnetic waves with periods between 1 s and 10 min which are of importance in understanding many magnetospheric phenomena. The wavelengths of these waves are comparable with the length scales of various regions in the magnetosphere. The nature of the boundary conditions becomes important. For this reason they cannot be treated by methods introduced in this book which is largely based on the theory of waves in infinite media. A full treatment of u. l. f. phenomena would require another book.

15.4 Waves in the Magnetotail, Magnetosheath and Solar Wind

The most complete understanding of naturally occurring plasma waves has been obtained in the nearer parts of the magnetosphere. We have restricted our examples largely to the large variety of waves in the plasmasphere and on the field lines mapping to the auroral zone. There are many reported observations of waves in more distant parts of the magnetosphere and also on other planets. A variety of phenomena occur in these regions. Examples are (i) "lion's roar" [95], a whistler mode wave generated by ion cyclotron resonance, (ii) bow shock plasma waves [40], turbulent bursts of whistler waves in the neighborhood of the bow shock, (iii) upstream whistler mode waves [40] caused by beams of ions and electrons streaming away from the bow shock and (iv) broad band electrostatic noise observed in the magnetotail [44].

15.5 Future Work

The theory of linear and quasi-linear waves in uniform or slowly varying plasmas is a mature field. Further advances in such areas are likely to be in the nooks and crannies of parameter space. New experimental observations will suggest treatments using particular mixes of particle species with different distribution functions. In different regions of space the morphology may give rise to propagation and coupling effects as complex as those occurring for ion cyclotron whistlers. The theory for studying such phenomena is available; what is missing is more detailed experimental observation of plasma composition and wave polarization in the many regions of the magnetosphere which have not yet been surveyed in this way.

Non-linear theory, on the other hand, is much less well developed. It is clear that wave–particle interactions play a key role in magnetospheric particle dynamics. In addition wave–particle and wave–wave interactions are the key to understanding the relaxation of non-equilibrium particle distribution functions to equilibrium. The microphysics of gross magnetospheric dynamics is the understanding of instabilities which depend on the understanding of the waves.

Apart from the exploration of more distant parts of our magnetosphere and the magnetospheres of other planets much experimental work is likely to apply our understanding of the nature of various kinds of waves in order to probe the magnetosphere. The outstanding early work of this type was the use of naturally occurring whistlers to infer the structure of the plasmasphere and plasmapause.

The study of wave phenomena is a field in which the marriage of theory and observations of natural phenomena has achieved a very high degree of success. It continues to provide a satisfying challenge to both theoretician and experimentalist.

Appendix A

The Essence of Cartesian Tensors

Cartesian tensors are often useful in the derivation of formulae. A set of Cartesian axes is labelled x_1, x_2 and x_3 corresponding to x, y and z. Vectors and higher rank tensors are then written in terms of their components in these coordinates e.g. A_j, T_{jk}, ϵ_{jkl}. The subscripts j, $k, \dots n$ are used to denote Cartesian tensors. We shall not use the subscript i for this purpose; it is reserved to denote the ith ion species. Some important results are summarized here. A detailed treatment can be found in [61] or [59].

1. It is understood that subscripts j, k, l etc. take values 1, 2, 3.

2. A_j represents the vector with x_1-component A_1, x_2-component A_2, and x_3-component A_3.

3. Higher rank tensors $B_{jk\dots l}$ can be defined (*e.g.* the elements of A_{jk} can be written as a 3×3 matrix)

4. The summation convention is *"sum over the repeated suffix"*, *e.g.*

$$A_j B_j \equiv A_1 B_1 + A_2 B_2 + A_3 B_3$$

5. The Kronecker delta is:

$$\delta_{jk} = \begin{cases} 1 & j = k \\ 0 & j \neq k \end{cases}$$

   It operates as a *substitution tensor*

$$\delta_{jk} A_k = A_j$$

6. The alternating unit tensor is:

$$\epsilon_{jkl} = \begin{cases} 1 & jkl = 123,\ 231,\ 312 \\ 0 & j = k \text{ or } k = l \text{ or } l = j \\ -1 & jkl = 132,\ 213,\ 321 \end{cases}$$

7. The relationship between ϵ_{jkl} and δ_{jk} is:

$$\epsilon_{jkl}\epsilon_{jmn} = \delta_{km}\delta_{ln} - \delta_{kn}\delta_{lm}$$

8. Whenever we have a product $\epsilon_{jkl}\epsilon_{jmn}$ we replace it by the expression involving the δ_{jk}'s and use the Kronecker delta as a substitution tensor.

9. Scalar product: $\mathbf{A}.\mathbf{B} = A_j B_j$

10. Vector product: $(\mathbf{A} \times \mathbf{B})_j = \epsilon_{jkl} A_k B_l$

Appendix B

Some Mathematical Results

B.1 Properties of Bessel Functions and Related Results

The classic reference on Bessel functions is the work by Watson [118]. A convenient summary of their properties and extensive tables are presented by Abramowitz and Stegun [1]. Only those results which are used in this book are listed here.

B.1.1 Bessel Functions

Bessel Function of the First Kind

1. The Bessel function of the first kind of integer order, $J_n(z)$, is one of the solutions of Bessel's equation,

$$z^2\frac{d^2w}{dz^2} + z\frac{dw}{dz} + (z^2 - n^2)w = 0 \tag{B.1}$$

2. For small z and $n > 0$ we have

$$J_n(z) \simeq (z/2)^n/n! \tag{B.2}$$

3. The derivative with respect to argument can be written

$$J_n'(z) = [J_{n-1}(z) - J_{n+1}(z)]/2 \tag{B.3}$$

4. A related recursion formula is

$$J_{n+1}(\zeta) + J_{n-1}(\zeta) = \frac{2n}{\zeta}J_n(\zeta) \tag{B.4}$$

Modified Bessel Function of the First Kind

1. The modified Bessel function of the first kind, $I_n(x)$, is related to $J_n(x)$ by

$$I_n(x) = (-i)^n J_n(ix) \tag{B.5}$$

2. It satisfies the differential equation

$$z^2\frac{d^2w}{dz^2} + z\frac{dw}{dz} - (z^2 - n^2)w = 0 \tag{B.6}$$

which is equivalent to

$$\frac{1}{z}\frac{d}{dz}\left(z\frac{dw}{dz}\right) - \left(1 - \frac{n^2}{z^2}\right)w = 0 \tag{B.7}$$

3. The derivative with respect to argument is

$$I_n'(z) = [I_{n-1}(z) + I_{n+1}(z)]/2 \tag{B.8}$$

In particular

$$I_0'(z) = I_1(z) \tag{B.9}$$

4. The series representation is

$$I_n(z) = \left(\frac{z}{2}\right)^2 \sum_{k=0}^{\infty} \frac{(z^2/4)^k}{k!\Gamma(n+k+1)} \tag{B.10}$$

The first few terms can be used when z is small.

5. The first term of the asymptotic approximation for large z is

$$I_n(z) \sim \frac{e^z}{\sqrt{2\pi z}} \tag{B.11}$$

B.1.2 Related Results

We derive some related results:

1. We define

$$\delta_{nm} = \begin{cases} 0 & n \neq m \\ 1 & n = m \end{cases}$$

Then

$$\int_0^{2\pi} e^{i(n-m)\theta} d\theta = 2\pi\delta_{nm} \tag{B.12}$$

and

$$\begin{aligned} &\int_0^{2\pi} \cos(\theta + \psi) e^{i(n-m)\theta} d\theta \\ &= \frac{e^{i\psi}}{2}\int_0^{2\pi} e^{i(n+1-m)\theta} d\theta + \frac{e^{-i\psi}}{2}\int_0^{2\pi} e^{i(n-1-m)\theta} d\theta \\ &= \pi\{e^{i\psi}\delta_{n+1,m} + e^{-i\psi}\delta_{n-1,m}\} \end{aligned} \tag{B.13}$$

2. An important relationship is

$$\exp(iz\sin\theta) = \sum_{n=-\infty}^{\infty} J_n(z)e^{in\theta} \tag{B.14}$$

3. Define

$$I_1 = \sum_{n=-\infty}^{\infty}\sum_{m=-\infty}^{\infty} J_n J_m e^{in\psi}\int_0^{2\pi} e^{i(n-m)\theta}\,d\theta \tag{B.15}$$

Then, using the results above,

$$\begin{aligned} I_1 &= 2\pi\sum_{n=-\infty}^{\infty}\sum_{m=-\infty}^{\infty} J_n J_m e^{in\psi}\delta_{nm} \\ &= 2\pi\sum_{n=-\infty}^{\infty} J_n^2 e^{in\psi} \end{aligned} \tag{B.16}$$

4. A related integral is

$$\begin{aligned} I_2 &= \sum_{n=-\infty}^{\infty}\sum_{m=-\infty}^{\infty} J_n J_m e^{in\psi}\int_0^{2\pi}\cos(\theta+\psi)e^{i(n-m)\theta}\,d\theta \\ &= \pi\sum_{n=-\infty}^{\infty} J_n e^{in\psi}\left\{\sum_{m=-\infty}^{\infty} J_m\left[e^{i\psi}\delta_{n+1,m} + e^{-i\psi}\delta_{n-1,m}\right]\right\} \\ &= \pi\sum_{n=-\infty}^{\infty} J_n\left\{J_{n+1}e^{i(n+1)\psi} + J_{n-1}e^{i(n-1)\psi}\right\} \end{aligned} \tag{B.17}$$

Now n is a dummy variable. In the first term replace $n+1$ by n, and in the second replace $n-1$ by n. Then, using the recursion formula (B.4) above

$$I_2 = 2\pi\sum_{n=-\infty}^{\infty}\frac{n}{\zeta}J_n^2 e^{i(n-1)\psi} \tag{B.18}$$

Two important integrals are Weber's first and second exponential integrals [118, Watson §§13.3, 13.31]:

$$\int_0^{\infty} J_0(at)\exp(-p^2t^2)t\,dt = \frac{1}{2p^2}\exp(-a^2/4p^2) \tag{B.19}$$

and

$$\int_0^{\infty} J_n(at)J_n(bt)\exp(-p^2t^2)t\,dt = \frac{1}{2p^2}\exp\left(-\frac{a^2+b^2}{4p^2}\right)I_n\left(\frac{ab}{2p^2}\right) \tag{B.20}$$

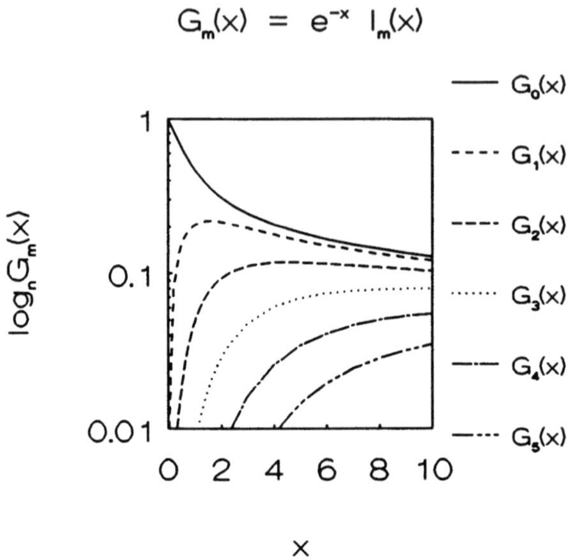

Figure B.1: The function $G_n(x) = e^{-x} I_n(x)$ where I_x is the modified Bessel function of the first kind.

where $I_n(x)$ is the modified Bessel function of the first kind. The special case for which $a = b$ is of importance for studying instabilities in plasmas. In this case Weber's second integral becomes

$$G_n\left(\frac{a^2}{2p^2}\right) = \frac{1}{2p^2} \exp\left(-\frac{a^2}{2p^2}\right) I_n\left(\frac{a^2}{2p^2}\right) \tag{B.21}$$

In Figure B.1 we show the function $G_n(z)$ for $n = 1, 2, 3, 4, 5$ and for real values of z

A useful integral representation is

$$\int_0^{2\pi} \cos(z \sin\theta - n\theta)\, d\theta = 2\pi J_n(z) \tag{B.22}$$

Clearly

$$\int_0^{2\pi} \sin(z \sin\theta - n\theta)\, d\theta = 0 \tag{B.23}$$

because the integrand has period 2π and is an odd function of θ.

B.2 The Plasma Dispersion Function

The plasma dispersion function $Z(\zeta)$ is defined by

$$Z(\zeta) = \frac{1}{\sqrt{\pi}} \int_C \frac{e^{-\xi^2}}{\xi - \zeta} d\xi \tag{B.24}$$

The contour C is that shown in Figure 6.1. An alternative definition involving integration along the real axis is thus:

$$Z(\zeta) = \begin{cases} \frac{1}{\sqrt{\pi}} \int_{-\infty}^{\infty} \frac{e^{-\xi^2}}{\xi-\zeta} d\xi & \Im(\zeta) > 0 \\ \frac{1}{\sqrt{\pi}} P \int_{-\infty}^{\infty} \frac{e^{-\xi^2}}{\xi-\zeta} d\xi + i\sqrt{\pi} e^{-\zeta^2} & \Im(\zeta) = 0 \\ \frac{1}{\sqrt{\pi}} \int_{-\infty}^{\infty} \frac{e^{-\xi^2}}{\xi-\zeta} d\xi + 2i\sqrt{\pi} e^{-\zeta^2} & \Im(\zeta) < 0 \end{cases} \tag{B.25}$$

where P denotes the Cauchy principal value of an integral.

The properties of the Z-function are discussed in detail by Fried and Conte [39].

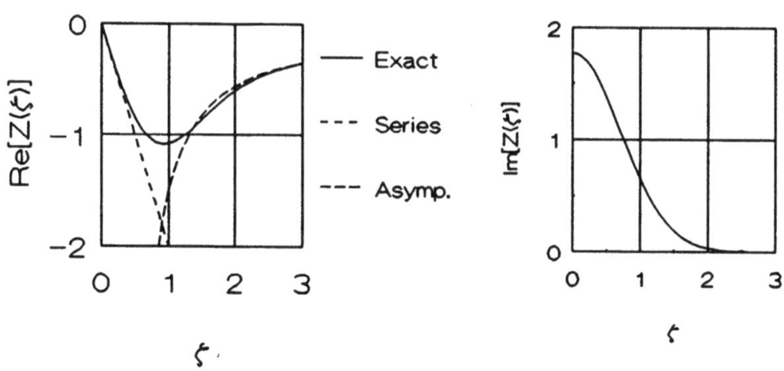

Figure B.2: The real and imaginary parts of the plasma dispersion function for real values of its argument compared with the two approximations. One term of the series expansion and two terms of the asymptotic approximation are shown. The approximations to the imaginary part are indistinguishable from the exact solution.

It commonly arises when kinetic theory is applied to plasmas whose zero order distribution functions can be represented by Maxwellians. It is related to the complex error function. Its properties have been tabulated for complex values of its argument [39].

The derivative with respect to its argument can be found by differentiating (B.24) with respect to ζ and evaluating the resulting integral by parts. The result is

$$Z'(\zeta) = -2[1 + \zeta Z(\zeta)] \tag{B.26}$$

Two approximations which are useful are

1. Series expansion:

$$\begin{aligned} Z(\zeta) &= i\sqrt{\pi}e^{-\zeta^2} - 2\zeta + \frac{4}{3}\zeta^3 - \frac{8}{15}\zeta^5 + \dots \\ &\dots + \frac{(-1)^{n+1} n!\, 2^{2n+1}}{(2n+1)!}\zeta^{2n+1} + \dots, \quad |\zeta| < 1 \end{aligned} \tag{B.27}$$

2. Asymptotic approximation for large $|\zeta|$:

$$\begin{aligned} Z(\zeta) &= i\sqrt{\pi}\left\{ \begin{array}{l} 0,\ \Im(\zeta) > 0 \\ 1,\ \Im(\zeta) = 0 \\ 2,\ \Im(\zeta) < 0 \end{array} \right\} e^{-\zeta^2} - \frac{1}{\zeta}\left\{1 + \frac{1}{2\zeta^2} + \frac{3}{4\zeta^4}\right. \\ &\left.\dots + \frac{(2n)!}{2^{2n} n!}\frac{1}{\zeta^{2n}} + \dots\right\} \end{aligned} \tag{B.28}$$

An important special case is that for real ζ. The real part of the function $Z(\zeta)$ is shown in Figure B.2 and compared with the approximate expressions.

The reason for the importance of the values for real argument is that, in cases of physical interest, the growth of an unstable wave or damping of a stable wave is often small. In such cases the imaginary part of ζ is small and positive for wave growth and small and negative for wave damping. In such cases an approximation due to Plemelj can be applied [31, p64]. If a pole is located above the real axis at $\zeta_r + i\zeta_i$ where ζ_i is small then, to zero order in ζ_i, for any function $f()$,

$$\int_{-\infty}^{\infty} \frac{f(x)}{x - \zeta_r - i\zeta_i}\,dx = P\int_{-\infty}^{\infty} \frac{f(x)}{x - \zeta_r}\,dx + i\pi f(\zeta_r) \tag{B.29}$$

In the same way, if the pole is located just below the real axis

$$\int_{-\infty}^{\infty} \frac{f(x)}{x - \zeta_r + i\zeta_i}\,dx = P\int_{-\infty}^{\infty} \frac{f(x)}{x - \zeta_r}\,dx - i\pi f(\zeta_r) \tag{B.30}$$

If these relations are applied to the plasma dispersion function we see that we can express values of $Z(\zeta)$ near the real axis in terms of values on the real axis. For all three possible cases

$$Z(\zeta) = \frac{1}{\sqrt{\pi}} P \int_{-\infty}^{\infty} \frac{e^{-\xi^2}}{\xi - \zeta} d\xi + i\sqrt{\pi} e^{-\zeta^2} \quad |\Im(\zeta)| \ll 1 \tag{B.31}$$

The real part of Z shown in Figure B.2 is just the Cauchy principal value. The values shown are appropriate both to cases of small growth and small damping.

Appendix C

Properties of the Earth's Dipole Field

The earth's magnetic field can be approximated by that of a dipole tilted with respect to the earth's axis. Some formulae relevant to a dipole field are presented here for reference.

The geomagnetic coordinate system is chosen to be a spherical polar system with the axis coinciding with the dipole axis. The radius is r. The geomagnetic longitude ϕ is measured positive eastwards. The usual polar angle is not used. Instead the geomagnetic latitude θ is measured from the equatorial plane with the positive direction northwards.

Let a be the radius of the earth. The dipole field is represented by a scalar potential V

$$V = -V_0 \left(\frac{a}{r}\right)^2 \sin\theta \tag{C.1}$$

The magnetic field can be found from $\mathbf{B} = -\nabla V$ getting

$$\mathbf{B} = -\hat{\mathbf{r}}\frac{2B_0 a^3}{r^3}\sin\theta + \hat{\boldsymbol{\theta}}\frac{B_0 a^3}{r^3}\cos\theta \tag{C.2}$$

Clearly

$$B = \frac{B_0 a^3}{r^3}\sqrt{4 - 3\cos^2\theta} \tag{C.3}$$

The equation of a field line can be found from

$$\frac{1}{r}\frac{dr}{d\theta} = \frac{B_r}{B_\theta} \tag{C.4}$$

which can be integrated to give

$$r = aL\cos^2\theta \tag{C.5}$$

L is the distance, measured in earth radii, at which the field line crosses the equatorial plane. A useful set of coordinates were introduced by McIlwain [77]. The parameter L forms one of these coordinates identifying a field line. His other coordinates are not used here. In a dipole field however another suitable coordinate is the potential V which effectively measures position on the field line. Equipotential surfaces are normal to the field lines. Thus the set L, V, ϕ forms a set of orthogonal curvilinear coordinates for a dipole field.

Appendix D

Definition of Symbols

Definitions of the more important symbols used in this book are listed here. In a work of this type it is not always possible to find a unique symbol for every quantity. Where a symbol is used in more than one sense the actual meaning should be clear from the context.

A: Coefficient of n^4 in formula for refractive index of a cold plasma.

A_m: Coefficient of n^m in formula for refractive index of a warm plasma.

a: Acceleration.

$\mathbf{B}$: Magnetic field in a wave.

$\mathbf{B}_0$: Steady background magnetic field.

B: Coefficient of n^2 in formula for refractive index of a cold plasma.

C: Constant term in formula for refractive index of a cold plasma.

c: Speed of electromagnetic wave in free space.

$\mathbf{D}$: Electric displacement field.

$D(\omega, k)$: Dispersion relation is $D(\omega, k) = 0$.

$\mathbf{E}$: Electric field in a wave.

E_R, E_L: Components of electric field in rotating coordinate system.

e: Magnitude of electron charge. 1.602×10^{-2} C.

f: Frequency.

$f(\mathbf{r}, \mathbf{v})$: Particle distribution function. Subscripts 0 and 1 indicate order of approximation.

$G_l(\xi)$: $e^{-\xi^2} I_l(\xi)$.

$I_l(\xi)$: Modified Bessel function of the first kind of order l.

$\Im(z)$: Imaginary part of z.

$\mathbf{J}$: Current density.

J_R, J_L: Components of current density in rotating coordinate system.

$J_l(\xi)$: Bessel function of the first kind of order l.

K: Boltzmann's constant.

$\mathbf{k}$: Wave vector.

k_0: ω/c.

L: Coordinate identifying a magnetic field line. In a dipole field it is the radius of the equatorial crossing of the field line measured in earth radii.

l_x, l_y, l_z: Direction cosines of the magnetic field.

m: Particle mass. Subscript may denote species.

N: Number density of particles. Subscript may denote species.

n: Refractive index vector. $c\mathbf{k}/\omega$.

n_x, n_y, n_z: Components of the refractive index vector.

n: Refractive index. c/V_P.

P: Electric polarization of medium.

P, P_0, p: Pressure.

q: Charge on a particle. Subscript may denote species.

R_E: Earth radius.

$\Re(z)$: Real part of z.

r: Position vector with components x, y and z.

T: Temperature.

t, t': Time.

U: Energy density in field *or* normalized hybrid sound speed $(\gamma_e T_e + \gamma_i T_i)/(m_e + m_i)c^2$ *or* $U = 1 - i\nu/\omega$. Subscript may denote particle species.

U_A: Normalized Alfvén speed. $B_0/c\sqrt{\mu_0 \rho_m}$.

U_S: Normalized sound speed, $c^{-1}\sqrt{\gamma KT/m}$.

V_A: Alfvén speed. $B_0/\sqrt{\mu_0 \rho_m}$.

V_G: Group velocity. $\nabla_k \omega$.

V_L: Component of group velocity in direction of wave normal. $d\omega/dk$.

V_P: Phase speed. ω/k.

V_R: Ray velocity. $V_P/\cos\alpha$.

V_S: Sound speed, $\sqrt{\gamma KT/m}$.

v: Particle velocity.

W: Power. Joule heating.

X: Square of normalized plasma frequency. ω_p^2/ω^2.

X_i: $(m_e/m_i)X$.

x, y, z, x', y', z': Cartesian coordinate systems. Usually $\mathbf{B}_0$ is parallel to z with $\mathbf{k}$ in the x–z plane and $\mathbf{k}$ is parallel to z' with $\mathbf{B}_0$ in the x'–y' plane.

Y: Normalized gyrofrequency. Ω/ω.

Y_i: $(m_e/m_i)Y$.

$Z(\xi)$: Plasma dispersion function.

α: Angle between wave normal and ray *or* coefficient of n_x^4 in Booker quartic.

β: Angle between ray and $\mathbf{B}_0$ *or* coefficient of n_x^3 in Booker quartic.

γ: Ratio of specific heats; c_P/c_V or $(n+2)/n$ where n is the number of degrees of freedom *or* coefficient of n_x^2 in Booker quartic.

δ_{jk}: Kronecker delta. $\delta_{jk} = 1$ if $j = k$, 0 otherwise.

δ: Coefficient of n_x in Booker quartic.

ϵ: Constant coefficient in Booker quartic.

ϵ_0: Permittivity of free space.

ϵ_{jkl}: Alternating unit tensor. $\epsilon_{jkl} = 1$ if the subscripts are an even permutation of 123, -1 if they are an odd permutation of 123 and 0 otherwise.

η: Polarization ratio $E_{z'}/E_{x'}$ in a wave.

η_i: Fractional abundance of ith ion species.

θ: Angle between wave normal and magnetic field.

κ_{jk}: Dielectric tensor.

κ_1, κ_2, κ_3: Non-zero elements of the dielectric tensor in rectangular coordinates.

κ_R, κ_L: Elements of the dielectric tensor in rotating coordinates.

λ_D: Debye length, $(\epsilon_0 KT/nq^2)^{1/2}$.

μ_0: Magnetic permeability of free space.

ν: Collision frequency.

χ_{jk}: Susceptibility tensor.

$\mathbf{\Pi}$: Poynting vector; $\mathbf{E} \times \mathbf{B}/\mu_0$.

ρ: Charge density *or* wave polarization *or* cyclotron radius mv/B_0e.

ρ_m: Mass density.

σ, σ_{jk}: Conductivity. Conductivity tensor.

τ: Time interval.

$\mathbf{\Phi}$: Electric potential *or* phase of wave in phase integral approximation.

$\mathbf{\Omega}$: Angular velocity of a particle about $\mathbf{B}_0$.

Ω: Gyrofrequency. $B_0 e/m$. Subscript may denote species.

ω: Angular frequency. $2\pi f$.

ω_p: Angular plasma frequency. $\sqrt{Ne^2/\epsilon_0 m}$.

$\perp, \|$: Subscripts denoting components perpendicular and parallel to $\mathbf{B}_0$.

An asterisk denotes a complex conjugate; X^* is the complex conjugate of X. A tilde denotes a Fourier transform; $\tilde{X}$ is the Fourier transform of X.

Bibliography

[1] Abramowitz, M., and Stegun, I. A., *Handbook of Mathematical Functions*, National Bureau of Standards, Washington D. C. 1964, republished by Dover Publications, New York 1965.

[2] Alfvén, H., On the existence of electromagnetic-hydrodynamic waves, *Nature, 150*, 405–406, 1942.

[3] Allis, W. P., Waves in a plasma, *MIT Research Lab. Electronics Quart. Prog. Rep. 54*, 1959.

[4] Angerami, J. J. and Thomas, J. O., Studies of planetary atmospheres 1 – The distribution of electrons and ions in the earth's exosphere, *J. Geophys. Res., 69*, 4537–4560, 1964.

[5] Appleton, E. V., Wireless studies of the ionosphere, *J. Instn Elect. Engrs, 71*, 642–650, 1932.

[6] Åström, E., On waves in an ionized gas, *Ark. Fys., 2*, 443–457, 1950.

[7] Ashour-Abdalla, M., and Kennel, C. F., Convective cold upper hybrid instabilities, in *Magnetospheric particles and fields*, ed. B. M. McCormac, 181–196, Reidel, Dordrecht 1976.

[8] Barkhausen, H., Zwei mit Hilfe der neuen Verstärker entdeckte Erscheinungen, *Physikalisch. Z., 20*, 401–403, 1919.

[9] Barrington, R. E., Belrose, J. S., and Mather, W. E., A helium whistler observed in the Canadian satellite Alouette II, *Nature, 210*, 80–81, 1966.

[10] Bernard, L. C., A new nose extension method for whistlers, *J. Atmos. Terr. Phys., 35*, 871–880, 1973.

[11] Bernstein, I. B., Waves in a plasma in a magnetic field, *Phys. Rev., 109*, 10–21, 1958.

[12] Berry, M. V., Waves and Thom's theorem, *Adv. Phys., 25*, 1-26, 1976.

[13] Booker, H. G., Some general properties of the formulae of magnetoionic theory, *Proc. Roy. Soc. Lond., 147*, 352–382, 1934.

[14] Booker, H. G., *Cold plasma waves*, Nijhoff, Dordrecht 1984.

[15] Brown, L. W., The galactic radio spectrum between 130 and 2600 kHz, *Astrophys. J., 180*, 335–350, 1973.

[16] Brice, N., Fundamentals of very low frequency generation mechanisms, *J. Geophys. Res., 69*, 4515–4522, 1964.

[17] Budden, K. G., *Radio waves in the ionosphere*, Cambridge University Press, Cambridge 1961.

[18] Budden, K. G., *The wave-guide mode theory of wave propagation*, Logos Press, London 1961.

[19] Budden, K. G., *The propagation of radio waves*, Cambridge University Press, Cambridge 1985.

[20] Calvert, W., and Hashimoto, K., The magnetoionic modes and propagation properties of auroral radio emissions, *J. Geophys. Res., 95*, 3943–3957, 1990.

[21] Carpenter, D. L., Whistler evidence of a "knee" in the magnetospheric ionization density profile, *J. Geophys. Res., 68*, 3727–3730, 1963.

[22] Carpenter, D. L., Stone, K., Siren, J. C., and Crystal, T. L., Magnetospheric electric fields deduced from drifting whistler paths, *J. Geophys. Res., 77*, 2819–2834, 1972.

[23] Chapman, S., and Cowling, T. G., *The mathematical theory of non-uniform gases*, 2nd ed., Cambridge University Press, Cambridge 1960.

[24] Chappell, C. R. and Gringauz, K. I. (eds.) *The physics of thermal plasma in the magnetosphere - Proceedings of symposium 9 of 26th meeting of COSPAR*, Published for COSPAR by Pergamon Press, Oxford, as Vol. 6, Number 3, of *Advances in Space Research*, 1987.

[25] Chen, F. F., *Introduction to plasma physics*, Plenum Press, New York 1974.

[26] Chiu, Y. T., and Schulz, M., Self-consistent particle and parallel electrostatic field distributions in the magnetospheric–ionospheric auroral region, *J. Geophys. Res., 83*, 629–642, 1978.

[27] Clemmow, P. C., and Dougherty, J. P., *Electrodynamics of particles and plasmas*, Addison-Wesley, Reading Mass. 1969.

[28] Clemmow, P. C. and Mullaly, R. F., The dependence of the refractive index in magneto-ionic theory on the direction of the wave normal, *The physics of the ionosphere*, 340–350, Physical Society, London 1955.

[29] Crawford, F. W., Cyclotron harmonic waves in warm plasmas,*J. Res. N. B. S. (Radio Sci.), 69D*, 789–806, 1965.

[30] Croley, D. R., Jr., Mizera, P. F., and Fennell, J. F., Signature of a parallel electric field in ion and electron distributions in velocity space, *J. Geophys. res., 83*, 2701–2705, 1978.

[31] Dennery, P., and Krzywicki, A., *Mathematics for physicists*, Harper and Row, New York 1967.

[32] Dowden, R. L., Doppler shifted cyclotron radiation from electrons: A theory of very low frequency emissions from the exosphere, *J. Geophys. Res., 67*, 1745–1750, 1962.

[33] Dowden, R. L., and Allcock, G. McK., Detarmination of nose frequency of non-nose whistlers, *J. Atmos. Terr. Phys., 33*, 1125–1129, 1971.

[34] Drummond, W. E., Malmberg, J. H., O'Neil, T. M., and Thompson, J. R., Nonlinear development of the beam-plasma instability, *Phys. Fluids, 13*, 2422–2425, 1970.

[35] Dungey, J. W., Waves and particles in the magnetosphere, in *Physics of the magnetosphere*, ed. R. L. Carovillano, 218–259, Reidel, Dordrecht 1968.

[36] Dusenbery, P. B., and Lyons, L. R., General concepts of the generation of auroral kilometric radiation, *J. Geophys. Res., 87*, 7467–7481, 1982.

[37] Egeland, A., Holter, Ø., and Omholt, A., *Cosmical geophysics*, Universitetsforlaget, Oslo 1973.

[38] Ellis, G. R. A., The *Z* propagation hole in the ionosphere, *J. Atmos. Terr. Phys., 8*, 43–54, 1956.

[39] Fried, B. D., and Conte, S. D., *The plasma dispersion function*, Academic Press, New York 1961.

[40] Greenstadt, E. W., and Fredricks, R. W., Plasma instability modes related to the earth's bow shock, in *Magnetospheric Physics*, ed. B. M. McCormac, 281–290, Reidel, Dordrecht 1974.

[41] Gurnett, D. A., The earth as a radio source: Terrestrial kilometric radiation, *J. geophys. Res., 79*, 4227–4238, 1974.

[42] Gurnett, D. A., The earth as a radio source: The nonthermal continuum, *J. geophys. Res., 80*, 2751–2736, 1975.

[43] Gurnett, D. A., Shawhan, S. D., Brice, N. M., and Smith, R. L., Ion cyclotron whistlers, *J. Geophys. Res., 70*, 1665–1688, 1965.

[44] Gurnett, D. A., Frank, L. A., and Lepping, R. P., Plasma waves in the distant magnetotail, *J. Geophys. Res., 81*, 6059–, 1976.

[45] Gurnett, D. A., and Inan, U. S., Plasma wave observations with the Dynamics Explorer 1 Spacecraft, *J. Geophys. Res., 26*, 285–316, 1988.

[46] Haerendel, G., Paschmann, G., Sckopke, N., and Hedgecock, P. C., The frontside boundary layer of the magnetosphere and the problem of reconnection, *J. Geophys. Res., 83*, 3195–3216, 1978.

[47] Hargreaves, J. K., *The upper atmosphere and solar-terrestrial relations: an introduction to the aerospace environment*, Van Nostrand Reinhold, New York 1979.

[48] Hartree, D. R., The propagation of electromagnetic waves in a refracting medium in a magnetic field, *Proc. Camb. Phil. Soc., 27*, 143–162, 1931.

[49] Hasegawa, A., *Plasma instabilities and nonlinear effects*, Springer-Verlag, Berlin 1975.

[50] Haselgrove, J., Ray theory and a new method for ray tracing, in *The physics of the ionosphere*, 355–364, Physical Society, London 1955.

[51] Hashimoto, K., A reconciliation of propagation modes of auroral kilometric radiation, *J. Geophys. Res., 89*, 7459–7466, 1984.

[52] Hashimoto, K., and Calvert, W., Observations of the Z mode with DE 1 and its analysis by three dimensional ray tracing, *J. Geophys. Res., 95*, 3933–3942, 1990.

[53] Helliwell, R. A., *Whistlers and related ionospheric phenomena*, Stanford University Press, Stanford 1965.

[54] Helliwell, R. A., A theory of discrete vlf emissions from the magnetosphere, *J. Geophys. Res., 72*, 4773–4790, 1967.

[55] Helliwell, R. A., Controlled vlf wave injection experiments in the magnetosphere, *Space Sci. Rev., 15*, 781–802, 1974.

[56] Helliwell, R. A., and Crystal, T. L., A feedback model of cyclotron interaction between whistler mode waves and energetic electrons in the magnetosphere, *J. Geophys. Res., 78*, 7357–7371, 1973.

[57] Helliwell, R. A., Katsufrakis, J. P., Bell, T. F., and Raghuram, R., VLF line radiation in the earth's magnetosphere and its association with power system radiation, J. Geophys., Res., 80, 4249–4258, 1975.

[58] Hines, C. O., The relation between hydromagnetic waves and the magneto-ionic theory, *International Series of Monographs on Electromagnetic Waves, 6*, 287–299, 1963; reprinted in *The upper atmosphere in motion* by C. O. Hines and colleagues, pp862–874, American Geophysical Union, Washington, 1974.

[59] Holt, E. H., and Haskell, R. E., *Foundations of Plasma Dynamics*, Macmillan, New York 1965.

[60] Isted, G. A., and Millington, G., The "dawn chorus" in radio observation, *Nature, 180*, 716, 1957.

[61] Jeffreys, H., and Jeffreys, B. S., *Methods of Mathematical Physics*, Cambridge University Press, Cambridge 1956.

[62] Jones D., Recent work on ion-cyclotron whistlers, *Ann. Géophys., 28*, 527–540, 1972.

[63] Jones, D., The second Z-propagation window, *Nature, 262*, 674–675, 1976.

[64] Jones, D., Latitudinal beaming of planetary radio emissions, *Nature, 288*, 225–229, 1980.

[65] Jones, D., Calvert, W., Gurnett, D. A., and Huff, R. L., Observed beaming of terrestrial myriametric radiation, *Nature, 328*, 391–395, 1987.

[66] Joselyn, J. A., and Lyons, L. R., Ion cyclotron wave growth calculated from observations of the proton ring current during storm recovery, *J. Geophys. Res., 81*, 2275–2282, 1976.

[67] Kennel, C. F. and Ashour-Abdalla, M., Electrostatic waves and the strong diffusion of magnetospheric electrons, in *Magnetospheric plasma physics* ed. A. Nishida, 245–344, Center for Academic Publications Japan, Tokyo 1982.

[68] Kennel, C. F., and Engelmann, F., Velocity space diffusion from weak plasma turbulence in a magnetic field, *Phys. Fluids, 9*, 2377–2388, 1966.

[69] Kennel, C. F., Lanzerotti, L. J. and Parker, E. H., (ed.) *Solar system plasma physics*, North-Holland, Amsterdam 1979.

[70] Kennel, C. F., and Petschek, H. E., Limit on stably trapped particle fluxes, *J. Geophys. Res., 71*, 1–28, 1966.

[71] Kennel, C. F. and Wong, H. V., Resonant particle instabilities in a uniform magnetic field, *J. Plasma Phys., 1*, 75–80, 1967.

[72] Laird, M. J., and Nunn, D., Full wave vlf modes in acylindrically symmetric enhancement of plasma density, *Planet. Space Sci., 23*, 1649–1658, 1975.

[73] Landau, L.D., On the Vibrations of the Electronic Plasma, *J. Phys. (USSR), 10*, 25–34, 1946.

[74] Lassen, H., Über den Einfluss des Erdmagnetfeldes auf die Fortpflanzung der elektrischen Wellen der drahtlosen Telegraphie in der Atmosphäre, *Elektrische Nachrichten-Technik, 4*, 324–334, 1927.

[75] Lewis, E. A., High frequency radio noise, in *CRC handbook of atmospherics*, ed. H. Volland, 179–250, CRC Press, Boca Raton, Florida 1982.

[76] Lyons, L. R., and Williams, D. J., *Quantitative aspects of magnetospheric physics*, Reidel, Dordrecht 1984.

[77] McIlwain, C. E., Coordinates for mapping the distribution of magnetically trapped particles, *J. Geophys. res., 66*, 3681–3691, 1961.

[78] Lyons, L. R., Thorne, R. M., and Kennel, C. F., Pitch-angle diffusion of radiation belt electrons within the plasmasphere, *J. Geophys. Res., 77*, 3455–3474 , 1972.

[79] Maeda, K., and Kimura, I., Calculation of the propagation path of the whistling atmospherics, *J. Atmos. Terr. Phys., 15*, 62–65, 1959.

[80] Nunn, D. A., A theory of vlf emissions, *Planet. Space Sci., 19*, 1141–1167, 1971.

[81] Nunn, D. A., A self-consistent theory of vlf emissions, *Planet. Space Sci., 22*, 349–378, 1974.

[82] Nunn, D. A., Comment on " A feedback model of cyclotron interaction between whistler mode waves and energetic electrons in the magnetosphere" by R. A. Helliwell and T. L. Crystal, *J. Geophys. Res., 80*, 4397–4398, 1975. Helliwell, R. A., and Crystal, T. L., Reply *J. Geophys. Res., 80*, 4399–4400, 1975.

[83] Paschmann, G., Haerendel, G., Sckopke, N., and Rosenbauer, H., Plasma and magnetic field characteristics of the distant polar cusp near local noon: The entry layer, *J. Geophys. Res., 81*, 2883–2899, 1976.

[84] Paschmann, G., Plasma structure of the magnetopause and boundary layer, in *Magnetospheric boundary layers*, ed. B. Battrick, 25–36, ESA SP–148, European Space Agency, Paris 1979.

[85] Persoon, A. M., Gurnett, D. A., and Shawhan, S. D., Polar cap electron densities from DE 1 plasma wave observations, *J. Geophys. Res., 88*, 10 123–10 136, 1983.

[86] Persoon, A. M., Gurnett, D. A., Peterson, W. K., Waite, J. H. Jr, Burch, J. L., and Green, J. L., Electron density depletions in the nightside auroral zone, *J. Geophys. Res., 93*, 1871–1895, 1988.

[87] Pitteway, M. L. V., and Jesperson, J. L., A numerical study of the excitation, internal reflection and limiting polarisation of whistler waves in the lower ionosphere, *J. Atmos. Terr. Phys., 28*, 17–43, 1966.

[88] Poeverlein, H., Strahlwege von Radiowellen in der Ionosphäre, *S. B. Bayer. Akad. Wiss., Math.-Nat. Klasse*, 175–201, 1948.

[89] Rabe, E., and Scourfield, M. W. J., Plasmasphere response to the onset of quiet magnetic conditions: plasma convection patterns, *Planet. Space Sci., 25*, 303–308, 1977.

[90] Roederer, J. G., *Dynamics of geomagnetically trapped radiation*, Springer-Verlag, Berlin 1970.

[91] Rönnmark, K., Borg, H., Christiansen, P. J., Gough, M. P., and Jones, D., Banded electron cyclotron instability – A first comparison of theory and experiment, *Space Sci. Rev., 22*, 401–417, 1978.

[92] Rostoker, G., Akasofu, S.-I., Foster, Y., Greenwald, R. A., Kamide, Y., Kawasaki, K., Lui, A. T. Y., McPherron, R. L. and Russell, C. T., Magnetospheric substorms: Definition and signatures, *J. Geophys. Res., 85*, 1663–1668, 1980.

[93] Scarf, F. L. and Chappell, C. R., An association of magnetospheric whistler dispersion characteristics with changes in local plasma density, *J. Geophys. Res., 78*, 1597–1602, 1973.

[94] Shawhan, S. D., Magnetospheric plasma waves, in *Solar system plasma physics* ed. C. F. Kennel, L. J. Lanzerotti, and E. H. Parker, 211–270, North-Holland 1979.

[95] Smith, E. J., and Tsurutani, B. T., Magnetosheath lion's roars, *J. Geophys. Res., 81*, 2261–2266, 1976.

[96] Smith, R. L., Brice, N. M., Katsufrakis, J., Gurnett, D. A., Shawhan, S. D., Belrose, J. S. and Barrington, R. E., An ion gyrofrequency phenomenon observed in satellites, *Nature, 204*, 274–275, 1964.

[97] Smith, R. L., Guiding of whistlers in a homogeneous medium, *J. Res. N. B. S. - D. Radio Propagation, 64D*, 505–508, 1960.

[98] Smith, R. L., Propagation characteristics of whistlers trapped in field aligned columns of enhanced ionization, *J. Geophys. Res., 66*, 3699–3707, 1961.

[99] Smith, R. L. and Angerami, J. J., Magnetospheric properties deduced from OGO1 observations of ducted and non-ducted whistlers, *J. Geophys. Res., 73*, 1–20, 1968.

[100] Stix, T. H., *The theory of plasma waves*, McGraw-Hill, New York 1962.

[101] Storey, L. R. O., An investigation of whistling atmospherics, *Phil. Trans. Roy. Soc. A, 246*, 113–141, 1953.

[102] Stott, G. F., Refractive index surfaces, *J. Atmos. Terr. Phys., 45*, 219–229, 1983.

[103] Strangeways, H. J. and Rycroft, M. J., Trapping of whistler waves through the side of ducts, *J. Atmos. Terr. Phys., 42*, 983–994, 1980.

[104] Strangeways, H. J., The effect of multi-duct structure on whistler-mode wave propagation, *J. Atmos. Terr. Phys., 44*, 901–912, 1982.

[105] Stringer, T. E., Low frequency waves in an unbounded plasma, *Plasma Phys. (J. nucl. Energy C), 5*, 89–107, 1963.

[106] Swanson, D. G., *Plasma Waves*, Academic Press, Boston 1989.

[107] Thorne, R. M., Church, S. R., and Gurney, D. J., On the origin of plasmaspheric hiss: The importance of wave propagation and the plasmapause, *J. Geophys.Res., 84*, 5241–5247, 1979.

[108] Volland, H., Low frequency radio noise, in *CRC handbook of atmospherics*, ed. H. Volland, 179–250, CRC Press, Boca Raton, Florida 1982.

[109] Wait, J. R., *Electromagnetic waves in stratified media*, Pergamon Press, Oxford 1962.

[110] Walker, A. D. M., The theory of guiding of radio waves in the exosphere – I. Guiding of whistlers, *J. Atmos. Terr. Phys., 28*, 807–822, 1966.

[111] Walker, A. D. M., The propagation of very low-frequency waves in ducts in the magnetosphere, *Proc. Roy. Soc. Lond. A, 321*, 69–93, 1971.

[112] Walker, A. D. M., The propagation of very low-frequency waves in ducts in the magnetosphere. II, *Proc. Roy. Soc. Lond. A, 329*, 219–231, 1972.

[113] Walker, A. D. M., Methods of scaling whistlers in the absence of the initiating sferic and nose frequency, *S. Afr. J. Antarctic Res., 4*, 27–30, 1974.

[114] Walker, A. D. M., The theory of whistler propagation, *Rev. Geophys. Space Phys., 14*, 629–637, 1976.

[115] Walker, A. D. M., The phase velocity, ray velocity, and group velocity surfaces for a magnetoionic medium, *J. Plasma Phys., 17*, 467–486, 1977.

[116] Walker, A. D. M., The ray velocity surface and the CMA diagram, *J. Plasma Phys., 18*, 339–346, 1977.

[117] Walker, A. D. M. and J. F. McKenzie, Properties of electromagnetic waves in ferrites, *Proc. Roy. Soc. Lond. A., 399*, 217–241, 1985.

[118] Watson, G. N., *A Treatise on the Theory of Bessel Functions, 2nd Edition*, Cambridge University Press, Cambridge 1944 (reprinted 1966).

[119] Wu, C. S., and Lee, L. C., A theory of the terrestrial kilometric radiation, *Astrophys. J.*, *230*, 621–626, 1979.

[120] Young, T. S. T., Callen, J. D., and McCune, J. E., High-frequency electrostatic waves in the magnetosphere, *J. Geophys. Res.*, *79*, 1082–1099, 1973.

[121] *Achievements of the International Magnetospheric Study*, ESA SP-217, European Space Agency, Paris 1984.

Subject Index

acceptance cone 227, 258
adiabatic equation of state 7
Airy function 243
AKR 292, 307
Alfvén modes 205
Alfvén speed 17
Alfvén wave 211
alternating unit tensor 321
ambipolar diffusion 197
angular spectrum of plane waves 47
Appleton-Hartree formula 27
Appleton-Lassen formula 27
angular frequency 11
anisotropic plasma 5
anisotropic particle distribution, conductivity for electrostatic waves 135
Åström speed 17
auroral hiss 289
auroral kilometric radiation 292, 307
auroral oval 289
Auroral whistler mode hiss 292

Bernstein modes 123, 124
Bessel function 323
  first kind 323
bi-Maxwellian distribution function 116
Birkeland currents 195
Boltzmann equation 9, 10
  collisionless 11
Boltzmann-Vlasov equation 11
Booker quartic equation 76, 235, 302
boundary layer 199
  entry layer 200
  high latitude 200
  low latitude 193, 201
bow shock 190, 198
bow shock plasma waves 318

Cartesian tensor 5
Čerenkov interaction 304
Čerenkov resonance 154, 170, 281, 292, 293
  non-linear 176
  velocity space diffusion 174
Chapman-Ferraro currents 189, 195
charge on electron 5
chorus 267
circular polarization 33
circularly polarized coordinates 20
cleft, polar 200
CMA diagram 55, 208
cold plasma approximation 3
collision frequency 6
collisionless Boltzmann equation 11
collisions, neutral particles 6, 24
complex frequency 24
complex notation for harmonic waves 11
composite distribution function 120
conduction current 22
conductivity electrostatic waves 135
conductivity relationship 4
conductivity tensor
  cold plasma 5, 18
  hot plasma 105
  hot plasma, parallel propagation 110
  warm plasma 85
configuration space 7
conservation of energy 11
constitutive relation 4, 18
  linear 5
continuity equation 6
contour of marginal stability 275
coordinates, magnetic 330
Coulomb collisions 9
coupling, effect of collisions on 260
coupling point 260
cross-over 39
cross-over frequency 37, 210
cross-tail current 195
current
  Birkeland 195
  Chapman-Ferraro 195
  cross-tail 195
  ionospheric system 195
  ring 195, 201
curvature of duct 245
cusp, polar 200
cut-off 37
  half gyrofrequency 243
  waveguide 225
cut-off boundary 55
cut-off frequency 37

cyclotron damping 123
 of whistlers 128
cyclotron frequency 17
cyclotron motion 102
cyclotron resonance 154
 Doppler shifted 129, 265, 268

D-layer 196
dawn chorus 267
Debye length 3, 198
Debye sphere 9
decay rate 114
depletion of electron density as duct 239
dielectric constant, hot plasma 105
dielectric tensor
 cold plasma 21, 22
 hot plasma, parallel propagation 110
 warm plasma 85
diffusion in velocity space 169, 272
diffusion, quasilinear 173
dipole magnetic field 330
dispersion 217
dispersion relation
 cold plasma 25, 29
 hot plasma electrostatic waves 107
 hot plasma, parallel propagation 111
 ion cyclotron whistler 251, 252
 particle beam 132
 warm plasma 86
displacement 4
distribution function 7, 8
 bi-Maxwellian 116
 composite 120
 evolution in time 9
 loss-cone 117
 modified Maxwellian 118
 subtracted Maxwellian 118
Doppler shifted cyclotron resonance 129, 265, 268
double hop whistler 215
duct
 curvature of 245
 field-aligned 215
 guiding of whistlers 236
 half gyrofrequency cut-off 243
 propagation of whistlers in 233

E-layer 196
earth–ionosphere cavity 214, 223
earth–ionosphere waveguide 215
electron, charge on 5
electron plasma 23, 34
electron plasma frequency 16
electron plasma waves 90
 resonance with beam 180
electron sound speed 84
electron whistler 249
electrostatic approximation 87, 101, 102
electrostatic oscillations in cold plasma 143
electrostatic wave 39, 88, 126
 conductivity, anisotropic particle distribution 135
 conductivity, loss-cone distribution 140
elliptical polarization 32
Ellis window 289, 297, 307
energy conservation 11
energy density
 in field 11
 time average 12
energy flux in wave 11, 40
energy flux, time average 12
energy transfer, wave-particle 112
energy transport by waves 11
enhancement of electron density as duct 237
entry layer 200
equation of motion 5
 of a species 6
equation of state, isothermal 7
equation of state, adiabatic 7
evanescence 113
evanescent wave 29
extraordinary wave 36

F-layer 196
F-wave 37
faller 267
fast Alfvén wave 96, 211
fast wave 37
field-aligned ducts 215
field energy density 11
fluid equations 6
fluid model 5
flux transfer events 193
fractional abundance 27
free space modes 205
frozen-in field lines 17

geomagnetic pulsations 318
group refractive index 253
 ion cyclotron whistler 253
group speed 46, 131

group velocity 45, 50, 52
  of ducted wave 241
group velocity surface 52, 63
growth rate 114
  for particle beam instability 133
guiding center 102
guiding center frame 152
gyrofrequency 17
gyroresonance 38, 154, 170
  and whistler mode 183
  velocity space diffusion 175

half gyrofrequency duct cut-off 243
harmonic oscillations 11
Harris instability 293
helium cyclotron wave 254
helium whistler 206, 211, 216, 251
Hermitian tensor 5
hiss 267
  auroral whistler mode 292
  whistler mode 293
hook 267, 284
hybrid resonance 38
hybrid sound speed 84
hydromagnetic medium 23
hydromagnetic waves 96

IGY 208
IMS 189
inhomogeneous wave 29, 243
interaction time 163
International Geophysical Year 208
International Magnetospheric Study 189
ion acoustic waves 94
ion cyclotron whistler 205, 206, 208, 211, 216, 249, 250
  cut-off, resonance and cross-over 256
  dispersion relation 251, 252
  generation 255
  group refractive index 253
ion cyclotron waves 283
ion plasma frequency 16
ion sound speed 84
ionosphere 195
  transmission of whistlers through 226
ionospheric layers 196
ISIS 2 250
isothermal equation of state 7

Jones window 289, 297
Joule heating 11

kinetic equation 7, 9
knee whistler 221
Kronecker delta 321

Landau contour 106, 130
Landau damping 84, 123
  linear 127, 178
  non-linear 178
  of electrostatic waves 126
left-handed polarization 33
lightning 216
  as source of whistlers 223
linear polarization 33
linearization 18
lion's roar 318
lobe 201
Lorentz force law 5, 9
loss-cone distribution function 117
  conductivity for electrostatic wave 140
low latitude boundary layer 193, 201
lower hybrid resonance 39, 209, 230

magnetic coordinates 330
magnetic merging 191
magnetic reconnection 191
magnetohydrodynamics 24
magnetopause 190, 199
magnetoplasma 3
magnetosheath 191, 198
magnetosonic waves 96
magnetosphere 189, 190
  closed 190
  formation 189
  open 192
magnetospheric substorm 193
magnetospherically reflected whistler 231
magnetotail 192
mantle 200
marginal stability, contour of 275
Maxwellian distribution function 114
Maxwell's equations 4
  for harmonic waves 11
mean values of plasma properties 8
merging, magnetic 191
modified Bessel function 323
modified Maxwellian distribution function 118
modified refractive index 244
motion, equation of 5
MR whistler 231
multihop whistlers 215, 266
multi-ion plasma 27

myriametric radiation, terrestrial 293

neutral particle collisions 6
nose frequency 218
nose whistler 267

O-wave 35
Ohm's law 4
optics convention for polarization 33
ordinary wave 35

particle beam 119, 130
    dispersion relation for 132
    growth rate for instability 133
particle orbit, unperturbed 101
particle resonance 108
particle trapping 176
permittivity 4
phase bunching 180, 286
phase integral solution 76
phase space 7
phase speed 25, 46, 131
phase velocity 52
phase velocity surface 52
pitch angle 102
pitch angle diffusion 278
    protons 283
plane waves 24
planes of constant amplitude 30
planes of constant phase 25, 29
plasma 3, 26
    anisotropic 5
    convection 191
    electron 34
    multi-ion 27
plasma dispersion function 327
plasma frame 152
plasma frequency 16
    as cut-off, resonance or cross-over 40
plasma mantle 200
plasma resonance 39
plasma sheet 201
plasmasphere 196
plasmaspheric hiss 267
Poeverlein construction 78, 233, 299
polar cap 198
polar cleft 200
polar cusp 200
polar wind 198
polarization
    circular 33
    elliptical 32
    left-handed 33
    linear 33
    medium 4, 30
    optics convention 33
    radio convention 33
    right-handed 33
    warm plasma 97
    wave 30
polarization current 21, 22
polarization label 31, 36
power line harmonics 317
Poynting vector 11
    in inhomogeneous wave 243
    time average 12
Poynting's theorem 11
pressure 6
principal resonance 38
principle of superposition 43
propagated wave 29
proton cyclotron wave 254
proton whistler 206, 211, 216, 249, 250
pulsations, geomagnetic 318

quasi-equilibrium 115
quasilinear diffusion 173
quasi-longitudinal propagation 210

radiation belts 201
radio convention for polarization 33
ratio of specific heats 7
ray 43, 46
ray direction 209
ray path 48
ray surface unfolding 55
ray tracing 79
    with ducts included 247
ray velocity 49
ray velocity surface 52, 63
reconnection, magnetic 191
reference frames 152
reflection, law of 74
refraction, law of, see Snell's law
refractive index
    cold plasma 26
    low frequency approximation 28
    method of modified 244
    warm plasma 86
refractive index surface 36, 49, 53, 63, 209
    proton whistler 211
    whistler 211
refractive index vector 25
relativistic resonance 154
resonance boundary 55
resonance condition 108

resonance cone 39, 53
resonance contour 312
resonance frequency 37
resonant diffusion, effect on velocity distribution 274
resonant particles, flux in velocity space 170
resonant wave-particle interaction 160
rest frame of wave 153
right-handed polarization 33
ring current 195, 201
riser 267

S-wave 37
saucer, v. l. f. 289
separable distribution function 116
signal front 52
single hop whistler 215
single wave characteristic 156, 312
  general case 168
  non-relativistic particles 157
  relativistic particles 158
slow Alfvén wave 96, 211
slow wave 37
Snell's law 74
solar wind 189, 198
solar wind pressure 190
sound speed 7, 84
specific heats, ratio of 7
speed, thermal 5
speed, sound 7
spheric 225
spitze 300
state, adiabatic equation of 7
stationary phase 44
steepest descents 44
Storey approximation 210, 217, 227
Storey cone 54
substitution tensor 321
substorm, magnetospheric 193
subtracted Maxwellian distribution function 118
superposition, principle of 43
susceptibility 4
  hot plasma 105
susceptibility tensor 21

tail lobe 201
temperature 6
tensor
  Cartesian 5, 321
  Hermitian 5
  substitution 321
terrestrial myriametric radiation 293, 304
test particle 162
thermal speed 5
time average of energy density and flux 12
TMR 293, 304
total internal reflection 30
triggered emission 268
tweek 225

u. l. f. pulsations 318
unducted whistlers 215, 227
unfolding 55
unit tensor, alternating 321
upper hybrid resonance 38
upper hybrid waves 145

v. l. f. 208
v. l. f. radiation 223
v. l. f. saucer 289
van Allen belts 265
velocity distribution function 8
very low frequency 208
velocity space 7
velocity space diffusion 169, 272
Vlasov equation 9

wave equation 13
wave frame 152
wave growth 113
wave injection experiment 317
wave normal 25
wave packet 43, 44, 50
wave polarization 30
wave vector 25
waveguide cut-off 214
waveguide, earth–ionosphere 215, 223
waveguide modes 214
Weber's integrals 325
whispering gallery 244
whistler 208
  amplification 283
  as magnetospheric probe 221
  double hop 215
  duct propagation 233
  effect of positive ions on propagation 230
  elementary theory 216
  fractional hop 249
  generation 214
  guiding by duct 236
  growth 269
  helium 216
  high latitude 218
  ion cyclotron 208, 211, 216

knee 221
low-latitude 217
magnetospherically reflected 231
mid-latitude 217
multihop 215, 266
near-transverse propagation 228
nose 267
observations 213
proton 216
refractive index surface 211
single hop 215
unducted 215
whistler growth 148
whistler mode 205, 208
upstream waves 318
whistler mode hiss, auroral 292, 293
whistler-mode noise 277
wind
polar 198
solar 189, 198
window 293
Ellis 289, 297, 307
Jones 289, 297
WKB solution 76

X-wave 36

Z-mode 206, 209
radiation 292, 294
resonance cone 293
window 292

MIX
Papier aus verantwortungsvollen Quellen
Paper from responsible sources
FSC® C105338

If you have any concerns about our products,
you can contact us on
ProductSafety@springernature.com

In case Publisher is established outside the EU,
the EU authorized representative is:
Springer Nature Customer Service Center GmbH
Europaplatz 3, 69115 Heidelberg, Germany

Printed by Libri Plureos GmbH
in Hamburg, Germany